Tarpons

Tarpons

Biology, Ecology, Fisheries

Stephen Spotte

Mote Marine Laboratory
Sarasota, Florida, USA

Registered Office
John Wiley & Sons, Ltd, The Atrium, Southern Gate, Chichester, West Sussex, PO19 8SQ, UK

Editorial Offices
9600 Garsington Road, Oxford, OX4 2DQ, UK
The Atrium, Southern Gate, Chichester, West Sussex, PO19 8SQ, UK
111 River Street, Hoboken, NJ 07030-5774, USA

For details of our global editorial offices, for customer services and for information about how to apply for permission to reuse the copyright material in this book please see our website at www.wiley.com/wiley-blackwell.

Library of Congress Cataloging-in-Publication Data

Names: Spotte, Stephen, author.
Title: Tarpons : biology, ecology, fisheries / Stephen Spotte.
Description: Chichester, UK ; Hoboken, NJ : John Wiley & Sons, 2016. | Includes bibliographical references and index.
Identifiers: LCCN 2016003051| ISBN 9781119185499 (cloth) | ISBN 9781119185703 (epub)
Subjects: LCSH: Tarpon. | Tarpon fisheries.
Classification: LCC QL638.M33 S66 2016 | DDC 597.5/7–dc23
LC record available at http://lccn.loc.gov/2016003051

A catalogue record for this book is available from the British Library.

Wiley also publishes its books in a variety of electronic formats. Some content that appears in print may not be available in electronic books.

Set in 9.5/13pt Meridien by SPi Global, Pondicherry, India
Printed and bound in Malaysia by Vivar Printing Sdn Bhd

1 2016

Contents

Preface

Two species of tarpons exist today, one in the Atlantic (*Megalops atlanticus*), the other *(M. cyprinoides*) in the Indo-West Pacific region. The name "tarpon," or "tarpom," is apparently of New World origin. The Englishman William Dampier encountered tarpons on his first voyage to the Bay of Campeche (which he called Campeachy), México, and his mention of the Atlantic tarpon is one of the earliest. Dampier wrote about the fish in his journals in 1675 and later included these entries in the account of his voyages around the world. The copies of *Dampier's Voyages* cited here are early twentieth-century editions edited by the poet John Masefield, but earlier versions were published in the seventeenth and early eighteenth centuries. In Volume II, Dampier (1906: 117–118) stated: "The Tarpom is a large scaly Fish, shaped much like a Salmon, but somewhat flatter. 'Tis of a dull Silver Colour, with Scales as big as a Half Crown. A large Tarpom will weigh 25 or 30 Pound." Because of their extensive distributions, both species have numerous other common names in many languages.[1] For simplicity, I refer to them as Atlantic and Pacific tarpons. The Pacific species is also called the Indo-Pacific tarpon and oxeye (or ox-eye), or sometimes oxeye (ox-eye) tarpon or herring.

Atlantic tarpons grow large, reaching 2.5 m and weighing 150 kg. The Pacific species is comparatively small, attaining only 0.6 m and 3 kg, although unsubstantiated reports exist of specimens three times this length (Seymour *et al.* 2008 and references). Despite the size disparity, their morphology, physiology, ecology, developmental biology, and other life-history features are so similar that I often found little justification for separate treatments, although I have separated them when possible for clarity. In some instances, such as discussion of distributions, I was handicapped by limited access to literature on the Pacific form.

Previous books have concentrated on just a few aspects of tarpon biology or restricted discussion to the Atlantic tarpon recreational fishery. My objective is to cover these and other topics without being too tiresome. The angling aspect presented (Chapter 8) is not about how to catch tarpons but how to conserve them and, if you must catch them, how best to do so with minimal stress to the fish and then release it in a manner offering the best chance of long-term survival.

My presentation of tarpon biology derives from a broad perspective, one in which I hope to assess the tarpon's unique life-history in terms of fishes generally.

---

[1]See, for example: http://www.catalogueoflife.org/col/details/species/id/17963572; http://eol.org/pages/339927/names/common_names; http://www.catalogueoflife.org/col/search/all/key/Megalops

Books like this are usually written by groups of specialists, the result being a series of chapters in which different aspects are partitioned, handed to separate authors, and subsequently treated in isolation. The result is often uneven, redundant, incompletely integrated, and fails to view the subjects themselves – tarpons in this case – as entities ruled by common natural forces for which data from more extensively studied species can sometimes apply just as well. Every biography is, in the end, a narrative of heritage and commonality.

My objective as a single author is to provide a cohesive picture of tarpon biology, ecology, and fisheries in which specialty aspects usually compartmentalized (e.g. physiology, larval development) blend at the edges and reinforce one another. I hope to accomplish this without loss of accuracy. For example, variations on the cube law used to practical advantage in fishery biology for predicting length and weight of individual fishes and assessing the condition of populations also apply theoretically to certain facets of water circulation in the buccal cavity (i.e. lamellar length scales isometrically with body weight). To receive full benefit of this integrative approach, chapters need to be read in sequence. This book, like my others, has been designed to be read, not consulted. Skipping through the text and examining sections out of sequence is guaranteed to be less satisfying. The reader has been duly advised, and I offer no apologies.

Symbols and abbreviations are generally defined at first use, but a roster of them is provided in the book's front matter. Background information necessary to understand certain concepts is given either superficially in the text or, if more detail is necessary, in occasional footnotes. The presentation overall assumes a certain advanced level of knowledge.

An early anonymous reviewer made the reasonable suggestion that I include a section on tarpon evolution. However, not being an ichthyologist I felt uncomfortable doing so. I therefore left this subject and certain other avenues of specialization (e.g. detailed aspects of tarpon skeletal anatomy) to the experts. I take a systems approach instead, integrating functional biology with ecology, and discussing both disciplines in terms of effects caused by humans in the recreational and commercial fisheries at both the individual and population level. Only a few reports exist on tarpon physiology, although other species can safely be used as proxies at the system and even cellular level, at which point any differences are of degree, not kind.

Tarpons are distributed widely throughout subtropical and tropical waters around the world. Appendix A at the end of the book comprises a partial list of countries from which both species have been recorded in the literature. Pusey *et al.* (2004) provided an outstanding short summary of the Pacific tarpon's natural history. Nothing I found on the Atlantic species in the recent literature matches it for brevity and completeness. Hildebrand's (1963) treatment came closest, but his information is outdated.

The Atlantic tarpon ranges north to Nova Scotia and south to Brazil. Some authors extend its southern range to the coast of Argentina (e.g. Castro-Aguirre *et al.* 1999: 89; Gill 1907: 36; Hildebrand 1963: 119), although I was unable to find a published record of its presence in either Uruguay or Argentina (e.g. Bouyat 1911 did not mention it). The warm, south-flowing Brazil Current stops at the mouth of the Río de la Plata, and the sea beyond, including off Patagonia, is temperate. It seems that any Atlantic tarpons found there could only be stragglers from Brazil.

Reports about the tarpon in the eastern Atlantic are uncommon in the refereed literature, aside from its inclusion in species lists or as notes mentioning its appearance in regional ichthyofauna. Tarpons in the eastern Atlantic range north to the Formigas, a group of small islands in the eastern Azores (Costa Pereira and Saldanha 1977) and the inshore waters of continental Europe including the Tagus River estuary of Portugal (Costa Pereira and Saldanha 1977), the Lee River of Cork County, Ireland (Twomey and Byrne 1985; Wheeler 1992), and the French Basque coast (Quero *et al.* 1982: 1022–1025). I could not find specific mention of Atlantic tarpons entering the Mediterranean, but surely they have.

Minimum water temperatures probably influence the distributions of Atlantic tarpons (Killam (1992: ix). Costa Pereira and Saldanha (1977) pointed out that north of Sénégal and south of Angola sea temperatures begin to cool and salinity rises, conditions they believed restrict the tarpon's latitudinal range in the eastern Atlantic. These regions are characterized by heightened rates of surface evaporation, lower seasonal rainfall, and weak fluvial flow into the Atlantic, factors that combine to keep coastal salinity values high and perhaps discourage inshore migration of metamorphosing tarpon larvae. Lower sea temperatures both to the north and south are the result of increasing latitude. The only elopiform listed by Penrith (1976) from Namibia and all of South Africa's western coast was the bonefish (*Albula vulpes*). Evidently this region falls outside the Atlantic tarpon's southern latitudinal range.

Costa Pereira and Saldanha (1977) did not mention that between Sénégal and Angola, where the "skull" of the African continent curves eastward, are sandwiched 13 countries with coastlines characterized by warm seas, high seasonal rainfall supporting tropical forests, mostly strong fluvial flow, an abundance of swamps and brackish lagoons, and other environments favorable for tarpons. The Atlantic tarpon's range is likely to be extended to southeast Asia at some future time now that specimens have been imported into Thailand and released into recreational fishing reservoirs (Chapter 7.2). Some individuals will inevitably escape into coastal waters or be released there. Atlantic tarpons are already established on the Pacific coasts of Panamá and Costa Rica,[2] having traversed the Panamá canal after its opening in 1914 (Anonymous 1975; Hildebrand 1939).

The Atlantic tarpon is essentially a straggler outside the subtropics, but the Pacific tarpon's normal range north and south seems broader, perhaps extending

[2] http://www.ticotimes.net/2011/07/06/tarpon-on-the-pacific-coast-you-betcha. Downloaded 18 July 2015.

Table 0.1 Length-length, length-weight, and otolith weight-age regressions for Atlantic tarpons from south Florida waters. Values of length in mm, weight (W) in kg, otolith weight (OW) in g, age in years (y). Length range for length-length regressions = 106–2045 mm FL; length range for length-weight regressions = 102–2045 mm FL; age range for OW weight-age regressions = 1–55 years (females) and 1–43 years (males). Source: Crabtree *et al.* (1995: 624 Table 2).

y	*x*	*n*	*a*	SE	*b*	SE	r^2
FL	SL	1342	10.8404	±0.6339	1.0423	±0.0007	0.999
FL	TL	1061	−10.8096	±0.8084	0.8967	±0.0007	0.999
SL	FL	1342	−9.9770	±0.6131	0.9588	±0.0007	0.999
SL	TL	1051	−21.1779	±1.0181	0.8606	±0.0009	0.999
TL	FL	1061	12.6345	±0.8937	1.114	±0.0009	0.999
TL	SL	1051	25.5839	±1.1622	1.1607	±0.0012	0.999
$\log_{10} W$	$\log_{10}$FL	1262	−7.9156	±0.0124	2.9838	±0.0045	0.997
$\log_{10} OW$ (females)	$\log_{10} Age$	193	−1.2083	±0.0199	0.5476	0.0152	0.872
$\log_{10} OW$ (males)	$\log_{10} Age$	106	−1.1734	0.0183	0.4614	0.0162	0.886

routinely into temperate waters. Wade (1962: 593) gave a latitudinal range in the western Pacific as Hamana Lake, Totomi Province, Japan (34.7°N, 137.6°E) to Victoria, Australia (≈37.0°S, 144.0°E). East to west, the species ranges halfway across the globe, from Tahiti in the Society Islands (17.7°S, 149.4°W) to Durban, South Africa (29.9°S, 31.0°E). Wade (1962: 594) gave its epicenter of abundance as the region encompassing "India, Ceylon, the Malay Archipelago, East Indies, southern Philippine Islands, and Polynesia." Ley (2008: 3) offered a similar range: 28°N (Japan) to 35°S (southern Australia and South Africa) and 25°E (east African coast) eastward to 171°W (Samoa). Most records of Pacific tarpons in the literature are from the Philippines and India (35% combined).

How the length of a fish is determined requires comment, because the terms defined in this paragraph are used throughout the book. A fish can be measured and its length expressed in several ways: *standard length,* SL (tip of the snout to end of the last vertebra); *fork length,* FL (tip of the snout to where the center-most rays of the caudal fin terminate); *total length,* TL (tip of the snout to end of the caudal fin, sometimes with the lobes compressed so they extend to maximum length); and *notochord length,* NL in early larvae, equivalent to standard length before skeletal development. If enough specimens of a species have been measured using SL, FL, and TL, any of these measures can be converted to the others, as shown for the Atlantic tarpon (Table 0.1). To Breder (1944: 219), TL was measured with the caudal fin lobes spread, and *overall length* was the term he used when they were compressed. I consider this to be TL too. In either case, TL includes the length of the caudal fin, and the literature is seldom specific about which method was used.

Acknowledgements

I thank the authors and publishers who allowed me to reproduce their figures and tables, and I thank Mote Marine Laboratory for use of facilities. I am grateful to Patricia E. Anyanwu of the Nigerian Institute for Oceanography and Marine Research for information on aquaculture of Atlantic tarpons, Michael Spotte for assisting with preparation of the figures, and Lucia Spotte for her patience and good humor while this work progressed. Special thanks is extended to Alyson Gamble, former Librarian at Mote's Arthur Vining Davis Library and Archives for her unwavering enthusiasm about all things academic, and for energetically tracking down obscure references, pursuing some of them literally to the ends of the Earth.

Symbols and abbreviations

≈	approximately
<	less than
≠	unequal
>	greater than
≤	less than or equal to
≥	greater than or equal to
°C	degree(s) Centigrade
µ	micro
µm	micrometer (micron)
µmol	micromole
A	age (days)
ABO	air-breathing organ (physostomous swim bladder)
AC(s)	accessory cell(s)
ASB	aquatic surface-breathing
bp	(years) before present
CC(s)	chloride cell(s)
CFTR	cystic fibrosis transmembrane conductance regulator
CI	confidence interval (statistics)
cm	centimeter(s)
COX2	cyclooxygenase type 2
d	day(s)
df	degrees of freedom (statistics)
DOM	dissolved organic matter
dph	day(s) post-hatch
F	fecundity
f_{ab}	air-breathing frequency (per unit time)
f_H	heart rate (beats per unit time)
fl	femtoliter(s) (10^{-15} l)
FL	fork length
g	gram(s)
G	growth (somatic)
GH	growth hormone
GSI	gonadal-somatic index
H	body height at its highest point
h	hour(s)
HA	H^+-ATPase
Hb	deoxygenated hemoglobin (g/l)

HbO_2	oxygenated hemoglobin (g/l)
Hct	hematocrit (%)
HL	head length
H_O	mean observed heterozygosity
H_S	genetic diversity value[1]
IGF-1	insulin-like growth factor 1
in.	inch(es)
kg	kilogram(s)
kPa	kilopascal(s)
L	body length (as FL, NL, SL, or TL)
L	liter(s)
lb	pound(s)
m	meter(s)
M	molar
MCH	mean cell hemoglobin (pg)
MCHC	mean corpuscular hemoglobin concentration (g/l)
MCV	mean corpuscular volume (fl)
min	minute(s)
mL	milliliter(s)
mm	millimeter(s)
mmol	millimole
$\dot{M}O_{2air}$	rate of oxygen uptake from air or ABO (mL/kg/min
$\dot{M}O_{2w}$	rate of oxygen uptake from water (mol/kg/min)
mol	mole
mOsm	milliosmole
MRC(s)	mitochondria-rich cell(s)
mRNA	messenger ribonucleic acid
ms	millisecond(s)
n	sample size
ng	nanogram
NHE	Na^+/H^+ exchanger
NKA	Na^+/K^+-ATPase
NKA1	secretory form of NKA
NKCC	$Na^+/K^+/2Cl^-$ co-transporter
NKCC1	secretory isoform of NKCC
NL	notochord length
Osm	osmole
OSTF1	osmotic transcription factor 1
O*W*	Otolith weight
oz	ounce
P	partial pressure
p	probability

[1] Nei's unbiased gene diversity across all loci.

P_{50}	PO_2 at which 50% of Hb is bound to O_2 (HbO_2)
P_a	partial pressure in arterial blood
Pa	pascal(s)
PAT(s)	pop-up archival transmitting tag(s)
pg	picogram (10^{-12} g)
pH_{cv}	caudal venous pH
ppt	parts per thousand
P_v	partial pressure in venous blood
PVC(s)	pavement cell(s)
P_w	partial pressure in water
$\dot{Q}$	cardiac output (ml/min/kg)
r^2	coefficient of determination
RBC	red blood cells (10^6/μl)
Rh(cg)	rhesus glycoproteins
Rhag	Rh-associated glycoprotein (ammonia transporter)
Rhbg	Rh-associated glycoprotein (ammonia transporter)
Rhcg	Rh-associated glycoprotein (ammonia transporter)
RSI	ram-suction index
S	salinity (mg/kg of seawater)
s	second(s)
S_A	absolute salinity scale (mg/kg of seawater)
SA	body surface area
SD	standard deviation
SE	standard error (slope of the regression)
SEM	standard error of the mean
SEM	standard error of the mean
SL	standard length
S_P	practical salinity scale (no units)
Su	Survivorship
TF2B	(osmotic) transcription factor 2B
TL	total length
TTPG	time to peak gape
U_{crit}	critical swimming speed (*l*/s)
V_{ab}	volume of a single breath of air
V_{ABO}	ABO volume (ml/kg)
vol%	vol/total vol × 100
V_R	(aquatic) ventilation rate (opercular pumps/min)
V_S	cardiac stroke volume (ml)
W	body weight, or mass (g)
W_r	relative weight (g)
y	year(s)
Σ	sum
$\overline{X}$	mean, or average
Z	Rayleigh's test of circular distributions

CHAPTER 1
Development

1.1 Introduction

The developmental biology of tarpons is so unusual that it seems a fitting subject for the opening chapter. I had originally intended to present the ontogeny of Atlantic and Pacific tarpons side by side, with salient features and timing emphasized at least partly in tabular form, but inconsistencies in the literature made it impossible. Specimens in the various reports were often measured at different lengths and unknown ages. Although captive rearing eliminates the age problem, it introduces confounding factors that can compromise normal rates of growth and development. Then, too, descriptions ranged in quality from detailed to superficial. Taxonomists sometimes favored particular characters, relegating others to lesser status or ignoring them. In short, I could not get comparative descriptions to line up without generalizing, which would have diluted the entire effort. What I present is therefore detailed, but in narrative form, with the two species treated separately.

Nonetheless, the pattern of their ontogeny is similar. The descriptions presented follow staging systems devised in the 1960s and 1970s by Brazilian and US scientists for Atlantic tarpons. Those for the Pacific species are less detailed and cohesive. The objective is to offer a detailed summary of tarpon ontogeny book-ended by separate sections devoted to just the leptocephalus larva.

1.2 The tarpon leptocephalus

There was a time when nobody knew what young tarpons looked like, but some still claimed to have seen them. Among the many tall tales is this whopper recorded in a letter from Charles H. Townsend to Mr. Grant and reproduced by Beebe (1928: 230). Townsend was traveling to the Galápagos Islands to capture

Tarpons: Biology, Ecology, Fisheries, First Edition. Stephen Spotte.

giant tortoises, probably for the New York Zoological Society's Bronx Zoo, when he penned this:

> *"In conversation with Mr. S. A. Venable of the Zone Police Force [Panamá Canal Zone Police], an experienced [Atlantic] tarpon fisherman, I was informed that the fish is* viviparous. *He has repeatedly observed the females seeking shallow water, generally less than 4 feet deep, where a continuous stream of young fish was poured from her vent, the young being apparently little more than ¼-inch long. The young immediately seek refuge in groups, under the large scales of the mother, each scale standing outward at an angle of probably 30°. The young clustered in these scale shelters as thickly as they could. Mr. Venable's many observations lead him to believe that the young shelter under the scales ten days or more, when they are ¾-inch long. The mother soon rids herself of the young by shaking herself and by leaping."*

Probably because the smallest tarpons he ever saw were juveniles, taxidermist and sportsman Victor Brown of Everglades City, Florida thought they hatch fully formed. In a letter to Kaplan (1937: 91), Brown wrote: "The newly spawned tarpons, 1 to 3 inches long, immediately commence to work their way entirely out of salt water into fresh water streams, into the multitude of small creeks and canals, some going as far inland as 25 miles from the Gulf [of Mexico]."

Contrary to these kinds of statements, baby tarpons do not emerge as miniature adults. They hatch from fertilized eggs as yolk-sac larvae before morphing into leptocephali, larval forms unique to relatively few species of fishes (Hulet and Robins 1989; Inoue *et al.* 2004; Wang *et al.* 2003). Greenwood *et al.* (1966) established the superorder Elopomorpha based on representatives of all its subgroups having leptocephalus larvae (Fig. 1.1). A *leptocephalus* is a bizarre shape-shifting creature, laterally compressed, transparent with a mucinous pouch, and described variously as ribbon-, band-, or leaf-shaped. Elopomorpha is a monophyletic group, the leptocephalus an elopomorph synapomorphy. Order Elopiformes (tarpons and ladyfishes) occupies the most basal place in elopomorph phylogeny, Albuliformes and a clade consisting of Anguilliformes and Saccopharyngiformes making up a sister group (Fig. 1.2). Smith (1989: 961–962) provided an abbreviated key to elopiform leptocephali occurring in the western North Atlantic.

What constitutes a "larval fish" has been standardized to some extent (e.g. Richards 2006). The traditional definition is the interval between hatching and absorption of the yolk sac, the post-larval stage extending from termination of the larval stage to appearance of juvenile characters. In Gopinath's (1946: 8) opinion, certain groups fail to conform with this progression. He listed specifically the bonefishes, ladyfishes, tarpons, left-eye flounders, and tonguefishes, "even though they are post-larvae according to the [accepted] definition", and termed them larvae instead, "since these [fishes] undergo a complete metamorphosis before the assumption of adolescent characters." In other words, by Gopinath's definition, a tarpon leptocephalus remains a larva to the moment it commences metamorphosis. Wade (1962: 548) considered the leptocephalus to the start of its metamorphosis as a post-larva, the larval period evidently restricted to the interval between hatching

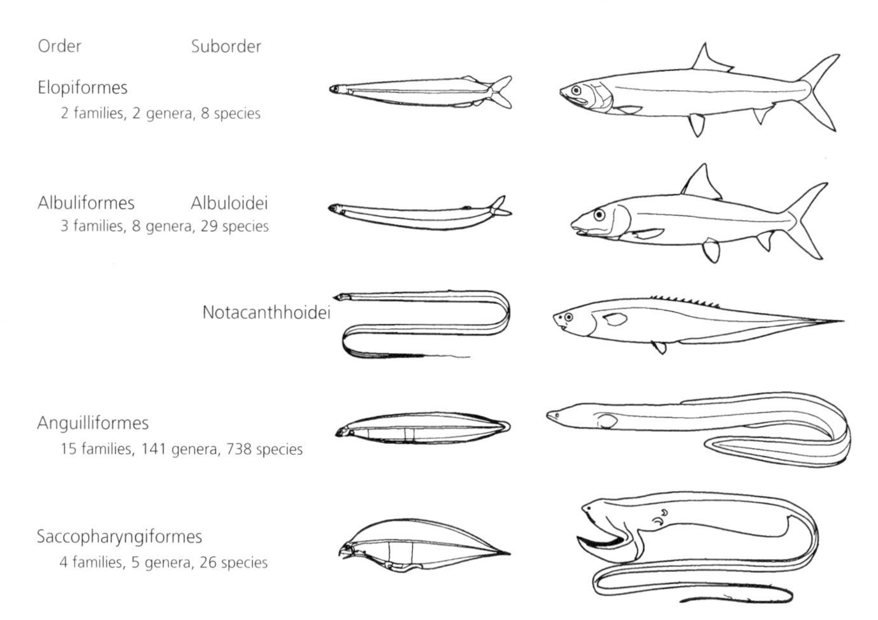

Fig. 1.1 Higher-level classification of orders in the Elopomorpha along with numbers of taxa presently included. Representative larval and adult body forms are illustrated for each group. The Elopiformes, to which the two extant species of tarpons (*Megalops atlanticus* and *M. cyprinoides*) belong, is represented by a ladyfish, of which six species exist (*Elops* spp.). Source: Inoue *et al.* (2004: 275 Fig. 1).

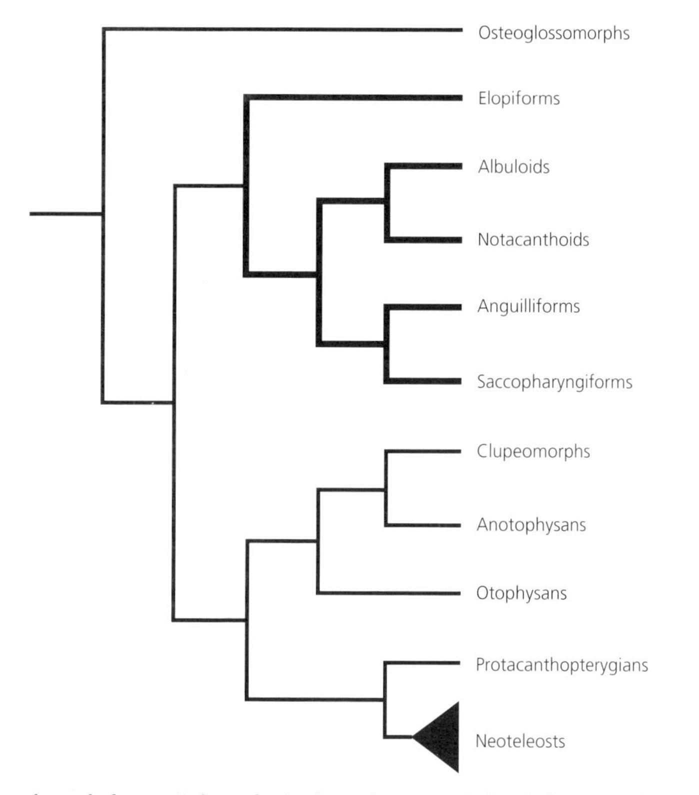

Fig. 1.2 A modern phylogenetic hypothesis about the monophyly of Elopomorpha. See source publication for history and details. Source: Inoue *et al.* (2004: 276 Fig. 2B).

and appearance of the leptocephalus (Stage 1, see below); that is, synonymous with the yolk-sac larva. So did Alikunhi and Rao (1951), although their terminology is less clear.

A more modern treatment of how a larval fish is defined (presumably a tarpon or any other) partitions the concept into four post-hatch stages in which *flexion* refers to when the notochord becomes flexible. These are: (1) yolk-sac; (2) pre-flexion (complete yolk-sac absorption and beginning of notochord flexion); (3) flexion (start of notochord flexion to its completion); and (4) post-flexion (end of notochord flexion and start of metamorphosis).[1] The last initiates post-larval transformation, or the metamorphic stage (start of metamorphosis to completion of fin-ray development and beginning of squamation), after which juvenile traits appear and development proceeds seamlessly to the adult form with eventual attainment of sexual maturity.

1.3 Staging tarpon ontogeny

To my knowledge, eggs and yolk-sac larvae of either species of tarpon have not been described. Anyanwu *et al.* (2010) of the Nigerian Institute for Oceanography and Marine Research purportedly obtained fertilized eggs collected in the wild, then hatched and reared them to the fry stage in laboratory aquariums (Chapter 8.6). Surviving fry were transferred to earthen ponds and grown to juveniles. Specific information was not provided. The ultimate goal was to develop these procedures so that a reliable source of fry could be available consistently to fish farmers.

The few details in this report are tantalizing and apparently unpublished formally, but if backed by adequate data would indicate that knowledge of early tarpon biology has advanced more quickly in western Africa than in the Western Hemisphere. For example, Anyanwu *et al.* (2010: 6) wrote, "The fertilized eggs are available in the coastal waters of Ondo State [Nigeria] which can be collected and hatched in the laboratory." They implied that fertilized eggs are recognized, collected, and cultured routinely by fish farmers (Chapter 8.6). The fertilized ova hatched after 24 hours, and early larvae measured 5.3–6.8 mm TL (5.0–5.7 mm SL). Plate 4 (p. 8) in their report is described as a photograph of the anterior half of a yolk-sac larva. Hatchlings experienced heavy mortality after 5 days, which Anyanwu and colleagues suggested could have resulted from a lack of appropriate food. This is doubtful, considering that evidence of feeding has been found only after metamorphosis (Section 1.6, Chapter 7.7, Appendix B).

In discussing subsequent larval stages of Atlantic tarpons, I rely mainly on descriptions of Jones *et al.* (1978: 53–62), which evidently were compiled from

[1] http://access.afsc.noaa.gov/ichthyo/StageDefPage.php. Downloaded 10 February 2015.

other sources, notably Mercado Silgado and Ciardelli (1972) and Wade (1962). Also see Mercado Silgado (1969, 1971) and Moffett and Randall (1957).

The protocol for staging tarpon leptocephali is clear through Stage 2, but Stage 3 can be confusing. To Mercado Silgado (1971) and later Cyr (1991: iv, 6), development of Atlantic tarpons comprises two larval stages. Cyr called them phases instead of stages. His Phase 1 is the equivalent of Stage 1 of other authors; his Phase 2 commences at the start of shrinkage (beginning of Stage 2) and continues until positive growth is resumed, or the beginning of Stage 3. This too conforms with the staging protocol adopted by other investigators. Wade (1962: 548), for example, wrote, "The period of initial length increase to the size at which shrinking begins is considered as *Stage I*. In *Stage II* the body gradually looses [*sic*] the 'leptocephalus' form while it is shrinking in length." To Gehringer (1958) the Atlantic tarpon larva consists of Stage 1 exclusively. However, he called Stage 2 a "metamorphic larva." Such terminology is barely useful if both are thought to be larvae, but such inconsistencies are unavoidable. Obvious interruptions in the developmental sequence are seldom clear, and at times my own descriptions of staging might seem equally vague or confusing.

Cyr cited Hardy (1978) as the source of his staging protocol, but the reference should be Jones *et al.* (1978), in which Hardy is listed as third author. They described what at first reading could be three larval stages, but of the four specimens depicted as representing Stage 3 (Jones *et al.* 1978: 60 Figure 28), the top two illustrations (Figure 28A, B) are labeled larvae, the bottom two (Figure 28C, D) juveniles. This is intentional, not an error or misprint. Under the heading "Larvae" (their p. 53), they defined Stage 3 as "a second period of length increase *which terminates with the onset of the juvenile stage* [emphasis added]." Thus they considered early Stage 3 tarpons – both Atlantic and Pacific – to still be "larvae," but the point at which the transition into juveniles happens is less exact. Wade's Stage IIIA for Pacific tarpons correlates directly with the top two Atlantic tarpons depicted by Jones *et al.* (1978: 60 Figure 28A, B), based on Harrington's (1958) work; that is, the transitional state during which body proportions switch abruptly from allometric to isometric growth (for a discussion of these terms, see Chapter 2.2).

Wade did not examine Pacific tarpons of what he called Stage IIIA (i.e. > 25 mm SL), writing (Wade 1962: 549): "Fish larger than 40 mm [SL] are without a full complement of scales, gill rakers, branchiostegal rays and the dorsal whip [the extended last dorsal ray] until they attain a length of about 130 mm [SL], but are easily distinguished as young tarpon." He considered these fish to be juveniles, designating them as Stage IIIB. Nor did he examine any Stage IIIB specimens of Pacific tarpons. As to the Atlantic tarpon, Wade (1962: 574) noted "a gradual change from allometry to isometry at the end of the Stage IIIA period." Harrington's (1958) and Wade's findings lead me to conclude that both Atlantic and Pacific tarpons commence the juvenile stage at ≈45 mm SL, the length at which growth in most body proportions (as percentage or fraction of SL) switches from allometric to isometric.

To Mercado Silgado (1969: 4; Table 1 1971: 12 and Figs. 1–4), Stage 3 represented "fry," and he cited Harrington (1958, 1966), Rickards (1968), and Wade (1962) as sources in his 1971 publication. In the 1969 report (his pp. 4–5), Mercado Silgado listed five stages, calling Stage 3 *Crecimiento Alevínivo* (i.e. Growth of Fry).[2] He termed Stage 4 *Crecimiento Juvenil* (Juvenile Growth). His Stage 5 described the adult. With few exceptions (e.g. Anyanwu and Kusemiju 2008), those writing in English have seldom applied the term "fry" to young tarpons, but its adoption might prove a useful descriptor for the stage immediately preceding the juvenile in both species if defined like this: *Tarpon fry have resumed growth at the start of Stage 3 (≈ 13.0 mm SL) and continue increasing in size until proportional morphometric growth shifts from allometric to isometric at ≈45.0 mm SL. A juvenile tarpon of either species is a Stage 4 fish ranging from ≈45.0 mm SL to onset of sexual maturity, throughout which proportional morphometric growth ceases to be allometric and becomes isometric.*

Harrington (1958: 3) investigated this division in the life history of Atlantic tarpons in detail using a large series of specimens generally classified as larvae and juveniles, noting that "The differential ... growth of body parts and regions clearly reveals a transitional period ...." He compared morphometric measurements of his fish with an earlier series of young Atlantic tarpons examined by Breder (1944) and found extreme allometry of body proportions in specimens of 16–19 mm SL that continued to ≈35–40 mm SL, "when it gradually resolves itself into what is essentially *incrementum in universum*." He continued: "The precise point at which allometry yields to isometry is not obvious, and if the latter is not complete, it is no less so than in Breder's 164 specimens, which ranged from 50 mm to 2030 mm in standard length, and in which growth was deemed only slightly heterogonic [i.e. allometric] ...."

The combined series covered a large range (Harrington's from 16.0–109 mm SL). Breder had taken 18 morphometric measurements (e.g. dorsal fin origin, pelvic fin length, head length, and so forth) and presented the values as a percentage of standard length. Harrington measured the same characters. Breder's data showed negative allometry in all proportions except the last dorsal fin ray, which was conspicuously positive. Harrington's were negatively allometric only in distances from snout to origins of the dorsal and anal fins up to 35–40 mm SL.

[2] *Alevín* is sometimes translated from Spanish as the "fry" of a fish, but the term is often not specific and can simply mean "young fish." Mercado Silgado (1969: 4 and Table 1) used *alevínivo*; the term applied by Mercado Silgado and Ciardelli (1972: 157 Table 1) was *alevínico*. However, *alevin* in English ordinarily refers to yolk-sac larvae of salmonids (e.g. Hasler *et al.*, 1978, Helfman *et al.*1997: 136, Moyle and Cech, 1982: 244, Varsamos *et al.* 2005). To Bond (1979: 421), *alevin* was more general: "If yolk-bearing larvae transform directly into a juvenile [*sic*], as is the case in many salmonids and certain sculpins, these larvae are called alevins." Still other writers (e.g. Alderdice, 1988: 175) simply referred to *alevin* without a definition, evidently assuming the reader knows what it means. Tarpon larvae are excluded in any case because of their intermediate leptocephalus stages.

He wrote (Harrington 1958: 4): "Thus in the earliest growth the majority of the obvious body proportions show extreme positive allometry with reference to standard length, all these proportions then becoming isometric at about 35–40 mm. standard length, and thereafter all but one of them showing slight but unmistakable negative allometry." A Stage 3 fish and one approaching Stage 4 are shown in Fig. 1.3. How these changes became incorporated into growth is illustrated here diagrammatically (Fig. 1.4). The larger fish shown was 36.8 mm SL, and a photograph of it can be seen in the bottom figure of Harrington's Plate II. A single row of scales had formed recently, and a second row was just becoming apparent.

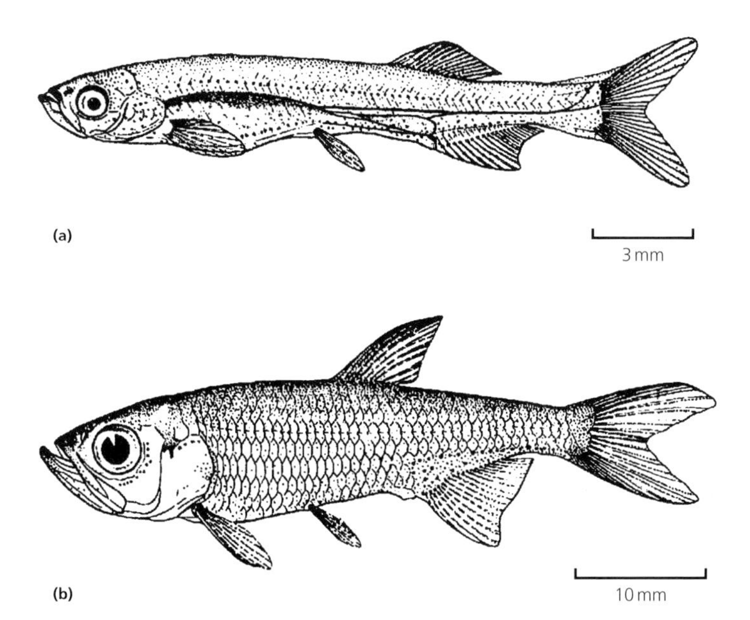

Fig. 1.3 **(a)** Atlantic tarpon fry, Stage 3 phase X (16.9 mm SL). **(b)** Atlantic tarpon in late Stage 3 (41.0 mm SL) approaching the end of allometric growth. Source: Mercado Silgado and Ciardelli (1972: 181, Fig. 10A, B).

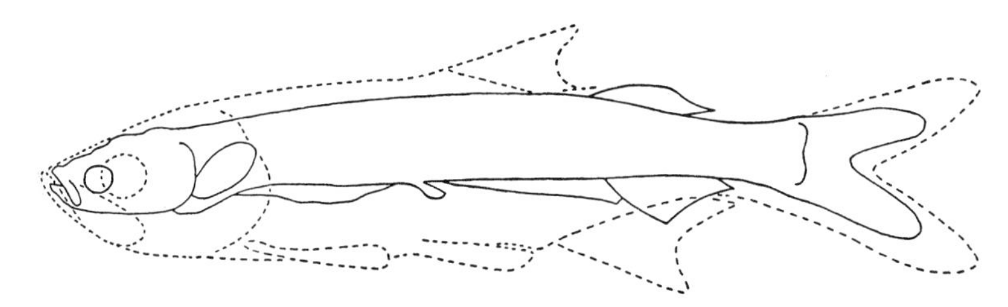

Fig. 1.4 Profiles of an Atlantic tarpon 36.8 mm SL (broken lines) and an earlier specimen of 16.0 mm SL (solid lines), the second superimposed onto the first and enlarged proportionately so that standard lengths of the two illustrations coincide. Source: Harrington (1958: 5 Fig. 2).

In discussing Stage 4, Mercado Silgado (1969: 4) wrote: "*Se observa en este Estado, el nuevo crecimiento y la verdadera morfología de un sábalo adulto. Es aquí donde se empieza a notar las escamas y la prolongación del último radio de la aleta dorsal, llegándose a observar la aparición de este radio claramente cuando el animal alcanza una longitude aproximada de 71 mm de longitud standard en el laboratorio.*" [This stage reveals the true morphology of an adult tarpon. It is here where you start to notice the scales and begin to clearly see the extension of the last ray of the dorsal fin, this ray becoming clearly evident when the animal reaches ≈71 mm SL in the laboratory.]

His description of Stage 4: "*Este Estado abarca los juveniles de sábalos en el momento en que aparece la prolongación del último radio de la aleta dorsal hasta una longitud aproximada a los 1000 mm de longitud standard que es cuando el sábalo pasa a ser adulto por llevarse acabo a esta longitud aproximadamente su primer desove.*" [This stage encompasses juvenile tarpons from the moment when prolongation of the last dorsal fin ray is apparent to ≈1000 mm SL; that is, when the tarpon becomes an adult, approximately the length at its first spawning.]

The last ray of the dorsal fin is diagnostic of adult tarpons of both species (e.g. Fig. 1.5). Mercado Silgado's Stages 3 and 4 are identified mainly by development of the last dorsal fin ray, and a fish in Stage 3, although still a juvenile, begins to resemble the adult. Keep in mind that for Stage 3 this conclusion is correct by his definition because Stage 3 is extended to 71 mm SL and based essentially on a single character (appearance and elongation of the last ray of the dorsal fin). These observations scarcely compare with Harrington's important finding that a tarpon longer than ≈45 mm SL ceases to grow allometrically. Consequently I would define Stage 3 as encompassing 13.0–45.0 mm SL instead of the range 13.0–71.0 mm SL recommended by Mercado Silgado and Ciardelli (1972), as reflected here in Table 1.1. An Atlantic tarpon between ~45 and ~1000 mm SL therefore can be considered a juvenile, unless evidence can be found that specimens within any part of this range are sexually mature. Length at maturity is much less for Pacific tarpons, perhaps as short as 300 mm FL (Chapter 3.6).

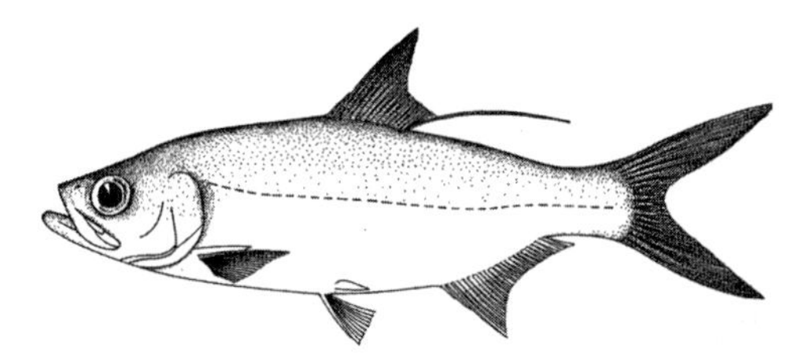

Fig. 1.5 Illustration of an adult Pacific tarpon showing the elongated last ray of the dorsal fin. Source: Food and Agriculture Organization of the United Nations. 1984. Bianchi (1984: 3). *Field Guide: Commercial Marine and Brackish Water Species of Pakistan* by G. Bianchi. FAO Species Identification Sheets for Fishery Purposes, Project UNDP/FAO Pak/77/033. Rome, Italy. Reproduced with permission.

Table 1.1 Growth stages and phases partitioned by length of Atlantic tarpon leptocephali (Stages 1 and 2) and fry (Stage 3). Stage 3 has been modified to 13.0–45.0 mm SL based on Harrington's (1958) finding that allometric growth ceases at ≈45 mm SL. See text for an explanation of phases. Source: Mercado Silgado and Ciardelli (1972: 157 Table 1).

Stage	Phase	Length (mm SL)
Stage 1	I and II	1.7–11.0
	III	11.0–17.5
	IV	17.5–24.0
	V	24.0–28.0
Stage 2	VI	28.0–25.0
	VII	25.0–20.0
	VIII	20.0–15.0
	IX	15.0–13.0
Stage 3	X	13.0–71.0 [13.0–45.0]

Mercado Silgado (1971) and Mercado Silgado and Ciardelli (1972) further partitioned the first three developmental stages into 10 phases identified by Roman numerals (Table 1.1): Stage 1 (phases I–V), Stage 2 (phases VI–IX), and Stage 3 (phase X). The second report, published in Spanish, is careful, detailed, and unavailable in English. I translated their descriptions of Atlantic tarpon developmental stages (see Appendix B). Their effort adds depth to the original staging systems of others.

Mercado Silgado (1971) did not mention a tenth phase or describe fry development. However, Mercado Silgado (1969: unnumbered page Table 1) listed Stage 4 (juvenile) as consisting of phase XI and the adult (Stage 5) as encompassing phase XII. These last two were eliminated by Mercado Silgado and Ciardelli (1972) in their final staging system.

Changes in morphology in the following sections are described as occurring at approximate body lengths. I emphasize that length alone is an unreliable predictor of the age of a leptocephalus and therefore not representative of its true ontogenetic status. Growth varies by individual, and so does the timing of metamorphosis. Increments formed on the *otoliths*, or "ear stones," are sometimes used to estimate the age of fishes (e.g. Chapters 2.6, 7.2). These are hard structures in the vestibular labyrinth consisting mainly of calcium carbonate and lesser concentrations of other elements embedded in a matrix of minor organic components.

Shenker *et al.* (2002), for example, examined otoliths of 41 Atlantic tarpon leptocephali caught at Sebastian Inlet, Florida as they entered Indian River Lagoon from the Atlantic Ocean during summer 1995, finding no correlation between length (15.5–22.1 mm SL) and age (15–26 days, $\bar{x}$ = 20.2 days). The oldest larvae (24–26 days) included both the shortest specimen and some of the longest (>20 mm SL). Thus the time when metamorphosis commences seems not to follow a particular pattern.

Findings of Tzeng *et al.* (1998: 182) for Pacific tarpon leptocephali were similar. Based on otolith counts, leptocephali entering Gongshytyan Brook estuary, northern Taiwan between 15 and 24 September 1995 were 20–39 days old ($\bar{x}$ = 28.5 days) and had already begun metamorphosis, meaning that some were twice as old as others. The authors noted that length on arrival was independent of age and that age related inversely to growth rate, implying differential rates of growth offshore through Stage 1 to onset of metamorphosis. The conclusion: "Slower-growing fish apparently metamorphosed later, and faster-growing fish arrived in the estuary earlier, than did [*sic*] slower-growing ones." If subsequent growth at inshore "nurseries" indeed offers survival advantages (Chapter 4.6), early penetration of lagoons and estuaries would appear to enhance fitness.

As touched on above, larval tarpons of both species undergo sequential growth stages (also called stanzas) during which organs and structures develop as metamorphosis proceeds. When experiencing metamorphosis, an organism advances to the next developmental stage through changes in shape and size. As mentioned, an Atlantic tarpon's early development encompasses three such stages encompassing radical changes. Because these occur along a continuum rather than abruptly, metamorphosis is like a time-lapse film as the animal shape-shifts, its appearance blending smoothly through one stage and into the next as certain features arise and others fade from view. Partitioning metamorphosis into stages is inevitably artificial and misleading. The depiction of leptocephali caught in a plankton tow (Fig. 1.6) or by some other means are snapshots in time, single frames extracted from a running film.

Hildebrand (1934) caught what he believed was a larval Atlantic tarpon – a leptocephalus – in transition to becoming a juvenile. The specimen, obtained at the mouth of Core Creek, Beaufort, North Carolina might have been the first found in the Western Hemisphere, but it was inadvertently destroyed before a drawing could be made. Only Hildebrand's cursory description remains.

1.4 Development of Atlantic tarpons

Some areas of the body where morphometric measurements of young Atlantic tarpons have been described are illustrated diagrammatically in Fig. 1.7.

Stage 1 – Growing leptocephalus

Stage 1 is a period of growth taking place offshore and characterized by transparency, a ribbon-like body, and large fang-shaped teeth. It encompasses specimens of 1.7–28.0 mm SL (Table 1.1). Cyr's (1991: 11) Stage 1 specimens (Phase 1 in his terminology), which included data from two Gulf of Mexico cruises, were 6.3–23.8 mm SL and 5.2–30 days old. Wade (1962: 555) reported fish of 11.0 and 11.7 mm SL. Spawning and early development into Stage 1 take place entirely offshore in full-strength saline waters (e.g. Cyr 1991: 17; Jones *et al.* 1978: 53; Smith 1980) and culminates in a completely formed leptocephalus.

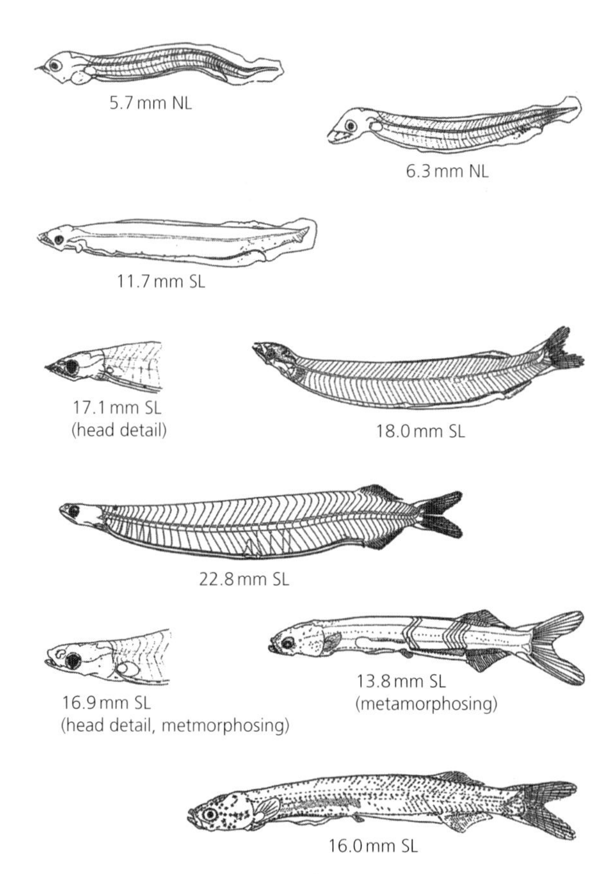

Fig. 1.6 Stages 1 and 2 Atlantic tarpon larvae. Source: Fahay (2007: 13). © Northwest Atlantic Fisheries Organization. Reproduced with permission.

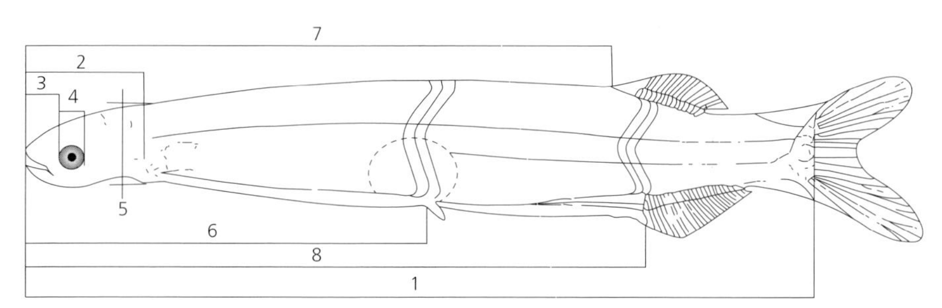

Fig. 1.7 Atlantic or Pacific tarpon larva, depicting where some (but not all) measurements are typically taken. Numbers indicate the following measurements (mm) or counts (9–15 not illustrated): 1 – *standard length* (SL); 2 – *head length* (HL) tip of snout to posterior fleshy margin of operculum; 3 – *snout length*, tip of snout to anterior edge of bony orbit; 4 – *eye diameter*, anterior inner edge of bony orbit to posterior inner edge of orbit; 5 – *depth*, angle of base of pelvic fin vertically to dorsal outline of body; 6 – *prepelvic length*, tip of snout to origin of pelvic fin; 7 – *predorsal length*, tip of snout to origin of dorsal fin (or dorsal fin fold); 8 – *preanal length*, tip of snout to origin of anal fin (or posterior edge of anus); 9 – *fin-ray counts*; 10 – *total myomere counts*, from anterior-most to last myomere in caudal area, these last becoming indistinct when hypural plate forms; 11 – *prepelvic myomere counts*, from anterior-most myomere to myomere the ventral extremity of which approximates origin of pelvic fin; 12 – *predorsal and preanal myomere counts*, same as 11 above; 13 – *lateral line scales*, counted from opercular flap to posterior scale of caudal fin; 14 – *teeth*, number on each side of upper and lower jaws; 15 – *gill rakers*, number (including rudiments) on upper and lower limbs of first gill arch on one side. Source: Wade (1962: 551 Fig. 1, 552, 615–616 Table 3, 619–622 Table 5).

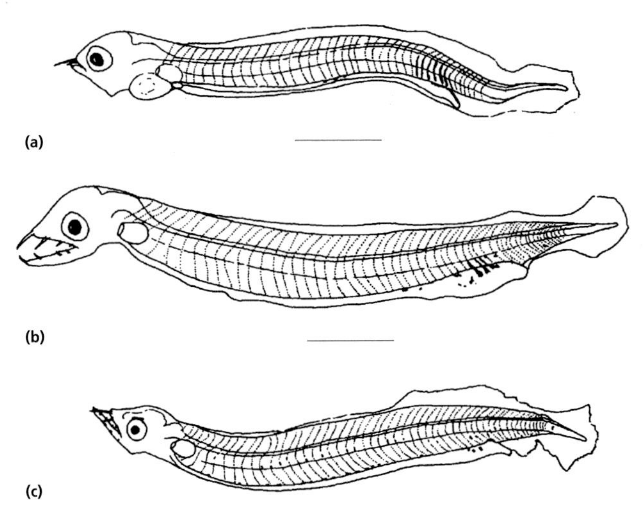

Fig. 1.8 Early Stage 1 Atlantic tarpons from the Yucatán Channel, Mexican Caribbean, and Gulf of Mexico (exact collection locations unclear). **(a)** 5.7 mm NL, **(b)** 6.3 mm NL, **(c)** 8.1 mm NL). Scale bars = 1 mm. Source: Smith (1980: 138 Fig. 2).

As mentioned, eggs and yolk-sac larvae (pre-Stage 1) have not been identified, but this statement might be true only for Atlantic tarpons in the Western Hemisphere. Floating masses of fertilized eggs have reportedly been collected and photographed off western Nigeria (Anyanwu and Kusemiju 2008: 120 Figure 9.4). Among the shortest tarpons so far recovered in the western Atlantic was a recently hatched specimen from the Gulf of Mexico (Smith 1980). It measured 5.7 mm NL and retained remnants of a yolk sac (Fig. 1.8a). The yolk sac evidently disappears by ≈6.0 mm NL (Smith 1980). Cyr (1991: 11) reported specimens of 8.1–23.2 mm SL (age 9.5–30 d, n = 29) and 6.3–23.8 mm SL (5.19–22.25 d, n = 103) for Stage 1 leptocephali caught during his cruises. Stage 1 is variable and lasts ≈30–40 d (Cyr 1991: iv), and growth is linear over days 7–24 post-hatch (Cyr 1991: 13). Cyr (1991: 15) speculated that if growth is asymptotic before subsequent metamorphosis, Stage 1 could be prolonged substantially.

Principal sources used in descriptions: Jones *et al.* (1978: 53–57) for leptocephali of 9.4–27.9 mm SL and the original descriptions of Harrington (1958, 1966); Mercado Silgado and Ciardelli (1972: 159–166); and Wade (1962: 555–559). Also see Chacón Chaverri and McLarney (1992, Appendix C); Gehringer (1958); Mercado Silgado (1969: 4, 1971: 9–10); and Smith (1980).

Meristic description: Fin rays: dorsal 12–13, anal 20–22, caudal 17 (at 11.7 mm SL, 19 at ≥17.5 mm SL). Myomeres (at 22.0–27.9 mm SL): total 54–57, predorsal fin 37–42, preanal fin 40–43, prepelvic fin 22–24, at swim bladder 21–25. Teeth (at 9.4–22.0 mm SL): upper 1 + 7 to 0 + 3, lower 1 + 6 to 1 + 3. Vertebrae (at 17.5 mm SL): 7 hypural plates.

Body proportions as percentage of SL: At 9.4–22.0 mm SL, height at pectoral fins 5.1–8.5, snout length 2.9–4.9, horizontal eye diameter 2.0–3.2; at 9.4–27.9 mm

SL, head length (HL) 8.2–14.5, preanal fin length 77.6–88.0; at 21.3–27.9 mm SL, preventral fin length 49.2–55.9.

Body proportions as percentage of HL: At 11.0–21.3 mm HL, snout length 24.4–31.3, horizontal eye diameter 17.7–23.5.

Narrative description: Body ribbon-like early in Stage 1, elongated, thin laterally, deep; head small, triangular, eel-like, wider than body in dorsal aspect; brain clearly visible; eye nearly round; snout sloping gently from top to tip; upper body height reduced at pectoral region by 17.5 mm SL; body compressed laterally at 23.0 mm SL, thicker along whole length and not ribbon-like; at 24.0–27.9 mm SL height greatest at pelvic fins and has decreased at caudal peduncle and pectoral fin region. Head still triangular when viewed dorsally and wider than body to at least 17.5 mm SL, shifting at 23.0 mm SL from eel-like to bullet-shaped, losing triangularity and now slightly wider than body when viewed in dorsal perspective; width nearly uniform except for bulge at eyes; snout rounded. Snout more pointed by 27.9 mm SL, cartilaginous structures evident in posterior operculum. Nostrils visible as shallow depressions at 17.5 mm SL, evidently still not bifurcate. Mouth large early in Stage 1, oblique and extending to pupil, lower jaw protruding at 11.7 mm SL, jaws equal at 17.5 mm SL. Mouth smaller by 23.0 mm SL, gape much shorter. First tooth in upper jaw fang-like; posterior teeth needle-like, uniform in diameter, in a single row extending to angle of gape; teeth of lower jaw thicker, anterior pair evidently not set in jaw; teeth absent by 23.0 mm SL, and cartilage developing in maxillary and mandible. Eye nearly round at 11.7 mm SL, oval at 17.7 mm SL. Gill filaments well-formed at 23.0 mm SL, but gill rakers absent. Dorsal finfold originating ≈66% of body length behind head; caudal finfold truncated, margin invaginated dorsally and ventrally anterior to urostyle at 11.7 mm SL. Finfold reduced to remnants anterior to caudal fin at 21.3 mm SL. At 11.7 mm SL, 8 probable ray bases in dorsal finfold visible opposite myomeres 41–44. At 13.4–14.0 mm SL, 8 incipient dorsal fin-ray buds appear, the fin rays first seen at 20.3 mm SL. By 23.0 mm SL, 12th dorsal fin ray splits, posterior half slightly elongated. At 24.0 mm SL, origin of dorsal fin apparent at myomere 42. At 11.7 mm SL, an opaque area is visible in the postanal area of the median finfold, perhaps indicating a developing anal fin; at 13.4–14.0 mm SL, 14–15 incipient anal fin-ray buds can be seen, and rays are obvious at 20.3 mm SL. At 27.9 mm SL, an incipient anal fin is visible underneath myomere 44. At 17.5 mm SL, caudal fin forked with unbranched rays, but start of branching apparent by 23.0 mm SL. Pectoral fin a convex bud at 11.7 mm SL, a little larger by 21.3 mm SL. Pelvic fin buds visible at 20.0 mm SL and present at myomere 24 by 23.0 mm SL. Developing vertebrae visible at 11.7 mm SL, and urostyle prominent and tipped slightly upward, the angle becoming steeper by 17.5 mm SL. Tubular gut extending ≈75% length of body at 11.7 mm SL, terminating at anus opposite myomeres 44–47; at 17.2 mm SL, gut is slightly looped, or indented, at ventral surface just anterior of vent. By 24.0 mm SL, heart located posteroventrally to pectoral fin in the shape of a figure eight, but nonfunctional. The swim bladder, which develops as an out-pocket of the esophagus, is apparent by 11.7 mm SL at myomeres 22–23, expanding slowly, and at 21.3 mm SL resembling a short cylindrical sac arising from the digestive tract at myomeres 23–24. Swim bladder extends dorso-caudally ≈ 33% of the distance to the central nerve cord, extending ≈66% of the distance by 23.0 mm SL and now stretched between myomeres 23 and 25. At 17.5 mm SL, kidney located dorsally to gut between myomeres 35 and 41, appearing larger by 23.0–27.9 mm SL, extending from myomeres 35–45 and now separated from posterior end of digestive tract.

Stage 1 larvae are lightly pigmented, possessing a few scattered melanophores on the posterior area of the gut dorsally to the central nerve cord at 11.7 mm SL. Also at this length, three chromatophores are evident on the ventral surface of the opercula, six on the dorsal border of the gut anterior to the swim bladder, and one on the swim bladder itself. From ≈ 13.4 mm SL to the end of Stage 1, dense, dark brown chromatophores show up as a fringed patch curving over the eyeball when viewed from a dorsal perspective, and small patches of chromatophores are occasionally evident on the fleshy margin below the eye. By 17.1 mm SL, a series of elongated chromatophores is visible along the dorsal edge of the intestine; a few others are scattered over the posterior part of the intestine and above the anus, and a series of elongated chromatophores can be seen on the myosepta below the midline. The caudal fin has a few chromatophores, and one exists below the pectoral fin. By 22.8 mm SL, about five lines of pigment are apparent below the lateral line on the caudal peduncle. By 24.0 mm SL, there is one stellate chromatophore on the lower head anterior to the heart, one on the heart, and 3–4 behind the heart. A row of elongated chromatophores extends along the dorsal surface of the gut to where the intestine and kidney separate; about four chromatophores can be seen above the kidney. A series of melanophores is visible at the base of the anal fin rays, as are four lines of melanin in dorsoventral alignment below the lateral line on the caudal peduncle. By 27.9 mm SL, dorsoventral lines are apparent on a minimum of five myomeres in a J-shaped pattern on the caudal peduncle.

Duration: Based on back-calculated hatch dates, Cyr (1991: 12, 26 Figure I-7) gave the estimated duration of Stage 1 as 33–51 days (95% CIs, $\bar{x}$ = 38 days, n = 29, 1981 cruise) and 27–29 days ($\bar{x}$ = 28 days, n = 103, 1989 cruise). Estimates based on counts of otolith increments: 15–32 days ($\bar{x}$ = 23.5 days ± 3.77 SD, n = 23). Smith (1980) had earlier proposed 60–90 days, but his sample size was small (n ≈25), and collections had been made at far-flung locations.

Stage 2 – Shrinking leptocephalus

Often called the "metamorphic stage," although changes that are obviously metamorphic continue through Stage 3 and early Stage 4. Growth stops drastically during the second larval stage, and tarpon larvae shrink, a startling example of what some have called "negative growth," an oxymoronic term. Stage 2 demonstrates that length alone is an unreliable diagnostic feature of tarpon ontogenesis. Wade (1962: 548) justified defining growth as change in morphology with age, not necessarily accompanied by increased size. To me, this stretches the definition beyond usefulness and its original descriptive intent (Chapter 2). I consider Stage 2 strictly a period of shrinkage accompanied by morphological change, but not "negative growth" or growth in any sense.

Stage 2 ordinarily involves specimens of 28.0–13.0 SL (Table 1.1). *Note in the descriptions below that when leptocephali shrink during Stage 2, length ranges are reversed and instead of increasing they diminish.* Stage 2 is thus notable for dramatic reduction in overall size and characterized by diminishing length, gradual loss of the ribbon-like form, and anterior shifting of the fins. Stage 2 occurs almost exclusively inshore (Cyr 1991: 17; Jones *et al.*, 1978: 53; Mercado Silgado 1971; Mercado Silgado and Ciardelli 1972; Smith 1980; Wade 1962; Zerbi *et al.* 2001). Exactly when the larval stage terminates and the next stage begins, as assessed from otolith increments, is often not obvious. Stage 2 is accompanied by a loss in length of > 14 mm SL (>40%) over two weeks. Richards (1969) described and illustrated the first Stage 2 Atlantic tarpon larva recovered from west Africa in the eastern Atlantic.

Principal sources used in the descriptions: Jones *et al.* (1978: 58–61) for leptocephali of 27.3–13.0 mm SL. Also see Mercado Silgado and Ciardelli (1972: 167–182) and Wade (1962: 559–561).

Meristic description: Fin rays: dorsal 12–13, anal 20–22, caudal 17. Myomeres: total 55–57, predorsal 42–36, preanal 43–38, prepelvic 24–21, at coelom 14–24. Vertebrae (at 17.0 mm SL): ≈6 hypurals.

Body proportions as percentage of SL: at 23.7–16.9 mm SL, height at pectoral fins 6.8–9.5, snout length 3.8–4.7, horizontal eye diameter 1.7–2.5; at 27.3–13.0 mm SL, head length 9.2–26.9, prepelvic fin length 53.0–48.4, preanal fin length 83.7–71.5, predorsal fin length 79.3–69.2.

Narrative description: Body height decreases from 27.3 to 25.0 mm SL. At 17.0 mm SL, height at pectoral fins increases markedly, and bottleneck-like appearance disappears. At 27.3 mm SL, snout's dorsal concavity is almost gone; the cranial bones are visible. At 20 mm SL, mouth has shifted dorsally, and jaws are longer. By 17 mm SL the lower jaw is a little longer than the upper, head larger in relative size, and a swelling is visible between developing mandibles in area of the future gular plate. By 15 mm SL, nostrils well developed, gular plate forming; teeth present. At 15–13 mm SL, eyes have become more rounded, upper and lower jaws well formed. By 17.0 mm SL, dorsal and anal fins have become longer and higher; some rays of caudal fin are branched; pectoral fins are larger and more pointed, their fleshy bases reduced. At 15 mm SL, lobes of the caudal fin are symmetrical. At 15.0 mm SL, gut has formed completely; scales still absent. At 16.9 mm SL, a slight loop, or indentation, still evident in ventral surface of the gut just anterior to vent.

At 27.3 mm SL, heart clearly visible; by 20 mm SL, circulatory system has become functional. The swim bladder, oval in shape, is apparent by 27.3 mm SL above myomeres 23–24, extending from myomeres 24–27 at 21.1 mm SL. At 17.0 mm SL, swim bladder has become more inflated and extends forward to myomeres 20–21.

Pigmentation at 27.3 mm SL similar to that of 27.9 mm SL specimens of Stage 1; two melanophores apparent above dorsal area of swim bladder. At 25 mm SL, dorsal part of caudal peduncle has two stellate chromatophores, ventral part with dorsoventral lines that in some specimens have lost the J-shape. Pigment has become denser on the head and extremes of upper and lower caudal lobes and is now visible over hypurals, on pectoral fins and central nerve cord, and ≈75% of kidney; pigment has merged over swim bladder forming a single patch. By 23.7–16.9 mm SL, a series of elongated chromatophores visible following dorsal edge of intestine. A series of small chromatophores is present on intestine above anus and another on myosepta at midline. A few chromatophores visible on myosepta above kidney and on anal fin, and small chromatophores apparent on dorsal surface of swim bladder. A few others have appeared on the anal, dorsal, and caudal fins, and one below the pectoral fin. Pigment can be seen above and on the eyeball and its fleshy margin below. One specimen had four subsurface and one surface chromatophores on or below the lower part of the brain. By 15 mm SL, body has become more opaque, its dorsal and lateral portions silvery. Pigment is concentrated along top of head and extends along dorsal surface of body; coelom densely pigmented. At 13 mm SL, eye is black.

Duration: Cyr (1991: 16) estimated Stage 2 to last ≈ 14 d, but it often has no discernible endpoint. Based solely on counts of otolith increments, Cyr (1991: 12–13) estimated 5–24 days ($\bar{x}$ = 14.2 ± 4.25 SD days, n = 23). However, note the extensive range and large standard deviation. He mentioned that in estimating ages of both Stage 1 and 2 larvae, checks (i.e. discontinuities; see Casselman 1983: 2) in otolith microstructures were difficult to interpret and therefore subject to error (Cyr 1991: 15).

Stage 3 – Fry

Principal sources used in the descriptions: Jones *et al.* (1978: 61–62) for leptocephali of 12.6 to ≈25 mm SL. Also see Mercado Silgado and Ciardelli (1972: 182–183); Wade (1962: 561–566).

Stage 3 is defined by renewed growth and characterized by increasing length and dramatic alteration of body form. To Mercado Silgado and Ciardelli (1972) it encompasses specimens of 13.0–71.0 mm SL (Table 1.1). I restrict it to the resumption of growth at ≤ 13.0 mm SL and ending at ≤ 45 mm SL, the point at which proportional changes in body features become isometric with growth in length (Section 1.3). Notable changes include increased body height at the pectoral fins, increased snout and head length, increased height of the dorsal and anal fins, and enlargement of the pectoral fins. Near the end of Stage 3 the body loses some of its transparency, becoming gradually opaque and often silvery.

Meristic description: Fins spineless, dorsal 12–17, anal 19–25 (including fin rays that merge with continued growth). Myomeres of predorsal fin 37–39, those of preanal fin 38–41. Upper teeth 0–6, lower teeth 0–8. Gill rakers 1 + 7 (at 13.1 mm SL), 2 + 13 (at 13.8 mm SL), 5 + 14 (at 15.9 mm SL), 8 + 21 (at 20.2 mm SL). Branchiostegal rays 7–15 (at 13.1–15.9 mm SL).

Body proportions as percentage of SL: at < 17.2 mm SL, height at pectoral fins 9.9–17.0, head length 20.7–28.6, predorsal fin length 61.8–76.0, preanal fin length 70.2–78.6.

Body proportions as percentage of HL: at < 17.2 mm SL, snout length 17.3–26.2, eye diameter 21.4–29.3.

Narrative description: At 13.8 mm SL, body height at pectoral fins has increased relative to SL, but decreased posteriorly. By 15.9 mm SL, shape of head irregular, mandible oblique and extending to a point in vertical alignment with pupil; posterior end of mandible distinctly flared. Teeth present on lower jaw throughout Stage 3 and at 13.9–14.1 mm SL are developing on upper jaw. The eye at 12.6 mm SL has become compressed dorsoventrally; the nares are bifurcating. At 13.8 mm SL, swim bladder is enlarged anteriorly, now extending to myomere 12 and characterized by a dorsal finger-like projection from the posterior region to halfway up lateral line. Origin of dorsal fin now at myomere 37, that of anal fin at myomere 39. Height of anal fin exceeds that of dorsal fin; last ray of anal fin has split. Fleshy bases of pectoral fins have been reduced, incipient rays evident on pelvic fins.

By 13.8 mm SL, chromatophores present on head and body and densely populate snout, opercula, regions above brain and below midline. Pigment apparent on bases of dorsal fin rays and anterior and posterior rays of dorsal and anal fins, respectively. Surface of swim bladder pigmented, as is the gut and area separating the gut and kidney. By 15.9 mm SL, chromatophores outline the myomeres and are developed on the body above the midline.

Stage 4 – Juvenile

Principal sources used in descriptions: Harrington (1966); Jones *et al.* (1978: 62); Pinto Paiva and Ferreira de Menezes (1963); and Wade (1962: 566–567) for specimens ≥25.2 mm SL; and Mercado Silgado (1969 Table 1) for specimens of 71.0–1000 mm SL. Also see Moffett and Randall (1957). I have allowed some overlap in lengths of early juveniles with larger Stage 3 specimens starting at ≈25 mm SL.

Meristic description: Fins spineless, dorsal fin 14–18, anal fin 24–28 (up to 59.9 mm SL). With growth, counts of rudimentary fin rays have become reduced by consolidation. Gill rakers 9 + 24 (at 25.2 mm SL), 16 + 34 (at 35.0 mm SL), 17 + 34–22 + 40 (at 51–271 mm SL). Branchiostegals 22–25 (at 51–271 mm SL).

Narrative description: Body torpedo-shaped at 25.2 mm SL, but has deepened considerably by 51.0 mm SL. At 25.2 mm SL, mouth is large, lower jaw projected, maxillary wide and reaching to posterior margin of eye. Full appearance of the adult has been attained by 194.1 mm SL, at which point the maxillary extends past the eye, the snout is obtusely conical; bands of villiform teeth are apparent on the jaws, tongue, vomer, palatines, pterygoids, and sphenoid. Incipient scales first appear along lateral line at ≈30–34 mm SL (Harrington 1966: 868 stated 32 mm SL at first scale formation), and actual scales by 36.8 mm SL, 1 above and 2 below lateral line. Pores in lateral line can be seen at 51.0 mm SL. Axial scales form by at least 78 mm SL. At 25.2 mm SL, fourth dorsal and fifth anal fin rays are the longest; anal fin is falcate, its origin slightly posterior to insertion of dorsal fin; pectoral fins have broadened, their central rays almost to the origin of pelvic fins, which are now about halfway between the snout and hypural plate. By 140 mm SL, two specialized scales cover the uppermost and lowest caudal fin rays. At 194.1 mm SL the dorsal fin's filamentous ray has a visible groove on its underside; anal-fin sheath is scaly, and its last ray appears; caudal fin scaly. At 25.2 mm SL, body is opaque, the internal organs now hidden.

Pigmentation mostly above lateral line. Gular plate heavily pigmented; opercula silvery. Pigment present on tip of mandible, snout, and occiput. Juveniles continue to darken dorsally with age. Moffett and Randall (1957: 5) described a juvenile of 33 mm (FL?) seined from an isolated pond in the Florida Keys, declaring it recently metamorphosed: "The body is translucent except for the region over the abdomen and a less marked band the length of the body at the level of the vertebral column which are silvery. There is a dusky area mid-anteriorly in the dorsal fin. *The caudal fin is emarginated, not forked* (emphasis added). The last dorsal ray is not longer than the preceding rays."

They also examined a specimen of 42.5 mm FL from the same location. It was "more silvery, the dusky spot on the dorsal is still distinct, the caudal is now forked, and the last dorsal ray has elongated slightly." Still another specimen, this one measuring 63 mm FL, "is almost completely silvery (only a region along the back and another at the base of the anal fin do not show metallic reflection), the spot on the dorsal is faint, and the last dorsal ray is relatively longer (it does not exceed the length of the long anterior rays of the dorsal fin until a fork length of about 130 mm. is attained)." The length at which the tail becomes obviously forked – a clearly visible character – seems to have been largely overlooked in other descriptions I read.

Stage 5 – Adult Atlantic tarpon

Principal sources used in descriptions: Jones *et al.* (1978: 53 and references); Mercado Silgado (1969: 5–6).

Meristic description: Fin rays: dorsal fin 13–16, anal fin 22–25, caudal fin 7 + 10 + 9 + 6–7, pectoral fin 13–14; pectoral and pelvic fins with axillary processes.

Vertebrae: precaudal 53–57, caudal 33–34. In their original description, Cuvier and Valenciennes (1846: 398)[3] gave counts of dorsal 13, anal 22, caudal 30, pectoral 13, branchiostegal 22–23, ventral 9. Gill rakers 19–22 + 36–40. Lateral line scales: 41–48 (counts based partly on some juvenile specimens).

Body proportions (as percentage of SL and based partly on some juveniles): body height 23.5–29.0, head length 25.0–31.0, snout length 4.5–6.2, eye diameter 5.3–9.5.

Fin-ray enumeration in adult Atlantic tarpons depends on whether fused rays are counted as joined or separate. Counts of rays in dorsal and anal fins given above, which are cited widely and uncritically in species descriptions, might be low by three or four. As Breder (1944: 224) pointed out, "In large fish the first four or five [rays] are consolidated into a solid leading edge, which have generally been counted as one ray. In the smallest sizes the separation of these rays is evident and doubtless, if small fish instead of large were generally available to taxonomists, the usage would have developed differently."

This raises the dorsal and anal ray counts to, respectively, minimums of 16 and 25. Table V of Breder (1944: 225) displays ray counts of these two fins from the literature, indicating that only Fowler (1936) gave "full" counts; that is, by including in lower-case Roman numerals the number of rays prior to their fusion. The counts then become iv–v + 10–11 (dorsal fin) and iv–v + 18–19 (anal fin).

The older literature contains some peculiar speculation about possible function of the extended last dorsal ray. Southworth (1888: unpaginated) referred to it as "an osseous bayonet, about nine inches long ....Whether this weapon – for such it surely must be – is for attack or defense, no one, as yet, seems able to determine." Breder (1929: 59) wrote: "The produced last dorsal ray functions in the tremendous leaps that the tarpon is famed for. It is concave below and adheres to the side of the fish, bending and securing the dorsal to the right or left, so determining the direction of the fall."

Babcock (1936: 61) did not believe him, and rightly so, writing: "There is nothing in the anatomical or muscular structure of the tarpon that lends color to the theory that the fish controls its jumps by manipulating its dorsal fin by the use of the ray and after observing hundreds of fish I am satisfied that this is not the case."

Depending on how rays of the dorsal and anal fins of the Pacific tarpon are counted, numbers for the two species potentially overlap, in which case the

[3]According to Bailey (1951), Valenciennes alone should be credited. *Histoire Naturelle des Poissons* comprises 22 volumes. The description of the Atlantic tarpon appeared in Vol. 19. Cuvier died 13 May 1832, and Valenciennes prepared the material for all volumes, starting with Vol. 10 published in 1835. Bailey's assessment might be correct in the sense that Valenciennes did the work alone, but when read in the original French it seems his intent was to sustain their collaboration even after Cuvier's death (see my text comments). Even today, a colleague who has died can still be listed as a coauthor.

characters would fail to be diagnostic. The original description of *M. cyprinoides* (Broussonet 1782: 62–65) omitted any mention of counting method, and perhaps this feature should be revisited for both *M. atlanticus* and *M. cyprinoides* using samples consisting of a range of body lengths.

Narrative description: Fins spineless, pectoral and pelvic fins with axillary processes. Body deep, compressed, eyelids adipose. Mouth large, oblique, maxillary extending far past eye in large specimens; gular plate elongated between rami of lower jaw; mandible projecting prominently, tail deeply forked. Scales cycloid, exceptionally large, firmly attached, borders crenulated and membranous. Lateral line complete, decurved anteriorly. Single high dorsal fin with last ray elongated and easily distinguished. Coloration silvery, darker above.

The habitat seems to temporarily influence color. Atlantic tarpons both large and small reportedly acquire a "golden," or "brassy," tint while spending time in freshwater or waters of low ionic strength, a phenomenon apparently confirmed by Breder (1944: 233–234) in aquarium experiments. Ferreira de Menezes and Pinto Paiva (1966: 85) wrote that Atlantic tarpons appearing along coastal Ceará State, Brazil during the last quarter of the year are brassy, but those arriving in June, July, and August (austral winter) in smaller aggregations are "fatty individuals, silvery-white colored, suggesting that they come from waters of high salinity … ." However, Moffett and Randall (1957: 5) captured juveniles in an isolated tidal pond in the Florida Keys, where the *practical salinity*, S_P, (explained in Chapter 6.1) ranged from ≈ 19–33, reporting they "had a definite bronze cast." They also noted, "Juvenile tarpon from ponds with dark brown water were colored distinctly darker on the back."

These observations indicate to me that tannic and humic acids might actually be the agents staining the fish brownish, although they do not explain Breder's aquarium experiments. Nor do they eliminate the possibility of selective dorsal darkening in pigmented waters as camouflage from aerial and terrestrial predators. Breder (1944) and Ferreira de Menezes and Pinto Paiva (1966) had suggested that the bright, silvery appearance of a tarpon indicated it had recently been living in high-salinity waters, as did Victor Brown, taxidermist and angler, in a letter to Kaplan (1937: 92): "Within 3 days, after they [adult females] run into the brackish water rivers from the open Gulf [of Mexico] waters, their scales change from turquoise blue to bronze." Another possibility? Inshore waters are commonly tinted yellow by refractory organic pigments (*gelbstoff*). A reflective silvery appearance has camouflage value in transparent offshore waters (Brady *et al.* 2015), but turning brassy might be advantageous inshore.

Gudger (1937) described a rare albino Atlantic tarpon. It weighed 12.2 kg (27 lb) when alive; the mounted form measured 114.3 cm TL (≈45 in.).

Cuvier and Valenciennes (1846: 398–399) wrote of the tarpon's color in more detail based on a drawing sent by one of his correspondents, presumably in contact with Valenciennes and not Cuvier, who had died almost exactly 15 years before publication of Volume 19. Note, however, that the drawing was sent to *nous*, not *je* (to *us*, not to *me*), indicating that Valenciennes perhaps still considered Cuvier his collaborator when describing the Atlantic tarpon. The description reads:

> *"La couleur, bleu plombé sur le dos, est d'un bel argenté sous le ventre, sur les joues et sur les opercules. Le bord membraneux de cet os n'a pas cette tache noire si caractéristique dans l'espèce précédente. Les nageoires dorsale et caudale sont plus ou moins grises; les ventrales sont jaunâtres. D'après un dessin qui nous a été transmis par M. L'Herminier, il y aurait, quelques teints jaunes dorés sur les écailles de la nuque et des taches rougeâtres sur le bord du préopercule et à l'angle de l'opercule. La dorsale, lisérée de bleu, serait verdâtre comme l'anale; la caudale et les ventrales plombées."*

I translate this as:

"The color, leaden blue on the back, a beautiful silver on the belly, cheeks, and gill covers. The membranous edge of this bone [the operculum] does not have the characteristic black spot of the other species. The dorsal and caudal fins are more or less gray; ventral [fin] yellowish. From a drawing that was sent to us by Monsieur L'Herminier, there would be some tints of golden yellow on the scales of the neck and reddish spots on the edge of the preopercle and the angle of the operculum. The dorsal [fin], edged with blue, and being green like [the] anal [fin]; caudal and ventral [fins] leaden."

How much these subtle hues were influenced by light reflected and refracted through guanine crystals in the scales, shifting wavelengths caused by time of day and sky conditions, and status of the fish (alive or dead) is impossible to know. The black spot on the operculum is not evident in Broussonet's illustration of the Pacific tarpon, *Megalops* (= *Clupea*) *cyprinoides*, nor did I find it mentioned specifically in his original description of the fish's head or notice it in modern photographs and illustrations (e.g. Fig. 1.5). Perhaps this character, if it exists, is restricted to certain regional populations. The specimen depicted in a color photograph by Bagnis *et al.* (1987: 272), for example, has a black patch on the body visible at the posterior edge of the operculum.

Jordan and Evermann (1896: 409) removed the Atlantic tarpon from *Megalops* and placed it in a new genus, *Tarpon*, based solely on one perceived character difference: "The posterior insertion of the dorsal fin distinguishes the single species of *Tarpon* from the East Indian *Megalops cyprinoides*, a fish of similar habit, in which the dorsal is inserted above the ventrals." After examining a series of both fishes in collections of the US National Museum, Hollister (1939: 450–451) declared this character not valid and questioned whether Jordan and Evermann actually made the requisite measurements. She wrote:

> "It is apparent that the dorsal fin is in the same position in the two species, that is, in the same relative distance from the snout. But the ventral fins in *Tarpon* are nearer the snout than in *Megalops*, giving the illusion of the dorsal being more posterior in position in *Tarpon* than in *Megalops*."

She showed this in an illustration; whether the drawings were to scale was not mentioned (Fig. 1.9).

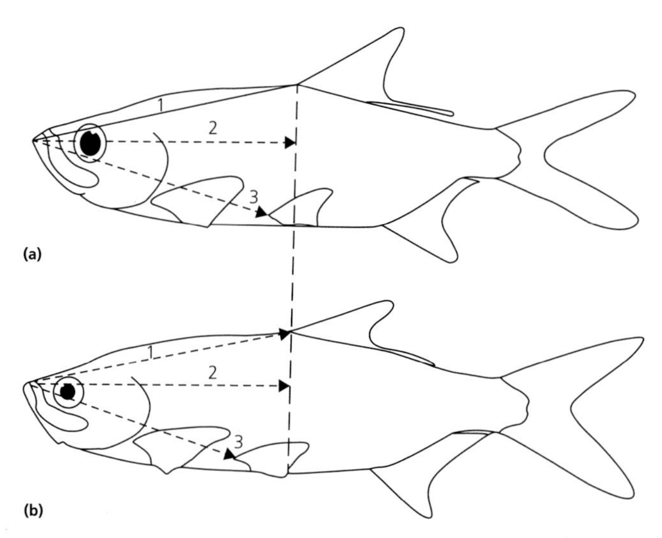

Fig. 1.9 **(a)** Pacific tarpon, **(b)** Atlantic tarpon. The distance in **(a)** represented by the line 3 is longer than that represented line 3 in **(b)**. The position of the ventrals is farther form the tip of the snout in **(a)** than the **(b)**. Lines 1 and 2 are equal in both, showing that position of the dorsal fins is identical in the two species. Source: Hollister (1939: 451 Text-fig. 1). Reproduced from G. Hollister: Young *Megalops cyprinoides* from Batavia, Dutch East Indies, including a study of the caudal skeleton and a comparison with the Atlantic species *Tarpon atlanticus*, with permission from the Wildlife Conservation Society Archives.

In addition to differences in placement of the ventral fins, Hollister noted variation in numbers of vertebrae (57 in the Atlantic tarpon, 68 in the Pacific species) and also in fin-ray counts, but whether she counted the rays as fused or separated is not stated (see meristic description above). The Atlantic tarpon supposedly has 12–15 rays in its dorsal fin compared with 19–21 in the Pacific tarpon. Counts of anal fin rays are 19–22 vs. 24–27. Details of the scales (Fig. 1.10) and caudal skeleton also seemed to Hollister to provide distinguishing features (see Hollister 1939: 460–467 and her accompanying Text-figs. 14–21).

In general appearance, the Atlantic tarpon is more slender (Gill 1907: 39) and has a smaller eye than its Pacific counterpart, which conflicts in one character with its original description. Cuvier and Valenciennes (1846: 398) wrote of the Atlantic species: "*Ce poisson la tête plus courte; le corps plus haut, plus trapu. L'oeil sensiblement plus petit.*" [This fish has a shorter head; the upper body, stockier. The eye significantly smaller.] Based simply on looking at photographs and illustrations, the Pacific tarpon appears to me the stockier species as an adult, which was also Hollister's conclusion, and seems evident in Fig. 1.9.

Maximum size and longevity: The current angling record is 130 kg (length unstated).

According to Heilner (1953: 220), "On August 6, 1912 ... native [commercial] fishermen at Hillsboro Inlet on the east coast of Florida caught a tarpon in their nets 8 feet, 2 inches long [≈249 cm] and estimated to weigh 360 pounds [≈163 kg]." Kaplan (1937: 93) gave its estimated weight as 352 lb [~157 kg].

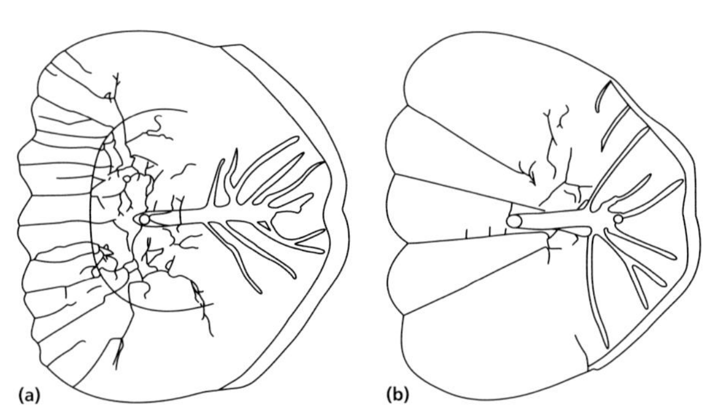

Fig. 1.10 Tarpon scales. **(a)** Pacific tarpon of 300 mm SL, 15th scale in the lateral line from below the anterior margin of the dorsal fin (see Text-fig. 9 of source publication). **(b)** Atlantic tarpon of 238 mm SL, 18th scale in the lateral line from below the anterior margin of the dorsal fin (see Text-fig. 11 of source publication). Source: Hollister (1939: 460 Text-fig. 12 and Text-fig. 13). Reproduced from G. Hollister: Young *Megalops cyprinoides* from Batavia, Dutch East Indies, including a study of the caudal skeleton and a comparison with the Atlantic species *Tarpon atlanticus*, with permission from the Wildlife Conservation Society Archives.

Norman and Fraser (1938: 97) also mentioned this fish, and in an example of how hearsay can eventually become fact, Beebe and Tee-Van (1928: 34), stated authoritatively: "The world's record for size, at this date, is 8 feet 2 inches long with an estimated weight of 350 pounds." According to Jordan and Evermann (1904: 85), "the largest taken with a harpoon weighed 383 pounds, if we may believe the record ...." Breder (1929: 60) wrote: "They reach a large size and records run as high as 8 feet 2 inches with an estimated weight of 350 pounds." Earlier, a fish like this might be considered medium-sized. Cuvier and Valenciennes (1846: 399), in describing the Atlantic tarpon, told us: "*M. L'Herminier nous en a envoyé un de quatre pieds un pouce, mais je trouve dans ses notes qu'on en pêche à la Guadeloupe qui ont jusqu'à seize pieds de longueur.*" [Monsieur L'Herminier sent us one [a specimen] four feet long, but I [presumably Valenciennes] find in his notes that [these] fish in Guadeloupe [presently part of the French West Indies] reach up to sixteen feet in length.] On the next page (p. 400): "*Marcgrave*[4] *en a déjà vu de onze à douze pieds de long et de la grosseur d'un homme* ...." [Marcgrave had already seen [specimens of] eleven to twelve feet in length and the size of a man ....] No doubt both reports are early fish tales.[5]

---

[4]Georg Marcgrave (1610–1644), German naturalist, astronomer, and cartographer whose writings on natural history impressed Cuvier. I assume Marcgrave's observations of tarpons were made during his explorations at the Dutch colony in Brazil, starting in 1638.

[5]Measurements in the old literature must be considered in historical context. In the mid-nineteenth century, one *pouce* (French inch) equaled 1.066 English inches, meaning that a 12-foot fish of 144 English inches examined by an Englishman would be ≈ 3657.6 mm, but longer (≈3899 mm) if a Frenchman measured it. The difference in this case (only 241.4 mm, or ≈ 6.6%) still leaves Marcgrave's reports looking fishy.

In a survey by Crabtree *et al.* (1995) of southern Florida tarpons, the oldest female was 55 years, the oldest male 43 years. Female Atlantic tarpons can live at least 64 years. A specimen captured in the Florida Keys in August 1935 died at Chicago's John G. Shedd Aquarium in October 1998 (Tim Binder, personal communication 5 January 2015). The husbandry record is incomplete. The fish was neither weighed nor measured when received, nor were any data recorded after its death. Apparently it was a juvenile when captured. Costa Rican Atlantic tarpons are also long-lived, some surviving to at least 48 years (Crabtree *et al.* 1997).

1.5 Development of Pacific tarpons

The Pacific tarpon's development has not been described in as much detail (e.g. Blanco 1955: 97–98; Hollister 1939; van Kampen 1909), and duplicating the above format of subsections is impossible. Species differences are slight (Wade 1962: 555) with exception of size at late juvenile and adult stages. According to Ellis (1956: 6), "Incubation lasts for about 20 hours, the young hatching out as prolarvae and quickly developing into small leptocephali." Neither a citation nor evidence to back this claim was presented. To my knowledge, initial development of the Pacific tarpon has not been reported.

Tsukamoto and Okiyama (1993: 379) considered the Pacific tarpon larva to have four early developmental stages, adding a "sluggish growth phase" between Stages 2 and 3, and noting that it seems unique to the Pacific tarpon in not having not been included in descriptions of the developmental stages of other Elopomorpha. I doubt its validity as a stand-alone feature and exclude it from the staging system used here, but incorporate the ontogenetic changes listed by Tsukamoto and Okiyama (1997) as occurring in early Stage 3 (here termed fry). The "sluggish growth phase" is probably an artifact of captivity. For example, captive Pacific tarpons reared by Chidambaram and Menon (1947) and maintained four weeks shrank progressively but never completed metamorphosis. Holstvoogd (1936: 4) claimed metamorphosis to take seven weeks in the laboratory prior to resumption of growth (i.e. initiation of Stage 3). Ellis (1956: 6) also gave seven weeks, but did not provide a citation. Others have reported metamorphosis as happening much faster (see Stage 2 description below). Alikunhi and Rao (1951:108), writing of grow-out of Pacific tarpons in aquaculture, said, "while marketable size was attained in the natural pond in the course of less than ten months, during the same period only less than half that size was attained in the aquaria [*sic*] and nursery tanks."

That captive conditions likely affect growth and development is illustrated by two examples. First is Holstvoogd's (1936: 4) description. Larvae captured when arriving at estuaries in the vicinity of Batavia (now Jakarta), Java, Indonesia were ≈23 mm SL (26 mm TL), and starting metamorphosis, when they shrank to 17 mm SL. During this process the usual shifting of structures occurred. Postanal myomeres increased from 17–20, for instance, and the anus shifted forward six myomeres. By completion of Stage 2 the anus had shifted forward a total of 13 myomeres. Growth in length (Stage 3) then commenced. At ≈24 mm SL the number of caudal vertebrae equaled the adult number, and except for shape of the dorsal fin, metamorphosis was considered finished.

In the second example, Alikunhi and Rao (1951: 105–109) described leptocephali captured in backwaters of the Adyar River near Madras (now Chennai), India on 6 October 1947. They measured 23.0–28.0 mm (presumably SL)[6] when acclimated to freshwater over ≤ 24 hours in the laboratory. Most had completed Stage 2 after 9 days (by 15 October), the shortest shrinking to 16.5 mm SL, and structures shifted as expected (i.e. at minimum standard length the dorsal fin was at myomere 36, the anal fin at myomere 42, the anus at myomere 41). Growth then resumed (Stage 3), reaching ≈20.5 mm SL over two weeks. During this time the dorsal and anal fins shifted forward, the dorsal beginning at myomere 27, the anus to myomere 38 (its adult position). Vertebrae numbered 38 (preanal) and 30 (postanal), and shifting of all structures was complete. *However, growth over the next four weeks reached 49 mm SL, more than twice that of Holstvoogd's fish in the same amount of time.*

The first tarpon leptocephalus described was identified as belonging to the Pacific species (Fig. 1.11). [van Kampen 1909: 10 unnumbered figure]. Beebe (1928: 228–229) provided an English translation of van Kampen's original article published in German (see below), along with a better inked reproduction of his illustration (Beebe 1928: 228–229, unnumbered figure following p. 228). Although van Kampen did not know the age of his specimen or note its length specifically, he seemed to be discussing a fish of ≈25 mm TL (≈22 mm SL) mentioned in his text. If so, larval development differs little from that of the Atlantic species. Because the leptocephalus came from an inshore location it was probably in Stage 2. Beebe's translation:

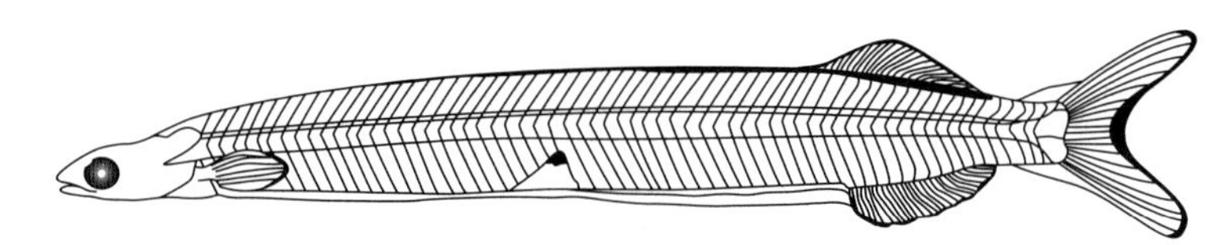

Fig. 1.11 First illustration of a Pacific tarpon leptocephalus (≈25 mm TL or ≈ 22 mm SL). The specimen is probably Stage 2 based on the presence of pelvic fins and pigment, dorsal and anal fins placed far back, and advanced development of median fins and swim bladder (Wade 1962: 594). The inshore capture site is also diagnostic. Source: (van Kampen 1909: 10 unnumbered figure).

[6]Alikunhi and Rao (1951) did not state whether their length measurements were FL, SL, or TL. I assume SL here.

In the month of January, fish larvae appear in the brackish water of the harbor canals of Batavia (see illustration) which are very similar to that of *Albula*, nevertheless it can not belong to this genus, but must be related to the *Megalops cyprinoides*, on later elucidated grounds. The total length of the larvae collected by me varies from 23 to 30 mm. yet all appear to be in the same stage of development. Older stages I have not yet found, and the development of the larvae is unknown.

The body is band-like. One animal of 25 mm total length (22 without the caudal) has a depth [height] of 3 mm. It is quite transparent; in life only the eyes and swim bladder are distinct.

The fins, with the exception of the ventrals, are already well developed, and much larger than in the youngest of Gill's stages (*Albula*), which otherwise correspond most closely with the Batavia larvae. Dorsal and anal fins lie far back, the first somewhat in front of, but for the most part over the latter. The dorsal has eighteen rays, and in front of these are plain evidences of one or more undeveloped rays; the last ray is somewhat larger than the others and this gives a hint of the filament of the adult *Megalops*. The anal fin possess [*sic*] twenty-seven rays, the caudal twenty, from each of which ten hypurals are formed. The small rim-rays of the tail fin are only visible in the first stage. The ventrals are the smallest of all, and lie in the middle of the body.

Of the inner organs the swim bladder is developed ahead of the rest. In life it appears as a small shimmering point. It lies above the ventrals.

There are about sixty-nine muscle bands evident, of which forty-eight or nine compose the body (between the ventrals and anus, 22), and twenty or twenty-one the tail.

On the ventral aspect of most of the myomeres is a chromatophore. Besides these, small chromatophores lie along the upper and lower caudal rays, and along the posterior edge of this fin. Above the swim bladder lies a pigmented cap.

A methylene blue preparation of [a] 30 mm larva, reveals the osseus beginnings of the cranium, the basal part of the pectorals and the fin-ray supports of all unpaired fins, but not yet the vertebral column. The ray supports of the dorsal and anal fin rest upon two pieces of which the distal (basiostegal) is small and round.

The great number of rays of the anal fin makes it unlikely that the larvae under consideration belongs to *Albula conorhynchus*, which possess only 9–10. In comparison the number agrees well with allied *Megalops cyprinoides* (anal rays 23–28), while in *Elops saurus* these are also much [*sic*] fewer (15–17). Other near related forms do not occur in the Archipelago. I dare assert with considerable assurance that this larva is that of some specimen of *Megalops*, especially as this genus is the only one which occurs commonly in the harbor canals of Batavia.

Stage 1 – Growing leptocephalus

Principal sources used in descriptions: Alikunhi and Rao (1951); Chen and Tzeng (2006); Chen *et al.* (2008); Chidambaram and Menon (1947); Delsman (1926); Gopinath (1946: 9); Tsukamoto and Okiyama (1997); Tzeng *et al.* (1998); and Wade (1962: 567–569).

At 16.1 mm SL, slope of snout flatter, eye diameter relatively larger than Atlantic tarpon, but still very similar in form and structure; eight teeth per side in both jaws, an increase of two in the upper and three in the lower, and two more teeth in each jaw than Atlantic tarpon of similar length, although arrangement, size, and shape of the teeth are the same. At 20–27 mm TL (Chidambaram and Menon 1947), head depressed, mouth with pointed teeth, alimentary canal straight, muscle fibers in myomeres arranged in parallel rows, caudal fin forked; HL 10% of SL, diameter of eye 25% of HL, body height 11% of TL; 69 myomeres total, 15 preanal, 18 postanal; swim bladder a club-shaped evagination in middle of alimentary canal between myomeres 27 and 31; pectoral fin fan-shaped with rounded margin; black pigment along middle of myomeres in a line and on latero-ventral aspect of myomeres in small patches.

Size range of specimens from Gongshytyan Brook estuary, northern Taiwan between 15 and 24 September 1995 were 20–39 days old ($\bar{x}$ = 28.5 d) and had already begun metamorphosis at 13.6 mm SL; similar in overall appearance to Atlantic tarpon leptocephali of comparable length: translucent with internal structures visible. Length at arrival inshore, 17.8–32.9 mm TL ($\bar{x}$ = 25.6 mm TL, n = 194). Age (A) vs. growth (G) relationship determined by Tzeng *et al.* (1998: 180):

$$A = 58.404e^{-.7886G} \tag{1.1}$$

Body proportions as percentage of SL: At 13.6 mm SL, compared with Atlantic tarpon, HL 21.8 vs. 31.3; eye diameter occupying greater proportion of head; differences of other body proportions (e.g. body height at pectoral region, predorsal, preanal) slight. By 24.1 mm SL, HL 10.4, just 0.4 greater than Atlantic tarpon of 21.3 mm SL; indices for body height equal at 6.2 (Pacific tarpon) and 6.1 (Atlantic tarpon). Snout length and horizontal eye diameter, which were 27.8 and 22.2, respectively, in Pacific tarpon larvae of 16.1 mm SL, decrease to 26.8 and 20.0 by 24.1 mm SL.

Narrative description: Body nearly identical to Atlantic tarpon's early in Stage 1; that is, ribbon-like, elongated, thin laterally, deep; head small, triangular, pointed, eel-like, wider than body in dorsal aspect; brain clearly visible; eye nearly round; snout sloping gently from top to tip. Central nerve and notochord well developed by 16.1 mm SL, urostyle turned up sharply, seven hypurals visible. Remnants of median finfold apparent in predorsal and postanal regions. Thirteen dorsal and 19 anal-ray bases visible, but fin rays absent. Predorsal and preanal myomere counts 48 and 52, total myomere count 65. Caudal rays 17, two more than Atlantic tarpon of comparable size. Pelvic fin absent, pectoral fin a transparent fleshy bud without rays. Development at 24.1 mm SL comparable to Atlantic tarpon's at ≈26.9 mm SL; position of dorsal fin unchanged, anal fin has moved slightly forward, outline of fins similar in both species. Caudal fin in Pacific tarpon more deeply forked than Atlantic tarpon, each with 19 principal rays. Development of pectoral fin static despite increase in body length of ≈8.0 mm SL; pelvic fin absent. Swim bladder a cylindrical sac arising from gut, pushing dorsally toward central nerve cord. Unidentified structure (kidney?) dorsal to gut extending from myomere 31 to anal area; kidney apparently developing later in Pacific than Atlantic tarpon. At maximum length a few melanophores along ventral abdomen. Stage 1 leptocephali in captivity tend to swim in the middle of the water column.

Head small and rounded when fully developed at ≈ 32 mm SL, body strongly compressed, dorsal and caudal fin rays starting to form; branched melanophores under eye, following along dorsal contour of abdominal cavity, dotting dorsal surface of swim bladder, between posterior-most myomeres; bud of pelvic fin just appearing. Nearly all elements of skull still cartilaginous. Flexion complete, but caudal complex poorly developed; cartilaginous buds of caudal skeleton appearing. Body filled with gelatinous matrix ("mucinous pouch"), which decreases quickly in Stage 2 (Section 1.6). Gills nonfunctional (filaments poorly developed, lamellae absent). Gut straight, esophagus and intestine divided by constriction at anterior ≈ 60% of SL; swim bladder connects immediately anterior to this constriction. At full length, gill filaments still poorly developed; no lamellae present.

Even at end of Stage 1, organismal development of the Pacific tarpon appears retarded compared with leptocephali of other genera (Tsukamoto and Okiyama 1997), and the same could be said for the Atlantic species. Visual and olfactory systems comparatively less advanced in Stage 1. Vision probably excludes formation of images, although differences in illumination are detectable. The nasal cavity is exposed and the anterior and posterior nostrils have not yet formed. Tsukamoto and Okiyama (1997: 31) wrote: "The development of those organs in fully grown leptocephali of Pacific tarpon is similar to that in other marine fish larvae at 2–3 days after hatching, when the yolk-sac is absorbed and the eye pigmented."

Duration: According to Tzeng *et al.* (1998), Stage 1 lasts 20–39 days ($\bar{x}$ = 28.5 days).

Stage 2 – Shrinking leptocephalus

Range ≈ 32–16 mm SL; body remains transparent until near end of Stage 2. Gill lamellae develop as length diminishes, and most components are present at termination of shrinkage. Length of alimentary canal decreases, especially the esophagus. Alimentary canal still straight, but intestine has thickened; stomach is developing. First uroneural and fifth and sixth hypurals ossify; most of caudal skeleton ossifies. At 27.4 mm SL, the Stage 2 Pacific tarpon closely resembles the Atlantic species of 23.0 mm SL. Head now bullet-shaped, HL 11.1, > 1.8 larger than Stage 1 larvae of 28.0 mm SL and equivalent to HL of Stage 2 Atlantic tarpon of 24.5 mm SL. Eye oval, elongated dorsoventrally and occupying a greater proportion of head compared with an Atlantic tarpon of 23.0 mm SL, becoming well developed as Stage 2 proceeds. Snout length as proportion of HL now decreased to 21.9 and also less than in the 23 mm SL Atlantic tarpon. Jaws larger than largest Stage 1 larva, gape now reduced, single nasal aperture, hindbrain anterior to dorsal half of first myomere. Position of fins similar to Atlantic tarpon. Dorsal fin rays 13–14, anal fin rays 22–24 compared to Atlantic tarpon's respective 12 and 20; branching apparent only for last ray of dorsal and anal fins; all but marginal rays of caudal fin are split. Predorsal and preanal myomere counts 50 and 51, same as largest Stage 1 larva of 24.1 mm SL. Pelvic fins starting to differentiate at myomere 29 (those of Atlantic tarpon develop in Stage 1). Swim bladder larger, rising straight up from gut at myomere 27. Kidney elongated, still attached posteriorly to gut (same as Atlantic tarpon), extending from myomeres 11–35. Pigmentation similar to Atlantic tarpon's, mainly a line of dashes on gut, kidney, base of anal fin, and appearing as scattered spots on caudal fin.

By 21.0 mm SL, percentages of body parts against SL of body are HL 13.4, body height at pectoral fin 8.4, prepelvic 51.6, predorsal 73.8, preanal 80.0. As percentage HL, snout length now 26.4, horizontal eye diameter 25.2. Myomere count 67, 27 prepelvic, 45 predorsal, 49 preanal.

These trends have continued by 15.6 mm SL. Slope of snout less steep, HL has increased to 21.2. Snout length now decreased to 19.7 of HL, eye diameter increased to 28.8 of HL. Body height at pectoral fin increased to 12.1 of body SL. Nares appear as single opaque areas on snout. At minimum size, viscous matter occupies almost half the body. Four gill arches, branchiostegals forming; pigment spots on upper ocular orbit. Body height now greater anterior of anus; median fins longer and higher, last ray of each branched. Dorsal fin origin has shifted posteriorly, its anterior insertion now increased by one myomere. Standard length has increased 3.8%. Insertion point of anal fin unchanged; pelvic fin has increased in size and shape, although change of positon negligible. Caudal peduncle now deeper, seven hypural plates visible; urostyle slender and pointed; lobes of caudal fin more elongated. Body becoming opaque, unlike Atlantic tarpon of comparable size; separation of gut and kidney still visible, beginning at myomere 32 and extending to anus. Swim bladder occupying large area mostly above and anterior to developing pelvic fins. Pigment spots apparent on dorsal surface of gut, ventral border of developing kidney, bases of dorsal and anal fins, caudal peduncle, along midline of body; one dark spot on every myomere.

The remarkable shrinking that occurs during Stage 2 is mediated by thyroid hormones, which in the Pacific tarpon are required for metamorphosis (Shiao and Hwang 2004, 2006). Thiorurea is an anti-thyroid hormone that inhibits production of thyroxine (T_4) and triiodothyronine (T_3). Leptocephali just entering Stage 2 treated with thyroxine or triiodothyronine showed slightly accelerated metamorphosis. However, those treated with thiorurea entered a metamorphic stasis lasting > 22 days, or a week beyond the normal interval of transformation to Stage 3. Although similar experiments have not been conducted with Atlantic tarpon leptocephali, there is little reason to believe the outcome would be qualitatively different. Time of development (i.e. rate of shrinkage) is unaffected by the environment's ionic strength. The timing until minimum length is attained correlates with completion of metamorphosis. However, otoliths grow continuously, indicating a decoupling of somatic from otolithic growth. Stage 2 leptocephali in captivity tend to swim lower in the water column compared with larvae in Stage 1, and closer to the bottom.

The larva depicted by van Kampen 1909: 10), presumably 22 mm SL (Fig. 1.11) is toothless and probably on the verge of shrinking. I count ≈68 myomeres on the drawing, the number confirmed in different specimens by others (e. g. Delsman 1926: 408 footnote 1, Hollister 1939).

Duration: Stage 2 lasts ≈10–14 days (Chen and Tzeng 2006; Chen *et al.* 2008; Shiao and Hwang 2006; Tzeng *et al.* 1998), or the same length of time required for metamorphosis by the Atlantic tarpon. According to Chen *et al.* (2008), its duration is unaffected by whether leptocephali are fed or starved. However, water temperatures that mimic winter (20°C) and summer (30°C), respectively, slow and accelerate metamorphosis slightly. The process generally takes ≈14 days at the optimal temperature of 25°C.

Stage 3 – Fry

Range 15–40 mm SL. Body length has briefly stabilized at start of Stage 3, but proportional changes in other characters as percentage of SL continue: HL, *H*, body width increase. Length of gill filaments and numbers of lamellae increase. Fin rays become well developed, last ray of dorsal fin elongates, and dorsal and anal fins move anteriorly. Pigments become denser. The intestine coils, and pyloric caeca start forming. Noticeable changes have occurred by 15.0 mm SL (fry, or Wade's Stage IIIA; Wade 1962: 571 Figure 6b, c; 572–573). Jawbones and most elements of hyoid and branchial regions have ossified. Wade described Pacific tarpon "fingerlings" as 36–72 mm SL, but mentioned the term only once and probably borrowed it from Alikunhi and Rao (1951: 107), who referred to fish exactly within this range from the Cooum River, Chennai (Madras), India as "fingerlings." This suggests a fish in Wade's Stage IIIB (equivalent of Stage 4 here), larger and more developed than a fry.

A fish entering the fry stage (Stage 3) has its greatest height anterior of the pelvic fins, its body tapering toward the anal fin and expanding again at the caudal peduncle. Between 15.6 and 15.0 mm SL, this change is dramatic, *H* now 18.7% of SL. The body is thicker than in stages 1 and 2 but still laterally compressed. Overall appearance more fish-like. Eye large and round and, 31.1% of HL, an increase of 5.2% over a Stage 2 specimen of 15.5 mm SL. Snout 3.3% of HL. Both changes as in Atlantic tarpon, but occurring less rapidly. Head length now 30.0% of SL, an increase of 8.8%. Mandible protrudes ventrally to below posterior edge of ocular orbit. Dentary and maxillary bear single rows of fine teeth. Nares not yet bifurcate, appearing as opaque areas on snout. Fins also showing striking changes with pectorals, pelvics displaying rays and increased surface area, pectorals extending two-thirds distance to pelvic, the pelvics extending two-thirds distance to anus. Origin of pelvic fins now slightly behind mid-point of SL. Outline of dorsal and anal fins similar, except posterior border of anal, which is falcate. The dorsal fin has shifted forward from myomere 46 to 34. Along with this anterior shift of the dorsal, the anal fin is now at myomere 44, having decreased 4.8% in distance from snout. Pigmented areas largely unchanged except for addition of spots on median rays of caudal fin.

By 24.2 mm SL (fry), or still Wade's Stage IIIA (Wade 1962: 571 Figure 6C, 573–574), the body is starting to look like that of a juvenile, notable for its increasing height anteriorly. In lateral view it resembles the torpedo-like shape of a 25.2-mm SL Atlantic tarpon. Snout and eye diameter as proportion of head length are also close to proportions of these measurements in Atlantic tarpons of similar length (1.7 and 0.4, respectively). The mouth is large, the maxillary extending down obliquely to where it aligns vertically with posterior border of pupil. In the Atlantic tarpon, this bone reaches to the level of the eye's posterior margin. Predorsal length has decreased 9.6% of SL, while prepelvic distance is now 10% greater. Anterior shifting of the dorsal fin has been become more obvious by the decrease in the predorsal myomere count to 28. Meanwhile the anal fin has also shifted. Pigment spots now found on every fin and scattered sparsely in a patternless manner over all regions of head and body. Scale formation is apparent below the lateral line from opercula to middle of the caudal peduncle. *Duration*: Not reported, to my knowledge.

Stage 4 – Juvenile

Meristic description: The information I found was too sketchy to assemble and report.

Body proportions as percentage of SL: At 13.6 mm SL, horizontal eye diameter larger (as percentage of HL) than Atlantic tarpon. *Body proportions as percentage of head length*: At 11.0–21.3 mm HL, snout length 21.8.

Narrative description: Stomach enlarges commensurate with decrease in relative length of esophagus. Squamation commences at ≈22 mm SL and spreads anteriorly with growth (Fig. 1.12). At ≈26 mm SL scales develop further,

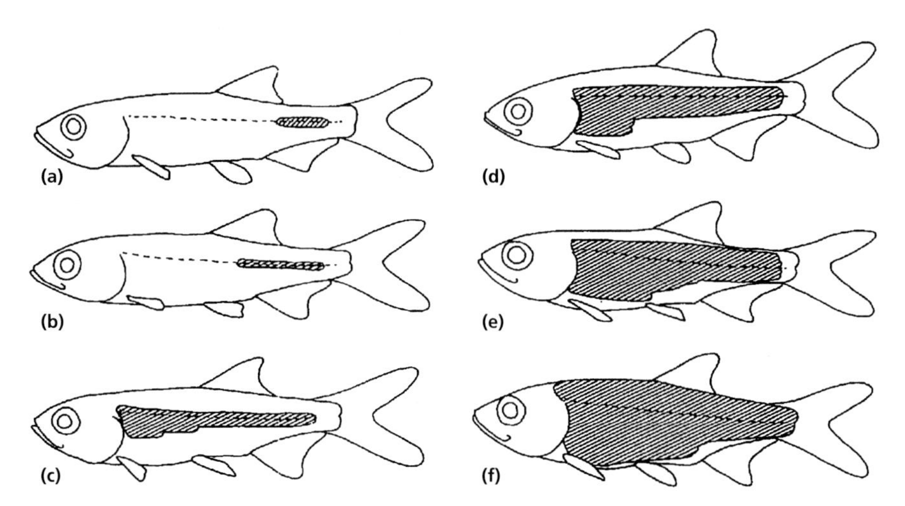

Fig. 1.12 Sequence of squamation development in Pacific tarpons during the juvenile growth phase. **(a)** 22.4 mm SL. **(b)** 24.3 mm SL. **(c)** 26.2 mm SL. **(d)** 27.0 mm SL. **(e)** 30.0 mm SL. **(f)** 33.5 mm SL. Source: Tsukamoto and Okiyama (1997: 26 Fig. 3).

dorsally and ventrally; by ≈28 mm SL half the body surface is scaled, and squamation is nearly complete at ≈33 mm SL and covering ventral surface by > 35 mm SL; all skull elements ossified by ≈35 mm SL. Shape of head bones as in adult at ≈50 mm SL, but ethmoid still cartilaginous. Most changes in body proportions are finished by ≈20 mm SL, (approximately half the length at which this occurs in Atlantic tarpons), when caudal elements have become ossified and gill formation is complete. Fish now have adult appearance except for incomplete squamation.

Duration: Not reported, to my knowledge.

Stage 5 – Adult Pacific tarpon

Principal sources used in descriptions: Boulenger (1909: 27–29); Merrick and Schmida (1984: 53); Pollard (1980: 53).

Meristic description: Fin rays: dorsal iv–v 14–16 (total 19–20); anal iii–iv 22–23 (25), total 26–27 (28); pectoral i 14–16; pelvic i 9–10. Gill rakers 14–16 + 29–33 (total 43–49 on first gill arch). Branchiostegal rays 24–27.

Body proportions as percentage of SL: Height 26.4–28.6, head length 28.2–30.0, snout length 4.7–7.7, horizontal eye diameter 7.2–7.9, post-orbital 12.6–14.0, inter-orbital 5.1–5.7, upper jaw 13.9–15.9, mandible 15.2–15.3, gular plate length 9.6–11.1, gular plate width 1.3–1.8, pectoral fin 17.1–20.2, pelvic fin 11.1–15.4, predorsal 52.3–57.0, prepelvic 52.3–57.0, preanal 72.4–79.0, last dorsal ray 27.0–30.4.

Narrative description: Fins spineless. Body oblong, compressed laterally, eyelids adipose, head moderately sized, snout sharply pointed, mouth large with strongly protruding mandible, tail deeply forked, maxillary extending past posterior border of orbit. Single high dorsal fin with last ray elongated and easily distinguished, about equal to or slightly longer than head length. Pectoral fin long and shallow, pectoral and pelvic fins with long fleshy axillary processes, axillary scales covering approximately two-thirds the length of both fins. Anal fin behind base of dorsal fin and without basal scaly sheath. Pseudobranch not exposed. Lateral line well developed with branched tubes. Scales exceptionally large, cycloid, firmly attached, lateral line distinct with 34–36 scales in lateral series, ≈4 more on caudal peduncle.

Color of back bluish-green to olive, head dark olive, sides silvery, ventral surface white, caudal yellowish, other fins greenish-yellow, dorsal fin with a dusky margin. Fresh specimens from different locations in eastern Africa have been described as having warmer colors: top of head dark brown, fins brownish, caudal and dorsal margins dark, pectoral and pelvic axillary scales speckled light brown. A dark patch on posterior edge of operculum has been reported by some observers, but not others.

Maximum size and longevity: According to Ley (2008: 8) the current angling record (weighed but apparently not measured) is a fish ≈611 mm FL (2.99 kg) caught near Gladstone, Queensland, but Ley also reported a 525-mm FL specimen from the Russell River estuary, Queensland and mentioned a specimen of 610 mm FL caught in the Calliope River near Gladstone. Barnard (1925: 105), without presenting evidence, listed Pacific tarpons as reaching 500 mm. Losse (1968: 81) reported the size of a female specimen from Tanga, Tanzania as 480 mm SL and 5.5 lb (≈2.5 kg). Bell-Cross and Minshull (1988: 42, 91) noted a specimen of 1.361 kg caught by angling at the confluence of the Save and Runde rivers, Zimbabwe, but did not provide a length, and they referred to another caught by an angler in Kenya that weighed 1.8 kg, again without stating its length. Rahman (1989: 236) mentioned adult lengths of Pacific tarpons in Bangladesh as 75–150 cm TL, but provided neither data nor a citation; 1500 mm TL is more than twice the maximum length typically stated. Without citing a source, Munro (1967: 41) wrote that Pacific tarpons reach "at least 40 inches," or ≈1016 mm. Pollard (1980: 53), without presenting evidence, claimed that lengths >1500 mm are attained, echoing Rahman and probably using him as an uncited source. Roughley (1953: 7) claimed that Pacific tarpons in Australian waters grew to 5 ft (≈1525 mm), but did not cite a source. Coates (1987) recorded 440 mm SL (1.5 kg) as the maximum size of specimens obtained from the Sepik River system, northern Papua New Guinea. The smallest was 103 mm (10 g). These lengths were repeated by Allen and Coates (1990: 52–53). Norman and Fraser (1938: 93–97) claimed that Pacific tarpons grow to 3 or 4 ft (≈914–1219 mm) without citing a source. Alikunhi and Rao (1951: 100) mentioned adult specimens >18 in. long (≈457 mm) in the Tamil Nadu region of southeastern India. Thomas (1887: 168) claimed to have seen Pacific tarpons in India measuring one cubit (≈460 mm). According to Shen *et al.* (2009), Pacific tarpons > 600 mm TL are rare in Taiwan. Kulkarni (1983, 1992) described growth of juvenile Pacific tarpons released into two freshwater Indian lakes in 1939. The specimens were not tagged, but intermittent seining over the years showed a near cessation of growth. Some fish caught in 1970 after 32 years were 650 mm TL (2.8 kg). In 1983, after 44 years, seined specimens were 670 mm TL (2.75–3.1 kg). A tarpon subsequently taken in 1991 after 53 years was 670 mm TL (3 kg), indicating no discernible growth in length. I could not find evidence that tarpons of either species living permanently in freshwaters are stunted, and according to Kulkarni (1992), the lakes from which these specimens came contained abundant food. Based largely on his information, I place the maximum length of Pacific tarpons at ≈ 700 mm TL.

1.6 Leptocephalus physiology

The leptocephalus is notable in many ways, but its physiology is truly remarkable. During the planktonic, or pre-metamorphic, phase it actually accumulates energy reserves as lipids and glycosaminoglycans, these last formerly called mucopolysaccharides (Pfeiler 1996). The unusual nature of this adaptation – perhaps unique among fish larvae – is hard to overstate. Although energy budgets of tarpon leptocephali have yet to be studied specifically, those of other species make illuminating proxies, with qualifications. Bishop and Torres (1999) evaluated how energy was partitioned between metabolism and excretion by leptocephali of four species of eels: margintail conger (*Paraconger caudilimbatus*), bandtooth conger (*Ariosoma balearicum*), honeycomb moray (*Gymnothorax saxicola*), and shrimp eel (*Ophichthus gomesii*).

All leptocephali have certain common features. Notable is the transparent, laterally compressed body consisting mainly of gelatinous material. The flattened shape provides a high surface-to-volume ratio, perhaps augmenting physiological exchange processes with the external environment (e.g. gas transfer, exchange of water and ions), and, in at least some species the possible uptake of dissolved organic matter, or DOM (Otake *et al.* 1990, 1993), common in many marine invertebrates (Pfeiler 1986). Uptake of DOM could thus be *per os*, by passive transfer across the integument, or through active carrier-mediated transfer linked with Na^+ (Fig. 1.13). A thin epidermis only a few cells thick possibly facilitates these functions (Pfeiler 1999 and references). Evidence of exogenous feeding, however, has been mostly inferred: in the Japanese eel (*Anguilla japonica*) leptocephalus, starvation after a certain amount of time appears to stimulate onset of metamorphosis (Okamura *et al.* 2012).

Fish larvae typically absorb the yolk sac soon after hatching and start to feed. Growth then continues to the juvenile stage. Alternatively, after yolk-sac absorption,

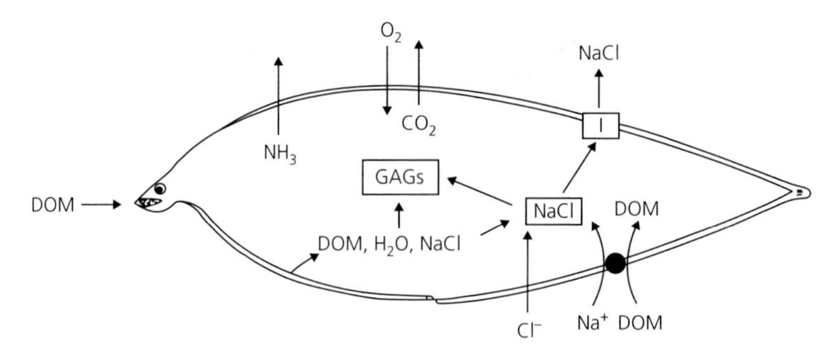

Fig. 1.13 Model postulating mechanisms of nutrient acquisition, Na^+ and Cl^- fluxes, and gas exchange in elopomorph Stage 1 leptocephali. The large, laterally-compressed body offers a high surface-to-volume ratio, favoring cutaneous respiration (the gills are not yet developed) and different pathways for possible uptake of dissolved organic matter (DOM). A portion of the Na^+ and Cl^- that enter passively by diffusion, including via Na^+-DOM cotransport involving specific carrier proteins (filled circles), and by intestinal absorption (along with water and ingested DOM, is bound to acidic glycosaminoglycans (GAGs). Some NaCl is probably transported actively out of the body by means of ionocytes (I) in the integument (see Chapter 6.3). Water and NaCl of leptocephali increase as GAG concentrations in the extracellular gelatinous body matrix rise during Stage 1 growth. The integument is also an important site of ammonia (designated here as NH_3) excretion. Source: Pfeiler (1999: 118 Fig. 2).

the larva grows unusually fast, often without an evident source of nutrients or sensory and anatomical systems sufficiently developed to locate and process food. Proposed energy sources include dissolved and particulate organic matter (Pfeiler 1986). Depending on species, leptocephali can remain in the plankton for days, months, or years, and they comprise the Albuliformes (bonefishes), Anguilliformes (eels), Elopiformes (tarpons and ladyfishes), Notacanthiformes (spiny eels), and Saccopharyngiformes (gulper eels). After absorbing the yolk sac the planktonic leptocephali of these five related orders, like other larvae, grow to species-specific, pre-metamorphic sizes. *However, unlike "conventional" fish larvae they accumulate energy reserves instead of expending them for immediate use.*

According to Pfeiler (1986), these reserves consist of a gelatinous matrix, or "mucinous pouch," extending almost the whole length of the body and comprising lipids and glycosaminoglycans. During Stage 1 the matrix serves as structural support in place of the absent vertebral column. During Stage 2 shrinkage, metamorphosis consumes these compounds, and they serve as the foundation for developing bones and muscles (Pfeiler 1984, 1986, 1996, 1999 and references) in a metabolic sequence apparently universal in leptocephali regardless of species (Deibel *et al.* 2012). The gelatinous matrix might also serve in buoyancy regulation during Stage 1, considering that the swim bladder does not become functional until metamorphosis (Pfeiler 1999 and references).

The normal situation is for organisms to experience a rise in whole-body respiration with increased body weight, but not leptocephali, for which no correlation is evident between mass and any index of metabolism, making them unique. Production of feces would seem unlikely for a life-stage that lacks the ingestion-digestion-excretion apparatus to process solid food. Most fishes are ammonotelic, excreting waste nitrogen as ammonia (Chapter 6), and leptocephali are no different (Bishop and Torres 1999; Pfeiler 1996). Bishop and Torres (1999) reported the rate of ammonia excretion to decline with increasing body weight and was highest in larvae ≤ 0.50 g ($n = 51$). Heavier larvae demonstrated excretion rates that were specific to wet weight and stayed more or less constant as weight increased.

Weight-specific consumption of oxygen (i.e. relative oxygen consumption rate, $\dot{M}O_{2w}$) fell steeply with increasing body weight when a power function was applied:

$$\dot{M}O_{2w} = aW^b \tag{1.2}$$

Here a is a scaling constant (i.e. the allometric coefficient, or intercept), W represents body weight, and b is the scaling exponent. Larvae < 0.20 g wet weight showed the greatest changes in $\dot{M}O_{2w}$. Depending on the value of b, relative metabolic rate as $\dot{M}O_{2w}$ regressed against body weight is either allometric and curved ($b \neq 1$) or isometric and linear ($b = 1$). In isometric relationships, metabolic rate and body weight scale in direct proportion; that is, in a straight line.

In conventional fish larvae, energy taken in is mostly expended with little being stored, making large and small larvae about equal in vulnerability to starvation. Overall, according to Bishop and Torres (1999: 2490), "a lower proportion of the mass of the leptocephalus is invested in metabolizing tissue than in other larval fish." This is certainly unusual, to which can be added that the energy reservoir (glycosaminoglycans) doubles as a pseudo-skeleton, permitting efficient propulsion in the absence of a bony scaffold, "and without appreciable metabolic costs other than that needed for acquiring and depositing the glycosaminoglycans." These authors found a sharp negative relation between increasing body weight and the weight-specific rate of O_2 consumption, excretion rate, and enzymatic activity in four species of eel leptocephali. The result was a substantial drop in metabolic rate with increasing size. They wrote (Bishop and Torres 1999: 2485): "The result suggests that the proportion of actively metabolizing tissue also declines with size, being replaced in large measure by the metabolically inert energy depot, the glycosaminoglycans." Their conclusion was that, "Leptocephali can thus grow to a large size with minimal metabolic penalty, which is an unusual and successful developmental strategy."

The combination of glycosaminoglycans and lipids is the impetus for rapid, low-budget growth, providing a fuel dump poised to fund the metabolic expense of impending metamorphosis. Mercado Silgado (1971: 14) speculated correctly when stating, "*Se cree pueda ser por ósmosis o reabsorción de tejidos, en todo caso, existe una disminución de tamaño, cuando no se ha formado su sistema digestivo, lo que hace pensar en el autoconsumo.*" [It (feeding) is thought to be by osmosis or reabsorption of tissues, although in any event size decreases before the digestive system has formed, indicating self-consumption.] Death by starvation is less likely when you *are* your own food, a situation that only gets better with increasing size when metabolism declines drastically instead of increasing as expected. For a leptocephalus, bigger is unquestionably better.

Atlantic tarpons in Stage 1 and early Stage 2 apparently do not feed exogenously (Dahl 1965; Harrington 1966). Feeding does not commence until Stage 2 (Phase VIII of Mercado Silgado 1971 and Mercado Silgado and Ciardelli 1972), or the approximate equivalent of Wade's (1962) Stage IIIA (also see Chapters 1.6, 7.7, Appendix B). This occurs at or near the end of metamorphosis during drastic shrinkage when the ribbon-like form becomes torpedo-shaped, simultaneously losing much of its surface area.

Stage 2 reveals enhanced development of organ, sensory, and structural systems; the epidermis thickens and becomes less permeable. The gelatinous matrix that served as structural support in Stage 1 is replaced by bony vertebrae and muscles; stiffening fin rays have attained their adult numbers. Stage 2 is relatively short, and during its progression the gelatinous matrix is quickly resorbed. Some leptocephali (those of the white-spotted conger eel, *Conger myriaster*, for example) convert the glycosaminoglycan component hyaluronan (which aids in control of water content) to glucose during metamorphosis, which might then be metabolized to glycogen and stored for use in further ontogenesis (Kawakami *et al.* 2009).

The matrix of bonefishes (*Albula* spp.) early in Stage 2 consists of sulfated keratan glycosaminoglycan (Pfeiler 1984, 1986, 1996; Pfeiler *et al.* 1991) in the form of repeating disaccharide chains displaying dominant anionic charge densities. Its complement in the Atlantic tarpon is heparan glycosaminoglycan and, in the ladyfish (*Elops saurus*), a form of chondroitin sulfate (Pfeiler *et al.* 1991). Glycosaminoglycans cause water retention and influence the distribution of ions, including sodium. Water content of some Stage 1 leptocephali is therefore >90% of total weight (Pfeiler 1986 and references). In bonefishes undergoing metamorphosis and not yet feeding, water and carbohydrate content both decrease ≈80% over 10 days or so in step with simultaneous loss of the gelatinous matrix (Pfeiler 1986, 1999). Although protein remains constant, the amount of collagen, lipid, and ash diminish by half. These events are indication of water accumulation during Stage 1. As Pfeiler (1986: 7) stated: "To argue otherwise would require that the recently hatched leptocephalus contain an amount of water equal to that of a fully developed Phase [stage] I leptocephalus ...." In other words, 10 times the whole-body wet weight at hatching.

Total glycosaminoglycans in bonefishes declines ≈87% during Stage 2 (Pfeiler 1984), which probably accounts for the substantial losses of water, sodium, and chloride. This reasoning assumes that glycosaminoglycans are synthesized during Stage 1, which they almost certainly are. As Pfeiler (1986: 7) wrote: "Again, to argue otherwise would require a high [glycosaminoglycan] content in recently hatched embryos which remains constant during the time when larvae form an extensive amount of gelatinous matrix."

Pfeiler (1986) outlined a hypothetical model. After the yolk sac has been consumed, Stage 1 leptocephali generate large quantities of gelatinous material, presumably a result of glycosaminoglycan synthesis. One consequence is water loading without altering the percentage of water to total wet weight (i.e. it remains balanced at ≈ 90%). Salt loading occurs simultaneously. Both are probably associated with the synthesis of polyanionic glycosaminoglycans (Pfeiler 1986, 1999). The strong anionic charge in the matrix drives the uptake and accumulation of sodium and chloride from the surrounding seawater. As Stage 2 ends and metamorphosis begins, glycosaminoglycans are catabolized, destabilizing tissue water and salt concentrations, which are then diminished through loss to the external environment.

But what about front-end loading? What serves as raw material for glycosaminoglycan synthesis, and where does it originate? As Pfeiler argued, that such large amounts of finished glycosaminoglycans could be retained in the yolk sac throughout Stage 1 until onset of Stage 2 is unlikely. As mentioned, DOM is the most likely raw material, its uptake enhanced by a high surface-to-volume ratio of the leptocephalus' flattened form. The conspicuous teeth, which Pfeiler (1986: 8) rightly called "enigmatic structures," warrant mention too. Their size and shape could easily mark them as the teeth of a planktonic predator, but this is apparently not the case. The teeth are resorbed or lost prior to metamorphosis when feeding on live prey starts. A function has not been identified (Pfeiler 1999).

The Stage 1 leptocephalus of tarpons has yet to acquire functional digestive and excretory systems, and identifiable food material that might properly be called "prey" (e.g. live plankton) has not, to my knowledge, been found in the gut of a first-stage leptocephalus of either species of tarpon. Leptocephali of some eels that remain in the plankton for extended periods are suspected of consuming "marine snow," including discarded appendicularian houses rich in microorganisms, zooplankton fecal pellets, and other forms of particulate organic matter (Deibel *et al.* 2012; Miller *et al.* 2011, 2013; Otake *et al.* 1990, 1993). In other words, those leptocephali that feed do so at a lower trophic level at which minimal energy is expended (Pfeiler 1999). Stage 1 leptocephali in general have low metabolism, meaning their demand for energy is low too (Pfeiler and Govoni 1993). Whatever their source of energy – dissolved or particulate organic matter, stored lipids or glycosaminoglycans – little is required to sustain metabolic functions. In any case, that bonefish, ladyfish, or tarpon Stage 1 leptocephali ingest exogenous matter is doubtful.

The contribution of free amino acids to energy reserves of larval bonefishes is minor (Pfeiler 1996). The combination of endogenous carbohydrate (principally keratan glycosaminoglycans) and lipid fuels metamorphosis during Stage 2, contributing, respectively, ≈20% and ≈ 80% of the overall energy budget (Pfeiler 1984, 1986, 1996). This adaptation diverges from the typical situation in which the yolk provides stored energy as lipid and protein until feeding begins, and carbohydrate contributes little (Pfeiler 1986). Growth in these other fishes is minimal through the yolk-sac phase, seemingly held in abeyance, commencing once feeding begins and proceeding uninterrupted to the juvenile stage. And DOM? Any contribution to metamorphosis is doubtful, at least to metamorphosing bonefish larvae. Unfed specimens kept in nutrient-free artificial seawater containing only inorganic compounds survived and developed normally (Pfeiler 1996). Whether dissolved organic compounds are taken up from the sea during Stage 1 and used as precursors of glycosaminoglycans or immediate energy sources is yet to be determined.

CHAPTER 2
Growth

2.1 Introduction

Millward (1995: 94) defined *growth* as "irreversible structural change ...." If we take this at face value – that is, in the absence of context – then the shrinkage of Stage 2 tarpon leptocephali could indeed be termed "negative growth." However, doing so would be misleading because vertebrate growth's principal feature involves interactions among the linear extension of bone, protein deposition in skeletal muscle, and dietary intake of protein (Millward 1995). In a less specific but more conventional sense, growth refers to increased size over time. *Size*, as I use the term, includes a fish's weight at the time its length is measured. Alternatively, body length or a proxy for it can be used to calculate weight, obviating the need of a weighing device. For example, girth or body height might be used instead of length, or length in combination with one of these other factors (Ault and Luo 2013; Jones *et al.* 1999). Any size determination must conform to statistical constraints of *accuracy* (degree of closeness of the measured value to its true value) and *precision* (the reproducibility of multiple measurements within an acceptable range of variation under the same conditions).

Unlike birds, mammals, insects, and many other creatures, fishes ordinarily continue to gain size after sexual maturity, a phenomenon called *indeterminate growth*. In fact, a fish typically grows until the moment of its death. In a sense, asking how *old* do fishes become is little different than asking how big do they get. Age-related growth rate in fishes can only be hypothetical, assuming a fish never quits growing in length, although to model the process requires setting asymptotic length to zero, in effect momentarily stopping the clock long enough to take a measurement. Actual growth (as opposed to modeled growth) has no endpoint; in the context of lifespan, age and growth are moving targets.

Because growth of the Pacific tarpon has not been well described, most of the species-specific information presented is for the Atlantic tarpon.

Tarpons: Biology, Ecology, Fisheries, First Edition. Stephen Spotte.

2.2 The cube law

The *cube law* in its various incarnations has persisted since 1638, when Galileo contended that the volume of an object increases as the cube of its linear dimensions (Froese 2006; Hirst 2012). Growth is of obvious theoretical interest, but it has a practical side too. As any angler knows, measuring a fish – especially a very large one – is usually faster and easier than weighing it. Assuming a fish's weight can be estimated with accuracy and precision from its length, weighing could be dispensed with, and the only equipment needed would be a simple measuring device. However, weight and length seldom increase in proportion over time. Instead, a healthy fish usually continues to gain weight and become plumper while its growth in length tapers off, or attenuates (i.e. tends toward the asymptotic). When applied in biology, in other words, the cube law encompasses a range of values, not simply a value cubed.

The relative proportions of mass and length in growing organisms (not only fishes) can still be described by a power law expressed arithmetically as:

$$W = aL^b \tag{2.1}$$

in which W = weight in grams (g) or kilograms (kg), L = length in millimeters (mm) or centimeters (cm), and a and b are constants. Parameter a is the arithmetic coefficient of equation 2.1 and parameter b the arithmetic exponent. Expressed logarithmically,

$$\log_{10} W = \log_{10} a + b(\log_{10} L) \tag{2.2}$$

in which a is the intercept of the regression and b the slope of the curve. Equation 2.2 shows that a decrease in the regression line's slope results in an increase in the intercept, and the reverse is also true. The cube law, as the name implies, exists in its idealized form when the power exponent b is exactly 3.0. Then W increases as the cube of L, and growth is *isometric*; that is, body proportions stay constant over time. If, in a regression plot, the 95% confidence intervals (CIs) include the value 3.0, growth of the species assessed is isometric. A fish that doubles in length while its shape remains isometric has a surface area that increases by a factor of 4 (i.e. 2^2) and a volume that increases 8 times (also expressed as 8×), or 2^3. Stated differently, when $b = 3.0$, small specimens have the same general shape and form as big ones (assuming all have been tested together), and body proportions remain unchanged.[1]

Values of $b < 3.0$ indicate more rapid growth in length than in other directions, as seen by a body shape that lengthens out of proportion with weight during

[1] Change in shape is actually possible when $b = 3.0$ because of changes in a, the proportionality constant (Jones *et al.* 1999).

ontogeny, and the fish becomes less rotund with increasing length. Values of $b > 3.0$ define more rapid growth in directions other than length (e.g. girth, thickness, height), and such specimens will appear plumper. In both cases growth is *allometric* because changes in the weight-length relationship are not proportional. A situation in which $b = 1.0$ describes a cylinder (e.g. a pipe of constant diameter) in the process of elongating.

Values of b outside the range 2.5–3.5 are rare for the thousands of fish species evaluated to date. In the 2005 version of FishBase used by Froese (3929 values for log W vs. log L covering 1773 species), the range of b was 2.7–3.4 with $\bar{x} = 3.03$ (see Froese 2006). Values of b <2.5 can arise when samples from which they originate have narrow ranges. Values both higher and lower than 3.0 can also result from small sample sizes, with closer clustering to the mean expected as the number of measurements increases.

It bears mentioning that although SL seems the most widely used measurement, TL is preferable to SL and FL for statistical and computational reasons, such as computing growth curves (Pardo *et al.* 2013). When applying the power law, for example, the type of length measurement used affects a but not b, the effect on a being TL < FL < SL (Froese 2006). Total length is also used to derive standard species-specific weight equations when relative weight is used to evaluate the general "condition" of a fish or the population to which it belongs (Section 2.4).

2.3 Sexually dimorphic growth

Atlantic tarpons exhibit *sexual dimorphism* (one sex grows larger than the other), and in this case the pattern is for females to outgrow males (Andrews *et al.* 2001; Chacón Chaverri 1993, translated in Appendix D; Crabtree *et al.* 1995, 1997; Cyr 1991: 54; Ferreira de Menezes 1967; Ferreira de Menezes and Pinto Paiva 1966). The reason is partly because females tend to live longer (Cyr 1991: 58, Ferreira de Menezes and Pinto Paiva 1966). In 1962, Ferreira de Menezes (1967) measured 535 tarpons (313 males, 222 females) taken from fish weirs and stationary nets set along the coast at Almofala (Acaraú, Ceará State) in northeastern Brazil. Males ranged from 725 mm FL (4.6 kg) to 1475 mm FL (39.0 kg), females from 825 mm FL (5.0 kg) to 1875 mm FL (67.0 kg). Application of equation 2.2 yielded:

$$\text{males: } \log_{10} W = -4.887 + 2.97(\log_{10} L) \tag{2.3}$$

$$\text{females: } \log_{10} W = -4.745 + 2.90(\log_{10} L) \tag{2.4}$$

The results are interesting for at least two reasons: in both sexes, values of b are <3.0; also, males are plumper than females. Mature females are probably thinner shortly after spawning, but whether females are generally thinner than males in this population is unknown. In any case, both sexes seem unusually thin. Values of $b > 3.0$ would be expected. Data were clustered in 50 mm ranges (males 700–1500 mm FL, females 750–1900 mm FL) instead of processed in a spreadsheet format using weight and length values of each fish. This blurred distinctions among individuals and reduced the sample size. In addition, breeding status was not considered, although the weirs and nets produced throughout the year. An earlier publication by this research group reported monthly gonadal maturation data for 10 052 Atlantic tarpons from 1962 through 1964 (Ferreira de Menezes and Pinto Paiva 1966), but we have no way of knowing which of these specimens were included in the later log *W* vs. log *L* relationship study of Ferreira de Menezes (1967).

Shape-shifting during ontogeny (Fig. 1.4) might make the *W* vs. *L* relationship in Stage 2 larvae and fry (Stage 3) of both species difficult to ascertain. Harrington (1958) collected young Atlantic tarpons in salt marshes of eastern Florida's Indian River Lagoon. They ranged from Stage 2 (16 mm SL) to early juvenile (Stage 4) at ≈45 mm SL (Fig. 2.1). Also see Rickards (1968: 230 Fig. 3), who provided a log-log plot of *W* vs. *L* for young tarpons from salt marshes at Sapelo Island, Georgia (range 19.6–273.5 mm SL, $n = 280$).

The *L* vs. *W* relationships between male and female Atlantic tarpons from southern Florida (sexed and unsexed fish combined, range 102–2045 mm FL, total $n = 1469$) did not differ significantly when the standard power equation was applied (Crabtree *et al.* 1995). However, females grew significantly longer than males (range 331–2045 mm FL vs. 203–1884 mm FL) and also lived longer (range 0–55 years, $n = 298$ vs. range 0–43 years, $n = 141$). The oldest female was 2045 mm FL, the oldest male 1710 mm FL. Fish of age one year (<400 mm), being juveniles, were difficult to sex. Data from some of these specimens ($n = 467$) were used in an earlier analysis by Cyr (1991: 44). The complete sample was admittedly biased, many fish having been obtained from recreational fishermen and taxidermists and thus more representative of trophy-sized specimens than a true range of size and age (Cyr 1991: 54). The length-frequency distribution with sexes combined was bimodal, revealing peaks at 500–700 mm SL and 1400–1600 mm SL (Cyr 1991: 49, 65 Fig. III-1A). Weight was distributed similarly, with peaks at 0–10 and 25–50 kg, again representing juveniles and specimens from the recreational fishery (Cyr 1991: 49, 66 Fig. III-2A). Separation of sexes produced a weakly bimodal pattern with females generally longer (Cyr 1991: 49, 65 Fig. III-1B, C) and heavier (Cyr 1991: 49, 66 Fig. III-2B, C).

The weight of male and female Atlantic tarpons increases slowly from 1–5 years, then more rapidly from 5–20 years in females (Cyr 1991: 52). Male growth curves are distinctly asymptotic at 15–20 years and 30–40 kg, implying determinate growth. Conclusive evidence of this remains to be found, and true determinate growth, in my opinion, is doubtful. Growth in females slows with

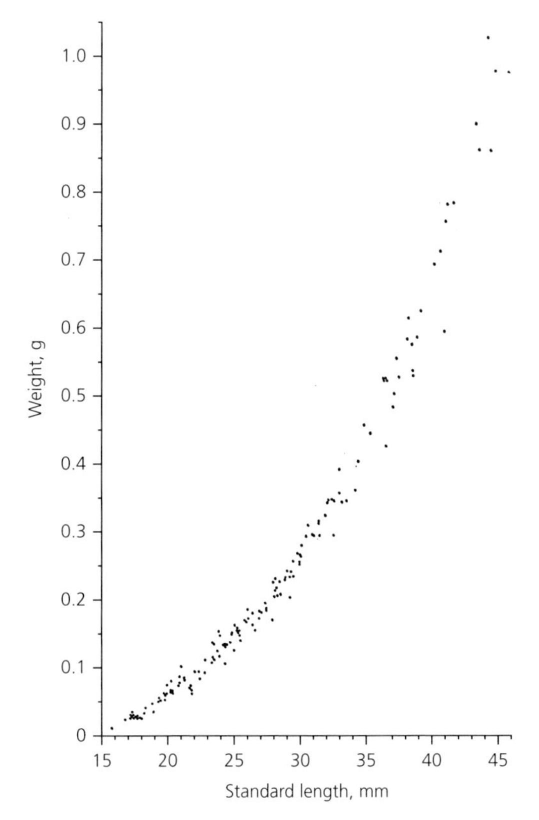

Fig. 2.1 Weight vs. length relationship for Stage 2 Atlantic tarpons from Indian River Lagoon, eastern Florida (range = 16.0–45.5 mm SL, *n* = 154). Source: Harrington (1958: 5 Fig. 3, plotted using equation 2.20). © American Society of Ichthyologists and Herpetologists. Reproduced with permission.

age, but unlike the case of males is clearly indeterminate. In small Atlantic tarpons, growth in length of both sexes outpaces gain in weight (Cyr 1991: 54). At 1000–1200 mm SL the pattern reverses: relative growth in length begins to decline, while the rate of weight gain rises.

Based on these observations, Cyr (1991: 54–55) concluded that weight is the preferable criterion for measuring growth, especially in older tarpons when length tends toward asymptotic. He considered length the better indicator in tarpons <1000–1200 mm SL. This remains problematic, once again in my opinion. Cyr (1991: 55) acknowledged that "Variability in weight at length increased with increasing fish size." He noted how this was especially true of large females in his sample, which in breeding condition had gonadal weights twice those of males ($\overline{x}$ = 2.4 kg vs. $\overline{x}$ = 1.22 kg). Many abiotic and biotic factors can affect weight in the short term. My opinion is that length, despite slowing or perhaps even stopping, still seems the truer measure (Section 2.4).

2.4 Condition

In context of the cube law a fish's "condition" is represented by parameter *a*, the intercept of the regression for a species relative to an established standard (Safran 1992). That the cube law does not apply perfectly to living organisms has been known for many years (Froese 2006; Fulton 1904). Its application also has built-in biases. Hayes *et al.* (1995) reported that estimates of mean weight at given length tend to be low; conversely, estimates of the intercept display a high bias. Still, the log *W* vs. log *L* relationship has two uses: first, to enable either weight or length to be predicted from the other factor; second, as an indicator, or "index," of general health and nutritional status. This latter function has been applied historically to assess both the individual condition of fishes sampled from populations and, by extrapolation, the health status of the populations themselves (e.g. Buchheister *et al.* 2006).

Several indices have been published for making such assessments (Bolger and Connolly 1989). Pope and Kruse (2007: 423) defined *condition* as "the well-being or robustness of an individual fish ...." They continued: "It has typically been estimated by comparing an individual fish weight to a standard weight for a given length and assuming that larger ratios (condition index) reflect a healthier physiological state .... Measures of condition are generally intended to be an indicator of tissue energy reserves, with the expectation that a fish in good condition should demonstrate faster growth rates, greater reproductive potential, and higher survival ...." *Standard weight* is the predicted weight at a given total length for a species of fish, and is derived from making numerous measurements and applying the power law.

Thus condition – in principle – offers fishery biologists a crude measure of commercial or ecological health, and plumper is thought to be better. In an evolutionary context, condition can be taken as an indirect indicator of fitness, assuming the parameters used to define it are empirically derived and capable of being tested separately from the data used to determine "condition." I wonder how often this distinction has actually been made in practice. If and when it has, any resultant model acquires putative strength by presenting a functional foundation as opposed to being purely phenomenological; that is, statistical. A positive correlation alone can never be evidence of function. This issue aside, measures of condition, whatever they are, should also be examined for insidious logical fallacies. When fishes heavier in proportion with length are considered fitter because of their "superior" condition, the reasoning teeters on the edge of circularity; the conclusion is contained in the premise, and heavier fish are fitter because they weigh more.

Despite its superficial simplicity, a truly accurate and precise method of determining condition under varied circumstances has proved elusive (Blackwell *et al.* 2000; Jones *et al.* 1999; Pope and Kruse 2007). Pope and Kruse (2007) rightly noted that any such attempt is contingent, affected by factors that impinge on growth (e.g. water temperature, season, spawning status). Condition is often expressed as a dimensionless ratio in which the quotient yields an "index," or coefficient, along a scale allowing rapid, arbitrary evaluation of an individual or population. Froese (2006) wrote that the earliest such attempt to devise a measure of condition seems to be Fulton's K (Fulton 1904), although he could not find mention of it in Fulton's report, and neither could I, no doubt because Fulton said nothing about it (Nash *et al.* 2006). Jones *et al.* (1999) cited Fulton (1911) as the original source, but this work is a history of English maritime law. Fulton's condition factor, $K = \frac{W}{L^3}$, therefore was never Fulton's. His putative formula, whatever its origin (Nash *et al.* 2006), is still used and can be written

$$K = \frac{W}{L^3\left(10^5\right)} \tag{2.5}$$

in which W is weight (g), L is length (mm) and 10^5 is a scaling constant, or multiplier, rounded to one decimal. This constant varies according the units used. Ezekiel and Abowei (2013) applied a scaling constant of 10^2, for example, which is appropriate for their units of measure (kg and cm), yielding a condition factor ranging from zero to unity, except the formula they presented was written incorrectly. Equation (2.5), which presumes that growth is isometric, can also be written

$$K = x\left(W\right)\left(L^{-3}\right) \tag{2.6}$$

in which x is a scaling constant varying with the units of measure (Jones *et al.* 1999). When growth obviously is allometric, as it almost always is, $b \neq 3.0$. As pointed out by Jones *et al.* (1999), if b is >3.0, K and L have a positive relationship, meaning that K and L increase together, although if b is <3.0, making the K-L relationship negative, K decreases with increasing L (Cone 1989).

The general condition of two populations of the same species can be assessed by examining the confidence intervals around the parameters (i.e. the values of a and b) calculated from equation 2.2. I shall use an example from Pope and Kruse (2007: 427–432 Box 10.1) for two populations, here labeled A and B. The slope of the regression, b, for population A = 3.069; for population B, b = 3.143. The 95% CIs for A show the range 2.977–3.160; for B the range is 2.934–3.352. Note that overlap of the two ranges is almost total, and that at least one (in this case both) CI envelops the slope of the other, indicating the values of b do not differ significantly (i.e. the weights of individuals sampled from the two populations change in lockstep). Intercepts (parameter a) in the example were similar too. Pope and Kruse (2007: 431) wrote: "Thus we conclude that although the respective transformed equations [equation 2.2 here] predict different average weights, neither population is significantly heavier or better conditioned than is [*sic*] the other…."

Pope and Kruse (2007: 432) gave an example of what to do in a situation when the CIs overlap but neither encompasses the slope of the other. In this case the standard error (SE) of the difference in the slopes b for populations C and D of the same species is:

$$\sqrt{SE_C^2 + SE_D^2} = \sqrt{0.2^2 + 0.15^2} = 0.25 \qquad (2.7)$$

and the CI for the difference is 0.6 (±1.98 × 0.25), with 0.6 representing the difference between b for the two populations and 1.98 is the t-value for $\alpha = 0.05$ (see Pope and Kruse 2007 for how t was derived). The CI does not include zero, evidence of statistically different values of b.

Blackwell *et al.* (2000) evaluated several indices for condition, concluding that relative weight (W_r) might be the best (but see Cone 1989):

$$W_r = \frac{W}{W_s(100)} \qquad (2.8)$$

in which W is the weight of the fish and W_s is the standard weight for a fish of the same length and species. Most of the published standard weight equations have been derived for North American freshwater fishes (Blackwell *et al.* 2000). I could not find one for the Atlantic tarpon. The misnamed Fulton's condition factor (equation 2.5) uses the simple cube law and assumes isometric growth, to which Fulton devoted much ink in his lengthy 1904 report refuting based on extensive data in which he showed that fishes grow allometrically and the cube law, if strictly applied, is inaccurate. Fulton (1904: 141) wrote:

> I have found that the law which governs the relation between the weight and dimensions of similarly-shaped bodies does not apply with precision to fishes. They increase in weight more than the increase in length would, according to the law, imply, and since the number of fishes in which the relation between the length and weight has been determined was large, viz. 5675, belonging to nineteen species, and in no case has the law been found to apply exactly, it appears to be well-established that on the assumption that the specific gravity[2] of the fishes does not change during growth they must increase

[2] Fulton recognized that the mass of a fish could be obtained directly by weighing it in air or calculating how much it weighed after first determining its volume. He assessed the specific gravity of trawled fishes because he was interested in whether volume might be more useful than direct weighing. In either case the independent variable would be length. *Density* is a measure of an object's mass per volume and represents an absolute quantity (kg/m^3 in SI units). *Specific gravity* is the ratio of the density of an object and the density of a fluid such as water, usually pure water at 4°C, which has a density of 1000 kg/m^3. The result is dimensionless (being a ratio of two densities the units cancel out). To determine the specific gravity of a solid the object is weighed in air and again while immersed in water. The difference in the two determinations, by Archimedes' principle, is the weight of the water displaced by the volume of the solid. Alternatively, the water displaced can be siphoned off and weighed, which is apparently what Fulton did (he mentioned the use of burettes for small fishes and a siphoning method for collecting the water displaced by big ones). His actual hands-on procedure is vague, and weight of the water must have been determined somehow. In any case he abandoned the volumetric method as too tedious and ended up simply measuring and weighing his specimens, presumably in air.

in some other of their dimensions, whether breadth or thickness, in greater proportion than they increase in length.

Beebe and Tee-Van (1928: 36) observed this in juvenile Atlantic tarpons, noting that proportional growth in various body measurements from juvenile to adult is "remarkably uniform" (i.e. tends toward isometric), as later demonstrated by Harrington (1958) and Wade (1962) and discussed in Chapter 1.3. However, "In weight, the relations ... are tremendously disproportionate." They pointed out that at 76 mm an Atlantic tarpon weighed 5 g, at 200 mm 65 g, but upon reaching 1020 mm was 16 300 g (16.3 kg). Harrington (1958: 6) noticed this too and was astonished, remarking that extrapolating his measurements of early growth revealed that "a tarpon 100 cm. long would weigh 6×10^{21} metric tons!"

The growth of fishes is obviously more complex than the a simple weight-to-length relationship, and using just these two variables yields a doubtful indicator (Jones *et al.* 1999). If a third variable (some measure of body shape) is introduced into a plot of log W vs. log L, the value of a could change considerably. The W vs. L relationship is affected by whether a species is compressed, fusiform, elongated, anguilliform, and so forth, and some investigators have taken this into account (e.g. Ault and Luo 2013; Jones *et al.* 1999; Kulbicki *et al.* 2005). Inserting shape into the mix as a categorical variable (Fig. 2.2) or another

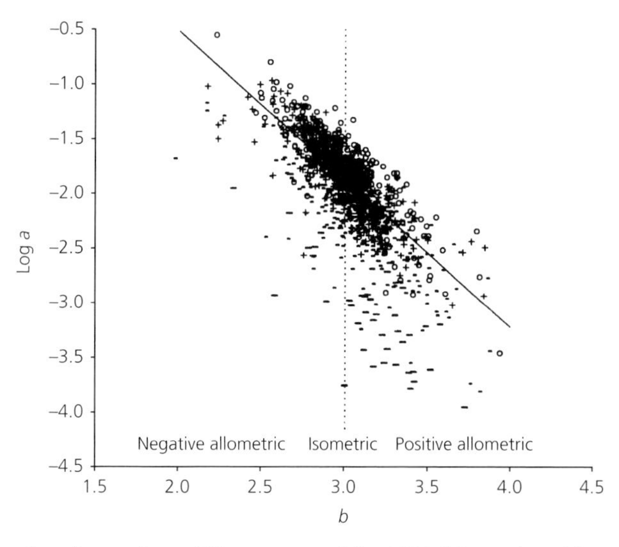

Fig. 2.2 Scatter plot of mean log a (TL) over mean b for 1223 fish species with accompanying body shape information: clear circles, short and deep; plus signs, fusiform; short horizontal bars, elongated; long horizontal bars, eel-like. Areas of negative allometric, isometric, and positive allometric change in W vs. L are shown. Regression line is based on a robust analysis of fusiform species ($n = 451$): intercept = 2.322 – 0.133 = 2.189, slope as in log $a = -1.358b + 2.322 - 1.137$ (1 if eel-like, otherwise 0), –0.3377 (1 if elongated otherwise 0), –1331 (1 if fusiform otherwise 0). Source: Froese (2006: 249 Fig. 9).

continuous variable derived from some combination of length and girth, thickness, or height is worth considering.

The W vs. L relationship alone is obviously an unreliable indicator of condition regardless of how the term is defined. Adding a third morphometric factor, as just noted, has value by bringing the condition concept into finer resolution. Jones *et al.* (1999) sought a method of assessing condition using just the proportionality constant. They proposed the model

$$W = BL^2H \tag{2.9}$$

and tested it against Fulton's condition factor, K, in the form $W = KL^3$, in which K was determined by linear regression of W against L^3. The other measure included for comparison was the standard power law ($W = aL^b$). Parameter B in equation 2.9 was obtained by regressing W against L^2H where H represents body height at its highest point. The intercept, a, was forced through the origin for determinations of B and K. Atlantic salmon (*Salmo salar*) and chinook salmon (*Oncorhynchus tshawytscha*) from hatcheries were the test fishes, but data from other species having a variety of shapes and sizes were also used to assess the model's sensitivity. Body height, H, was selected as the additional morphometric variable because measuring girth or body thickness requires longer handling times (Chapter 8.3); determining H requires only a quick measurement on one side of the fish. Furthermore, preliminary measurements of both girth and thickness induced greater variation in the regression parameters compared with H, and thickness, as the smallest of these whole-body assessments, is therefore accompanied by the highest probability of error.

Equation 2.9 proved superior to the others. Scatter plots indicated that values for the coefficients of determination (r^2) were tighter, and so was accuracy. The relationship between W and B was substantially better in both salmon species, with B seemingly species-specific. Fulton's K in particular became less accurate as the fishes increased in size. Unlike K, B varied little with age, offering a way of comparing a fish's condition over time. Parameter B was superior to K over a variety of morphological shapes (e.g. fusiform, anguilliform). By encompassing two dimensions, this model also captured body thickness. If a girth measurement is required, it can be obtained from B and L (Fig. 2.3).

The salient point so far is that fishes usually grow allometrically, not isometrically, meaning that weight and length do not increase proportionately (i.e. $b <$ or > 3.0). Rickards (1968) measured and weighed 280 juvenile Atlantic tarpons of 19.6–273.5 mm SL and derived the regression equation

$$\log W = -6.78247 + 3.18689(\log L) \tag{2.10}$$

As seen, slope of the growth curve, b, is >3.0, demonstrating that weight increases out of proportion with length as a juvenile tarpon grows. This is typically true of fishes. As mentioned, Fulton (1904) found the same trend in an

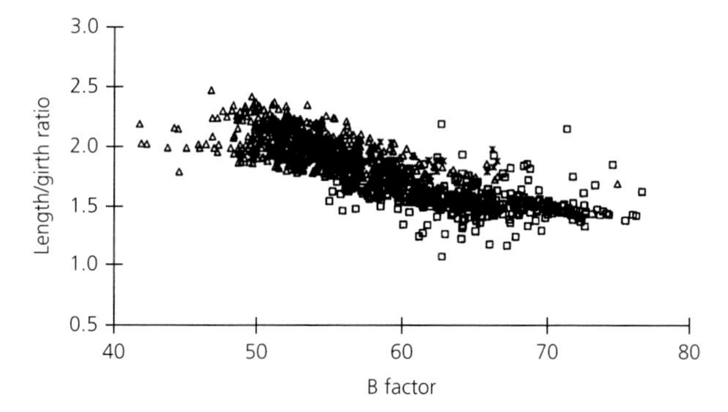

Fig. 2.3 Change in B with length/girth ratio in salmons. Clear triangles, Atlantic salmons; clear squares, chinook salmons; x, smolts. See text for explanation of the B factor. Source: Jones *et al.* (1999: 274 Fig. 6).

early analysis in which he tested the cube law using data from 5675 adult specimens of 19 species of fishes caught commercially in the Scottish North Sea. He also found the log W vs. log L ratio to vary by location, season, and physiological status. Nonetheless, he wrote (Fulton 1904: 142), "It is obvious ... that the true criterion of growth is the increase in the mass of the fish ...." This statement is not entirely true. Change *in length* is more reliable despite indeterminate growth because a fish's weight is influenced by sex, spawning and nutritional status, water temperature, and other factors. Were fishery biologists restricted to determining just one aspect of growth, it would have to be length. Male Atlantic tarpons might be exceptional (Section 2.3), although this has not been confirmed. In a comprehensive analysis of how the external body parts of fishes grow in length, Martin (1949), for example, reported values of b shifting with developmental stage and was able to change them experimentally by altering water temperature and by starvation. The log W vs. log L relationship is obviously dynamic in fishes for both body length and the lengths of their separate parts.

2.5 Growth rate

As discussed in Chapter 1, having moved inshore the subsequent ontogeny and growth of young tarpons are rapid (Fig. 2.4). As illustrated, from strange leptocephalus forms emerge recognizable tarpons, in this example within 50 days.

Atlantic tarpon juveniles at Sapelo Island, Georgia grew ≈30 mm SL/month (≈1.0 mm SL/d) through summer and autumn (Rickards 1968), remarkably close to the growth rate of captive juvenile Pacific tarpon larvae of 31.4 mm SL/month reared in freshwater ponds by Alikunhi and Rao (1951: 107).[3] The largest

[3] Alikunhi and Rao (1951) did not state whether their length measurements were FL, SL, or TL. I am assuming SL.

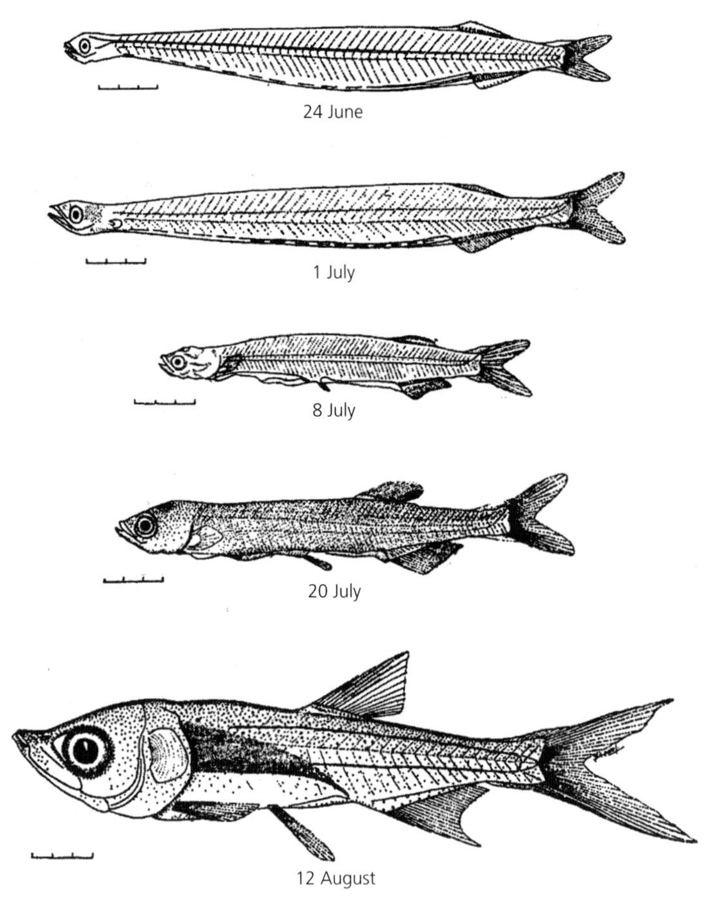

Fig. 2.4 Growth sequence over 50 days of Atlantic tarpons from inshore locations of Golfo de Morrosquillo, Colombian Caribbean. Scale bars in mm (0–3). Source: Dahl (1965: 11).

grew to 387.35 mm SL ($\bar{x}$ = 283.66 mm SL) in 289 d. According to Chacko (1948: 210) and Job and Chacko (1947), Pacific tarpons in freshwater pond culture at Madras (now Chennai), India grew to 14 in. (≈356 mm) and 0.68 kg the first year, or a rate of 0.97 mm/d. Juvenile Atlantic tarpons in an isolated pond in the Florida Keys grew at ≈1.43 mm/d in September when water temperature stayed >33.5°C, then the rate slowed by half in October when water temperature declined to a low of 29.5°C (Moffett and Randall 1957). Harrington (1966) noticed a similar pattern in specimens caught farther north in eastern Florida's Indian River Lagoon: growth was rapid until late November, then minimal until mid-April during the time of coldest temperatures and shortest day lengths. Water temperatures were recorded in early afternoon on sunny days; unrecorded nighttime values were undoubtedly lower.

In mid-August 1957, Moffett and Randall (1957) seined 30 juvenile tarpons of 238–388 mm FL from drainage ditches on Florida farmland and measured and tagged five, which were released at the capture site. In mid-October the fish were recaptured and measured again. Growth rate over 2 months was $\bar{x}$ = 1.0 mm FL/d (*n* = 5), similar to fish in Georgia at about the same time of year. Such rapid increase in size could easily account for the large range of lengths reported (e.g. 16.0–109.0 mm SL in Indian River Lagoon by Harrington 1958). However, hatch and recruitment dates also factor in, considering that larvae hatch on different dates over several months, moving inshore on propitious tides or during overwash events (Chapter 4.4, 4.6).

Water temperature in the subtropics is always an influential factor. Cyr (1991: 33–35, 42 Fig. II-3) estimated the growth of juveniles at two southern Florida sites (Collier-Seminole and Jack Island) to be seasonal, slowing to asymptotic from October through March and accelerating in spring and summer. Average first-year growth at Jack Island, based on length-frequency distributions, was 230 mm SL, or 0.50% SL/d (April-September) and 0.11% SL/d (September–February) of which 79% occurred in 5 months (May–September), when growth was exponential. Harrington (1966) also reported seasonal growth in captive Atlantic tarpons kept outdoors in southern Florida. Growth slowed 85% from 22 September to 1 November. None of this is surprising. Although Florida is within the Atlantic tarpon's normal reproductive range, it lies at the northern extent. Juveniles collected by Harrington (1966) caught in Indian River Lagoon and kept in outdoor tanks grew at 2.66% SL/d during warm weather, slowing to 0.58% SL/d in November. Breder (1944: 230 Table VII), who monitored Atlantic tarpons of 345–390 SL on Florida's southwestern coast, saw growth vary from not discernable to 0.05% SL/d, depending on time of year.

Fish growth rate can be estimated indirectly from length-frequency distributions or directly from tag-release-recapture studies. The first method loses both accuracy and precision from continuing recruitment, emigration and immigration, mortality, variations in individual growth rates, and sampling bias (Zerbi *et al.* 1999, 2005). These difficulties often make tag-release-recapture in nature a more meaningful technique by providing real-time data on such crucial factors as growth rate and residency times in the habitat, although the stress of being handled and carrying a tag introduces confounding variables (e.g. possible higher mortality from predation, incidence of infection at the tag site, reduced ability to capture food).

Using a tag-release-recapture protocol, Zerbi *et al.* (1999) monitored recruitment of Atlantic tarpons onto flooded mudflats (water depth ≈ 30 cm) located landward of mangroves bordering Rincón Lagoon, part of Boquerón Bay in southwestern Puerto Rico. Daily growth rate and duration of residency were assessed over 2 y. Examination of otoliths (microzonation; see Casselman 1983: 3) from unmarked juveniles and those injected a single time with oxytetracycline laid down one microannulus daily (Zerbi *et al.* 2005).

Late larvae and early juveniles appeared from June to February, and recruitment strength varied substantially from year to year (Zerbi *et al.* 1999). Juvenile tarpons recruited mostly in August and September and grew at 1.1% SL/d, attaining ≈200 mm SL between October and December. Residency of juveniles was unpredictable, lasting from June through February when the mudflats and lagoon retain a connection. During this period, tarpons and other fishes collected in vernal pools along the flats, which could be seined easily. The flats dried out near the end of February, by which time most of the tarpons had presumably emigrated to the lagoon. Maximum residency on the mudflats (the two years combined) was 126 d ($\bar{x}$ = 20 ±1.8 SEM d). Maximum size attained before emigration over years 1 and 2 were, respectively, 180 mm SL and 230 mm SL. Breder (1944: 230 Table VII, 231) mentioned three aquarium-reared juveniles captured at Key West, Florida growing from < 500 mm TL to a maximum of 1220 mm TL over 5 y, a rate of ≈212 mm TL/y, or ≈0.58 mm TL/d. Zerbi *et al.* (2005) reported that growth of juvenile tarpons was unaffected by abiotic habitat factors (e.g. temperature, ionic strength, dissolved oxygen, turbidity) despite extreme daily fluctuations. At their study location near Boquerón, Puerto Rico (18.0°N, 67.2°W), temperature is unlikely to be limiting, despite minor site variations.

Mateos-Molina *et al.* (2013) tagged and released tarpons (216–886 mm FL, $\bar{x}$ = 490 mm FL, n = 438) at six locations within Boquerón Wildlife Refuge, including continuously flooded canals, the Rincón Lagoon mentioned above, and adjacent lagoons. Maximum residency in the refuge was 337 d. One fish had a growth rate of $\bar{x}$ = 0.17 mm FL/d (size at capture unstated). Another of 474 mm FL (presumably at first capture) grew at 0.8 mm FL/d. Internal lagoons of the refuge yielded 287 tarpons, 95% of which were 400–600 mm FL and their estimated ages 1–3 y, assuming growth data from Florida are comparable (see Crabtree *et al.* 1995).

2.6 Modeling growth

Models of growth in fishery science are important because they help in assessing such factors as general health and nutritional status of populations, providing information about age at maturity, survivorship, lifespan, and effect of exploitation. Several models of fish growth have been developed and applied, but among the most important and widely used is one devised by Karl Ludwig von Bertalanffy (1938). The von Bertalanffy function, although not perfect, remains among the best and most popular descriptors of growth in fishes. The equation requires three parameters: L_∞, k, and t_0 (or l_0). Metabolism comprises two opposing factors, anabolism and catabolism. For modeling purposes Bertalanffy considered growth (defined previously as the increase in size over time) to be their net difference. Knowing that growth rate declines with size, and incorporating the cube law (the presumption of isometric growth and thus the proportionality of weight and length), he defined growth in length to be

$$\frac{dl}{dt} = k\left(L_\infty - L\right) \tag{2.11}$$

in which L is body length (cm), t is time, k is growth rate (i.e. the coefficient of annual growth), and L_∞ (called "L infinity") the asymptotic length (in centimeters) at which growth equals zero. Of course, in animals having indeterminate growth the value of L_∞ can only be hypothetical, a trend rather than a fixed point. After integration, equation 2.11 becomes

$$L_t = L_\infty\left(1 - e^{-k(t-t_0)}\right) \tag{2.12}$$

Alternatively,

$$L_t = L_\infty - \left(L_\infty - L_0\right)e^{-kt} \tag{2.13}$$

Expressed either way, L_t represents body length at age (time t in years), L_∞ is asymptotic size, and k the von Bertalanffy growth coefficient. Parameter t_0 adjusts equation 2.12 to the starting length of the target organism; that is, to the age at which its length equals zero. Its complement, L (equation 2.13), is body length at age zero. Again, because growth in most fishes is indeterminate, maximum length is never reached and k actually represents the rate at which growth approaches the asymptote – in other words, the point where the curve breaks and becomes more or less a horizontal straight line when growth is plotted against time, the equivalent of indeterminate. Thus $\ln\frac{2}{k}$ units of time pass as a fish grows halfway to L_∞ (Pardo *et al.* 2013).

Several ways of estimating either L vs. W relationship of tarpons have been devised. Seymour *et al.* (2007) provided a linear relationship for Pacific tarpon juveniles ($r^2 = 0.99$) in which FL = 0.49W + 145. This was not the objective of their research, and they gave no details of its derivation. Harrison (2001) provided an equation based on Pacific tarpons from South African estuaries:

$$W = 1.432 \times 10^{-5}\left(L^{3.052}\right) \tag{2.14}$$

in which L = mm SL and W = weight in g (range 160–478 mm SL, $n = 43$, $r^2 = 0.995$).

Bishop *et al.* (2001: 19) examined 155 Pacific tarpons (137–410 mm FL) gill-netted in the Alligator Rivers area, Northern Territory, Australia about midway through the wet season of 1978–1979 and determined this association to be:

$$W = 0.0242\left(L^{2.83}\right) \tag{2.15}$$

in which W = weight (g) and L = length (FL in cm); $r^2 = 0.70$. In the Sepik River system of northern Papua New Guinea, Coates (1987) found the W vs. L relationship to be:

$$W = 9.96 \times 10^{-6} \left(L^{-3.1}\right) \tag{2.16}$$

where W = weight (g) and L = mm TL.

Ley (2008: 11–15) reported the W vs. L relationship of 162 Pacific tarpons (190–520 mm FL) from 11 Queensland estuaries over a broad latitudinal range (19°50′S, 147°45′E to 10°48′S, 142°33′E) during February and June 1996, and March 1998 to March 2000. Sampling encompassed parts of both the rainy and dry seasons. The relationship:

$$W = 0.0152\left(L^{2.83}\right) \tag{2.17}$$

Kulbicki *et al.* (2005) provided this W vs. L equation based on 35 Pacific tarpons (170–470 mm FL) from New Caledonia, Micronesia:

$$W = 0.0122\left(L^{3.03}\right) \tag{2.18}$$

Ley (2008: 13) attributed discrepancies in some of these results to differential growth rates among populations of Pacific tarpons, but listed other factors too (e.g. genetics, foods, seasonality). Sampling bias and sample size are also possible sources of differences, the former often a consequence of where nets are set and their mesh sizes. Ley was careful to note that of the 305 specimens measured in her Queensland study, sizes ranged from 175–530 mm FL, and that 93% of the specimens were caught in a net of 102 mm mesh: "Thus, the size range present in the 11 estuaries along the main channels and creeks where the research nets were deployed was apparently well represented ... with a median of 430 mm FL ...."

Crabtree *et al.* (1995) caught Atlantic tarpons on Florida's east coast for age validation and presented regression equations for describing the otolith weight vs. age relationship (Table 1.1). As a control, some fish were measured, tagged, injected with 200 mg oxytetracycline (OTC), and their weights estimated from a power equation based on length. Afterward, they were held 13–50 months in captivity, then killed and their OTC-marked otoliths examined ($n = 18$). Growth rates ranged from 95 mm FL in 20 months to 235 mm FL in 21 months. Otoliths of 12 specimens yielded ages of 4–9 y and macrozonation (see Casselman 1983: 3) revealed that 1 annulus/y forms between December and May.[4] Temperature might affect the laying down of annulus layers because some fish kept at a facility with heated water could not be evaluated. In any case, otolith

---

[4] Note that *annulus* means "concentric ring." As emphasized by Casselman (1983: 2), "There is no connotation of yearly in the Latin definition of annulus ...."

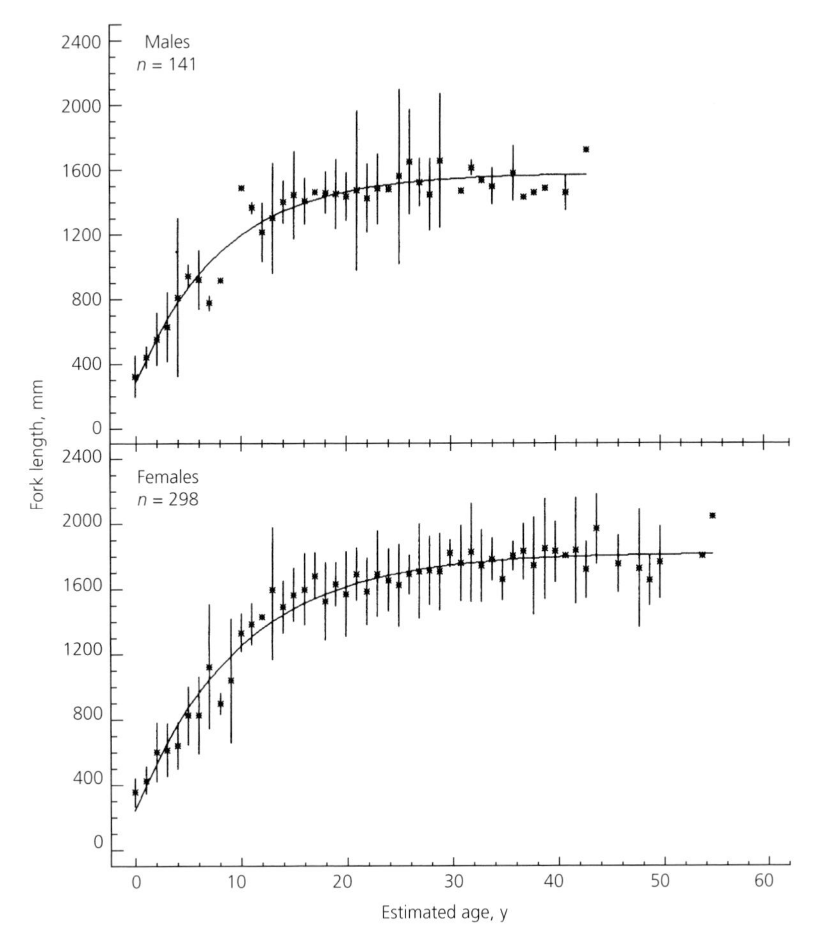

Fig. 2.5 Observed mean lengths (±2 SD) and predicted lengths from the von Bertalanffy growth model for male and female Atlantic tarpons. Source: Crabtree *et al.* (1995: 625 Fig. 3).

deposition by Atlantic tarpons in Florida waters is annual with formation occurring from late February to early March (Cyr 1991: 53). Ferreira de Menezes and Pinto Paiva (1966) concluded that annual checks in scales form during the last quarter of the year in Atlantic tarpons in coastal waters of Ceará State, Brazil.

Moffett and Randall (1957: 14) speculated that "The irregularity of the part of Breder's curve beyond age 12 is probably due to insufficient data to average out individual variation." The curve referred to is shown in Breder (1944: 231 Fig. 3). What it depicts is actually the rudimentary start of an asymptote. Additional data accumulated since reveal that Atlantic tarpons grow quickly until ≈ 12 years, when growth becomes clearly asymptotic (Fig. 2.5).

Moffett and Randall (1957: 15 Table 5) compared their annuli counts with those of Breder (1944) based on TL data of tarpons from Florida's west coast. The combined sets of data were presented in tabular form but not analyzed. I did this and found no significant difference.[5] They also compared the growth of tarpons from the east and the west coasts of Florida using number of annuli vs. fork length based on samples collected in 1955 and 1956 (Moffett and Randall 1957: 16 Table 6). These data were also presented in tabular form but not analyzed. I did this too and found no significant difference.[6]

Cyr (1991: 51, 63 Table III-3) reported the mean age of tarpons taken in the Florida recreational fishery to be 25.16 y based on examination of otoliths. The mean age of females was 25.8 y, that of males 21.1 y. In contrast, Ferreira de Menezes and Pinto Paiva (1966) reported respective mean ages of 8 and 10 y for male and female Atlantic tarpons of the same length obtained from weir fisheries in the northeastern Brazil. Cyr (1991: 58) speculated that the disparity was a result of their use of scale annuli instead of otolith increments to determine age, the latter generally being considered more accurate and precise for aging long-lived fishes, scale annuli usually underestimating age (e.g. Beamish and McFarlane 1983; Casselman 1983). However, as Casselman (1983: 3) warned, interpreting hard tissue provides an estimated osteological age, which might not correlate directly with a fish's chronological, or calendar, age, and Beamish and McFarlane (1983: 29) pointed out that "proving correct interpretation of an annulus for younger fish does not imply validation for older ages." Validation of a technique must therefore be demonstrated for all age groups of a population. The same applies to nullification. Until the findings of Ferreira de Menezes and Pinto Paiva (1966) are validated or refuted, the more rapid growth and earlier maturity of Atlantic tarpons in tropical waters (relative the subtropics) remain viable possibilities.

Determining the ages of Atlantic tarpons is difficult and confusing (Andrews *et al.* 2001; Crabtree *et al.* 1995) and especially so with older specimens. Their otolith increments become increasingly crowded and less distinct with time, and older ones might not even deposit of macrolayers annually (Andrews *et al.* 2001). Like the growth rings of a tree, environmental factors affect the width and regularity of deposition. A tree exerts no control over its habitat, but a tarpon ranges through places where extreme changes in abiotic and biotic conditions are routine and the length of time spent at a given location is unpredictable.

No aging method devised so far is sufficiently accurate and precise to determine how old a fish actually is with certainty, especially old specimens of long-lived species. Radiometric age validation using Atlantic tarpon otoliths has been applied

---

[5] I used a *t*-test of independent sample means comparing TL for 1–15 annuli. For data from Moffett and Randall (1957), $\bar{X}$ = 119.6 (±44.01 SD) cm TL (n = 15); for data from Breder (1944), $\bar{X}$ = 121.03 (±43.13 SD) cm TL (n = 15). Results: t = –0.089, df = 28, p = 0.929.

[6] I again used a *t*-test of independent sample means comparing FL for 1–15 annuli. Data for two Florida east coast fish with 16 annuli were omitted because paired values from Florida's west coast were unavailable. West coast: $\bar{X}$ = 109.77 (±41.28 SD) cm FL (n = 15). East coast: $\bar{X}$ = 99.13 (±36.22 SD) cm TL (n = 15). Results: t = 0.750, df = 28, p = 0.459.

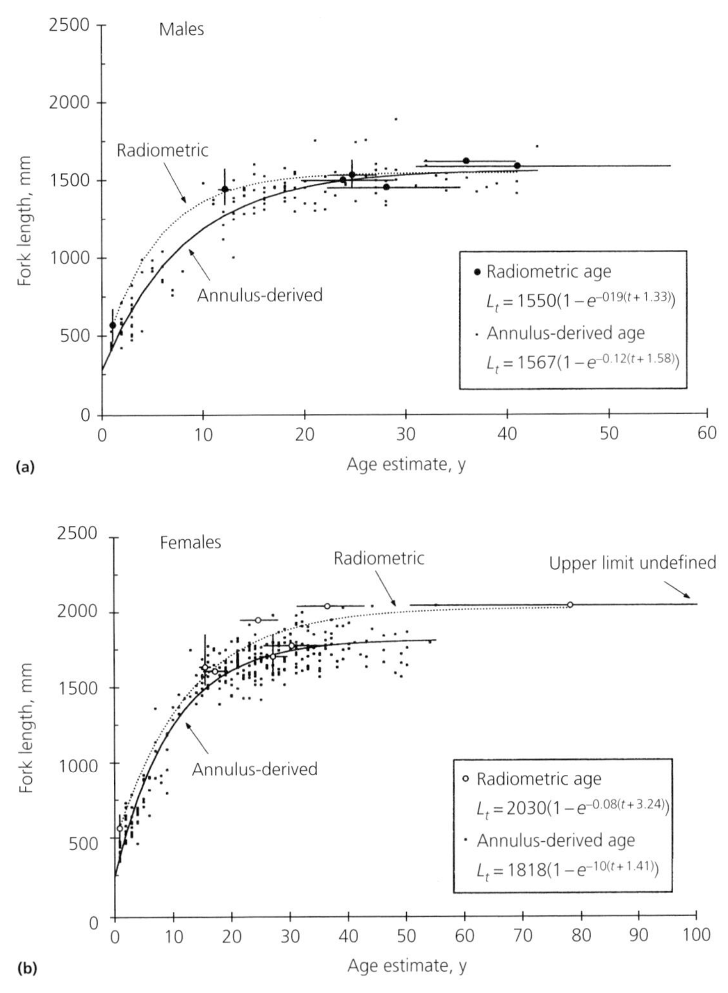

Fig. 2.6 Von Bertalanffy growth curves for male **(a)** and female **(b)** Atlantic tarpons showing fork length plotted against estimated radiometric age and compared with annulus-derived age. Vertical bars represent FL range for age groups. Horizontal bars represent low and high radiometric age estimates. Source: Andrews *et al.* (2001 395 Fig. 4 and Fig. 5).

more recently in an attempt to validate earlier age estimates based on the otolith annulus-derived method (Andrews *et al.* 2001). Agreement was inconsistent, and the coefficient of variation (used as a measure of precision) failed to explain the disparities (Fig. 2.6). The radiometric reading of a female of 2045 mm FL produced an age of 78 y; her annulus-derived age had been 55 y. The radiometric error overlapped the 55-y estimate by ≈4 y. The oldest male (1588 mm FL) had a radiometric age of 41.0 y, the error encompassing the annulus-derived age of 32 y by ≈ 1 y.

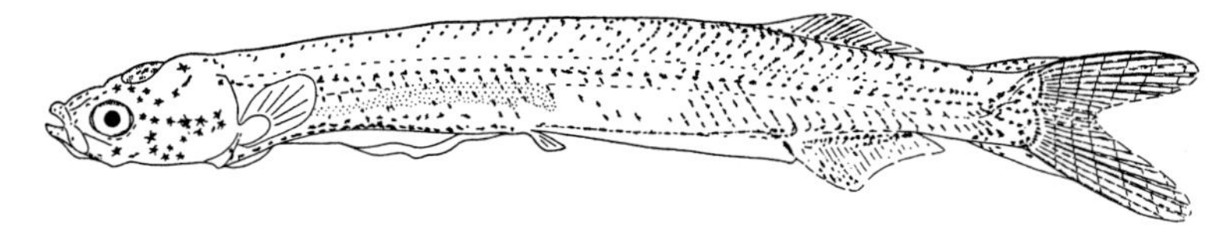

Fig. 2.7 The smallest post-larval Atlantic tarpon recorded by Harrington (1958). It measured 16.0 mm SL (18.8 mm TL). Source: Harrington (1958: 6 Fig. 5). © American Society of Ichthyologists and Herpetologists. Reproduced with permission.

2.7 Tarpon larvae

Evident so far is the profound change in shape and size of tarpons undergoing metamorphosis, which is not unusual. Larval fishes often produce extreme values of b at some point in the developmental sequence (Froese 2006). The growth of Stage 2 Atlantic tarpons is clearly allometric with weight increasing well out of proportion with length, and the cube law (equation 2.1) therefore is inapplicable (Fig. 2.1). Harrington (1958) provided the growth function

$$W = a + b\left(c^L\right) \tag{2.19}$$

Yielding:

$$W = -150 + 55.141.069^L \tag{2.20}$$

A staging system of tarpon development had not yet been devised when Harrington's 1958 report was published. That came later (Wade 1962). I presume Harrington's larvae had metamorphosed based on their sizes, appearance, and inshore capture locations. His smallest were evidently at the end of Stage 2 or entering Stage 3. The illustration of the smallest specimen is reproduced here (Fig. 2.7). Also see his Plate 1. In the caption of his Fig. 5 he referred to this specimen as representing "a post-larval tarpon" and "a transitional stage," which it certainly is. In addition, Harrington listed specimens of 56–109 mm SL omitted from his growth calculations, which had included only fish of 16.0–45.5 mm SL. Harrington (1958: 6) speculated: "Additional specimens of larger size were measured, and there is reason to suspect that the growth stanza within which the above weight-length relations holds ends at about 45 mm SL, but the additional measurements were too few to constitute more than an indication."

Mercado Silgado and Ciardelli (1972) considered the general range of Stage 2 Atlantic tarpons to be 28.0–13.0 mm SL, and Stage 3 as 13.0–71.0 mm SL (Table 1.2). This means that Harrington's smallest fish could straddle both late Stage 2 and early Stage 3 – what Mercado Silgado (1969, 1971) termed fry. Harrington's largest could only have been juveniles.

CHAPTER 3
Spawning

3.1 Introduction

Tarpons are gonochoristic, and fertilization is external; otherwise we know little about where tarpons spawn and what a spawning event looks like. Both species are *broadcast spawners*, adults gathering at specific locations and spewing gametes into the space around them, an evolutionary mechanism that relies on enough ova and sperm cells meeting in the water column to sustain populations.

Details of tarpon spawning mechanics and behavior are woefully thin. To my knowledge no biologist has yet witnessed a spawning event, and what becomes of the eggs once released and fertilized is anyone's guess. It would be enlightening to know the depth at which they drift, for example, the abiotic factors that impinge on development, the duration of fertilization to hatching, and the length of time spent in the yolk-sac stage, but such information is lacking completely. My only choice is to insert what little knowledge we have into this largely empty mosaic and fill the remaining voids with proxies.

3.2 Fecundity and early survival

The progeny of broadcast spawners experience high rates of mortality. Species using this method of reproduction are typically fecund, releasing thousands – even millions – of gametes in a single spawning event. *Fecundity* can be defined several ways. The definition I shall use is the number of potential offspring a female fish can produce during a breeding season, as determined by counting her mature eggs.

Little original information exists on the Atlantic tarpon's fecundity and still less on that of the Pacific species. Ovaries of the Atlantic tarpon are large (Fig. 3.1). Nichols (1929: 199) described its eggs as "exceedingly small and exceedingly numerous" and estimated that a female of 142 lb (≈64 kg) and

Tarpons: Biology, Ecology, Fisheries, First Edition. Stephen Spotte.

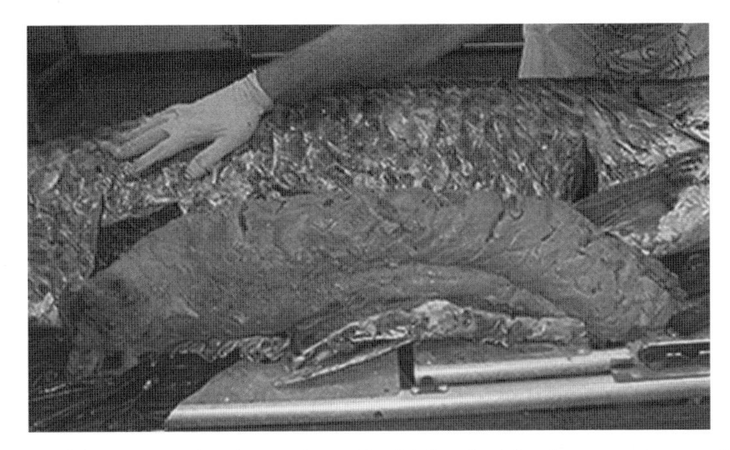

Fig. 3.1 One of the two ovaries of a 100 kg (≈220-lb) female Atlantic tarpon. Source: Florida Fish and Wildlife Conservation Commission.

measuring 6 ft 8 in. (≈2030 mm) contained ≈12 million, a number repeated many times in the literature without a formal citation (e.g. Babcock 1936: 41; Breder 1929: 60; Kaplan 1937: 90; Okoro *et al.* 2010: 18; Nichols and Breder 1927: 33; Villa 1982: 78), probably because it was never formally published. This estimate was derived as follows.

According to Moffett and Randall (1957: 4), "Babcock sent the ovaries [of this fish] to Mr. J. T. Nichols, who after counting the eggs in one ounce of ovary and making a correction for weight of ovarian tissue, estimated that the ovary contained over 12 million eggs." The eggs examined were not yet mature. Nichols found them to be 0.6–0.75 mm diameter, information once again repeated often and without a citation (e.g. Babcock 1936: 42; Hollister 1939: 454, Planquette *et al.* 1996: 54). Beebe and Tee-Van (1928: 34) reported that an Atlantic tarpon they examined of 40 in. (≈1016 mm, misstated by them as 1060 mm) and 36 lb (≈16.3 kg) contained 891 000 eggs.

Cyr (1991: 91–92) listed estimated fecundities of 1 081 330 for a 30 kg Atlantic tarpon and 19 518 400 for one of 54.09 kg. Weight-specific fecundity averaged an estimated 204 019 (±82,729) oocytes/kg of adult body weight. Fecundity also demonstrated a positive linear correlation with age (Cyr 1991: 92, 116 Fig. IV-16-B):

$$F = 1.616 + 3.72540 \times 10^5 (y) \tag{3.1}$$

in which F = fecundity and y = age in years ($r^2 = 0.51$, $n = 15$, $p < 0.0298$). The relationship, however, proved less robust than the association between fecundity and total body weight ($r^2 = 0.74$, $n = 15$, $p < 0.001$), but then aging any fish is a far less accurate and precise determination than weighing it. Crabtree *et al.* (1997) reported the estimated fecundity of Atlantic tarpons from southern

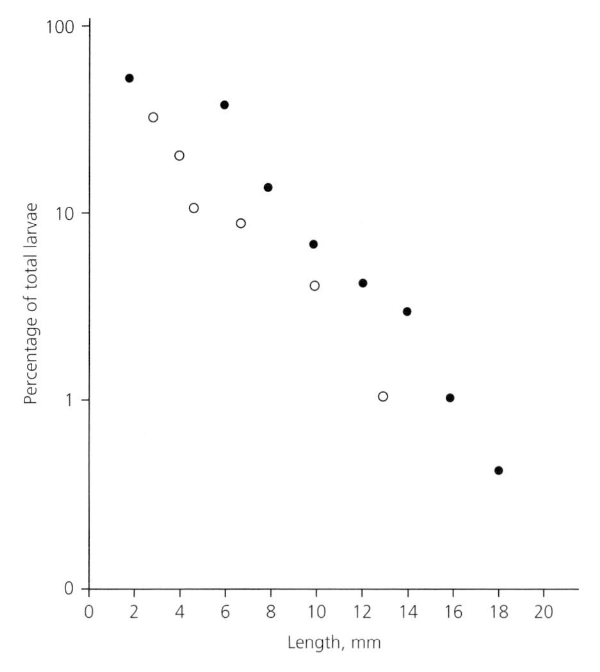

Fig. 3.2 Partial survival curves for larvae of Adriatic Sea pilchards, *Sardina pilchardus*, (clear circles) and northern anchovies, *Engraulis mordax* (solid circles) as percentage total larvae vs. length (presumably TL). The association is negative, indicating decreasing percentage mortality with increasing length. Source: Dahlberg (1979: 8 Fig. 5).

Florida as 4.5–20.7 million oocytes. The fecundity of Pacific tarpons has apparently not been assessed, Bishop *et al.* (2001: 24) stating just that it "is most likely high ...." For a female of either species to carry so many gametes necessitates they be tiny, mere specks adrift in a vast ocean once released.

From small eggs emerge small larvae. We know from studying other fishes that size matters where survival is concerned (Chapter 4). Bigger is better, often substantially so (Fig. 3.2). Nielsen and Munk (2004), for example, compared size-at-age survival of Atlantic cod (*Gadus morhua*) larvae, finding the relative survival probability of the largest to be 10 times that of the smallest. We know too that fast-growing larvae capable of shifting into sequential developmental stages more quickly have higher survival probabilities (e.g. Hare and Cowen 1997; Kaufmann 1990).

Based on a carefully constructed multi-variable model, Letcher *et al.* (1996) implied that the capacity to grow rapidly represents the *sine qua non* of larval survival. Proportional changes in prey availability, along with an extensive array of other variables, exerted lesser effects. Of intrinsic factors, capacity for growth (e.g. efficiency at which nutrients are taken up and assimilated) – not foraging capability or resistance to starvation – explained the most variance in survival of planktonic fish larvae. Among extrinsic factors, predator length accounted for

83% of the variance, requiring variability in prey density to increase threefold to equal its effect. Food density, historically thought to be crucial (e.g. Sirois and Dodson 2000 and references), had minimal effect on survival, explaining only 5% of the variance. How larval fishes died (e.g. starvation, predation) depended more on intrinsic factors associated with metabolism, resistance to starvation, and density of the smallest type of prey.

Evidence clearly points toward rapid growth as a critical survival mechanism, and outgrowing predators as soon as possible is perhaps the most effective adaptation. By developing rapidly a larval fish improves its ecological status. It becomes a faster, stronger swimmer, enabling it to capture prey more effectively and escape from predators by out-maneuvering or out-swimming them. Its gape increases, permitting ingestion of larger food items; its sensory systems become more advanced and finely tuned. On balance, these developments are advantageous. On the downside, the larva itself becomes more noticeable to predators that hunt visually (Chapter 4), larger size becoming a liability in clear, calm waters, but beneficial in waters that are turbid or turbulent, unless the predator is a cnidarian or other tactile hunter (Bailey and Batty 1984; Ohata *et al.* 2011), in which case degree of visibility becomes irrelevant.

I found no evidence of stage 1 tarpon leptocephali feeding in a conventional sense; that is, of purposely ingesting food *per os* (Chapter 1.6). In Atlantic tarpons, anatomical structures necessary for ingestion, digestion, and excretion are not sufficiently developed until Stage 2, phase VIII when active feeding commences (Mercado Silgado 1971; Mercado Silgado and Ciardelli 1972; Chapters 1.6, 7.7, Appendix B), or the approximate equivalent of Wade's (1962) Stage IIIA, and probably at a parallel level of development in the Pacific species. Larvae have moved inshore by then and are no longer part of the offshore plankton. This means that rapid growth *in size alone* – as opposed to the simultaneous development of requisite feeding and sensory systems – becomes the variable of greatest urgency to Stage 1 Atlantic and Pacific tarpons. Without the need to identify, capture, and process food, straightforward somatic growth directed toward increased size is the utmost priority. From tiny fertilized eggs, both species reach the end of Stage 1 within a month or less at ≈30 mm long, the culmination of a period of growth accompanied by retarded structural and sensory development compared with most fishes (Chapter 1.5).

A broadcast spawner's progeny from the newly released unfertilized egg forward are at a disadvantage, having been abandoned without parental care in waters rife with predators. Shine (1978) generated a simple theoretical model to test why offspring of organisms receiving parental care are generally larger. For convenience he combined an organism's life-history from the egg through its larval stages to juvenile into a single stage, the *propagule*. A propagule's size determines how long it remains at this arbitrary "stage" because large eggs take longer to hatch, and large larvae often develop more slowly than small ones. As a result, time spent as juveniles is typically reduced.

Instantaneous survivorship can be represented by l_p during a propagule's combined egg and larval stages, and by l_j on its becoming a juvenile. The "optimum" time spent as propagule vs. juvenile is represented by *t*. *Survivorship* (*Su*) is a measure of survival rate (i.e. survival vs. time), in this case to time *T*,

$$\left(Su_T = (l)_p^t\right)\left(l_j^{T-1}\right) = l_p^T\left(\frac{l_p}{l_j}\right)_t \tag{3.2}$$

Stated differently, survivorship represents how many individuals of a cohort remain alive over time. Defined narrowly, a *cohort* can be the offspring of a single spawning event by one female, and in tarpons could include millions of gametes. For purposes of his model, Shine associated fitness of the female parent directly with her fecundity (*F*), defined in this case as the number of eggs she produces during a single spawning event multiplied by the survivorship of her offspring. Fecundity ordinarily relates inversely to propagule size, which is proportional to the duration that propagules need to develop. Therefore,

$$C = \frac{k}{t} \tag{3.3}$$

where *C* represents "clutch size" (i.e. mass of ova released in a single spawning event), *k* is a constant, and:

$$F = l_j^T\left(\frac{l_p}{l_j}\right)_t\left(\frac{k}{t}\right) = \left(\frac{k^*}{t}\right)_t\left(\frac{l_p}{l_j}\right)_t \tag{3.4}$$

$$\frac{dF}{dt} = \frac{k^*}{t}\ln\left(\frac{l_p}{l_j}\right)\left(\frac{l_p}{l_j}\right)_t + \left(\frac{l_p}{l_j}\right)_t + \left(\frac{l_p}{l_j}\right)_t\frac{k^*}{t^2} \tag{3.5}$$

Slope of the line for fecundity (or fitness) is zero when $\frac{dF}{dt} = 0$. Thus:

$$\frac{\ln l_p}{l_j} = \frac{-k^*}{t} \tag{3.6}$$

As Shine (1978: 420–421) emphasized, his model illustrates a minimum – not a maximum – because the second derivative is positive. "Hence, parental fitness is maximized at the extremes of *t* – that is, by having either very small propagules or very large propagules." As he noted, what matters is the effect of the ratio of propagule to juvenile survivorship, $\frac{l_p}{l_j}$, acting on optimal propagule size. The advantage leans toward "large" as the ratio increases, which reduces how long the juvenile stage persists.

3.3 Where tarpons spawn

Beebe (1928: 74) wrote, "There seems to be a general agreement that the young are hatched or live up the rivers or in ponds of fresh or brackish water." Moffett and Randall (1957) thought Atlantic tarpons spawn in estuaries or nearshore waters, and Dahl (1965) reported seeing adult Atlantic tarpons in pairs and trios putatively spawning on April nights in water ≤ 1 m deep over sandy banks in Golfo de Morrosquillo, Colombia. Breder (1944: 226) had initially accepted similar conjecture as true, having been told by local citizens of southwestern Florida that spawning "has been repeatedly observed in shallow water between the numerous islands ...." If so, he reasoned, eggs and larvae ought to be "recoverable in tow-nets." He might also have read about it in popular books on recreational fishing (e.g. Kaplan 1937). In a personal letter to Kaplan, taxidermist and angler Victor Brown of Everglades City, Florida had stated with certainty that Atlantic tarpons spawn inshore in brackish waters (Kaplan 1937: 91): "It is an established fact that the tarpon spawns on Florida's lower West Coast from May to July, in such places as Woods River (between Fackahatchee River and Royal Palm Hammock), in the Fackahatchee River north of Everglades City; also in Turners, Lopez and all the other rivers south of Everglades City to Cape Sable Canal."

Five years of plankton sampling both summer and winter at benthic and surface inshore locations without recovering an egg or larva convinced Breder that reproduction and early development take place in the open sea. He proposed that if tarpons produce pelagic eggs, "at least some reproductive activity" must occur offshore (Breder 1944: 228). Wade (1962), writing later about both species, agreed. The likelihood of offshore spawning, however, had been deduced years earlier. In reference to Atlantic and Pacific tarpons, Meek (1916: 65) stated:

> *It is probable that these more oceanic allies of the herring [sic] have pelagic eggs, and that spawning takes place out at sea, but this does not preclude an anadromous migration taking place. The spawning migrants appear to migrate towards the coast before spawning takes place, but the spawning region is sufficiently far from the coast to demand a denatant drift of the eggs and larvae to the coast where the early life is spent. After metamorphosis the young stages are passed in relatively shallow water.*

According to Breder (1944: 228), reproduction and early development inshore would be successful only "if some morphologic, developmental or behavioristic habit insures the protection of the young until they are too large to be preyed upon by planktonic elements." He noted that other fishes breeding near the southwestern Florida coast either produce large, well-advanced young or provide parental care and protection from planktonic predators, especially larval crustaceans. With this in mind, and noting the Atlantic tarpon's tiny egg (capable of hatching only a tiny larva), he speculated that tarpons spawn in offshore blue waters where predators are more sparsely distributed and the young could delay moving inshore, also hinting that the larva is likely a leptocephalus. Breder was right. We know now that adults indeed move into pelagic waters to spawn at locations yet to be found. As Storey (1937: 10) explained in a slightly different context, "It is notably easier to establish the absence of a fish than it is to prove where it has gone."

Breder's conjecture about offshore spawning was preceded by proponents other than Meek. Having found juvenile Atlantic tarpons in a land-locked pool in Haiti, and citing a similar observation in Puerto Rico (Evermann and Marsh 1902: 80), Beebe (1927: 144) and Beebe and Tee-Van (1928: 36) wrote: "On account of the shallowness and the muddiness of the Haitian lagoon, it would be impossible for an adult tarpon of any size to enter it and deposit her eggs." The conclusion: "We can only surmise that the eggs are scattered out in the waters of the gulf ... and that the young, either as leptocephalids or as very small larval fish, make their way into the lagoon before the season when it becomes cut off from the gulf." This, of course, is essentially what happens. Atlantic tarpons along Florida's west coast aggregate inshore in late spring and summer before moving offshore into the Gulf of Mexico to spawn (Crabtree 1995; Crabtree *et al.* 1992). Further indirect evidence of offshore spawning was obtained when a 5.5 mm SL leptocephalus was captured over the continental slope ≈250 km west of Florida. Its estimated age of three days (Crabtree *et al.* 1992) was too young for it to have drifted so far seaward from a bayou or estuary.

The distance Atlantic tarpons travel to spawn is unknown, but at some locations in the eastern Atlantic and perhaps elsewhere it might not be far. Anyanwu and Kusemiju (2008: 121) wrote that in Nigerian waters of the eastern Atlantic, "Fertilized egg masses were found attached to floating vegetation and could be obtained at sea, as well as along the shorelines of the study area." Unfortunately the vegetation was not identified. If benthic it might indicate that spawning is benthic too or that the eggs sink to the bottom, sticking to plants that subsequently detach. However, it could also suggest that the eggs tend to coalesce and form masses, but also attach to floating objects such as detached grasses. In any case, the eggs are apparently not sticky (Breder 1944: 218).

Ferreira de Menezes and Pinto Paiva (1966) monitored fish weirs at Ceará State in northeastern Brazil, finding the tarpon catch to be highest during the dry season (October and November) and leading them to speculate that reproduction occurs when coastal waters of that region are saltiest. However, Carvalho Collyer and Alves Aguiar (1972), who also provided catch data for fishes and sea turtles taken from some of these weirs during 1968–1970, reported the highest catches of tarpons to occur from November–March, which encompasses part of the rainy season (January–May).

A mechanism-based hypothesis for the requirement of high salinity has not been proposed, although one implication could be that eggs and Stage 1 leptocephali (Chapter 1) are stenohaline, perhaps requiring full-strength seawater as suggested by Pfeiler (1986) based on salt-loading by bonefish (*Albula* sp.)

leptocephali and subsequent unloading as they migrate to less salty waters inshore (Chapter 6.8). Smith (1980: 140) speculated: "Premetamorphic leptocephali in general are not found in areas of low or fluctuating salinities and the tarpon appear to be typical in this regard." He continued: "Probably because of their need for constant high salinities, tarpon larvae, like all leptocephali, are normally found well offshore." Crabtree *et al.* (1992) cited Smith and seemed to agree.

Ellis (1956: 6) believed Pacific tarpons to be inshore or nearshore spawners, stating with neither a citation nor evidence, "In *M. cyprinoides* spawning occurs in shallow coastal areas where fresh and saltwater meet." He reaffirmed this opinion later in the same document (p. 9): "It appears that spawning normally occurs in salt or brackish water close to shore." Merrick and Schmida (1984: 54) wrote that "Adults with ripe eggs have been collected in estuaries in southern New Guinea, and adults school along the same coast late in the year." Documentation of these statements was not provided.

Coates (1987) perhaps came closer to the truth when he speculated that Pacific tarpons >400 mm SL in the Sepik River system of northern New Guinea return downstream to estuaries and coastal areas for reproduction, then remain permanently in saline waters. Allen and Coates (1990: 53) stated, "Adults spawn in sea [*sic*] and estuaries." Merrick and Schmida (1984: 54) reached the same conclusion, and Herbert and Peeters (1995: 21) wrote of Pacific tarpons, "They are thought to spawn in estuaries or shallow coastal waters around the onset of the wet season." Most likely, Pacific tarpons everywhere spawn offshore.

Strengthening this notion is lack of direct evidence that tarpons of either species reproduce in brackish or fresh waters. Whether survival of early larvae is even possible except in full-strength seawater could be tested quickly and crudely just by placing undamaged Stage 1 leptocephali in freshwater or aliquots of seawater diluted to predetermined ionic strength.

The case for freshwater spawning of Atlantic tarpons is equally weak. Some authors over the years have speculated that spawning occurs in Lake Nicaragua and Florida's Deep Lake, although leptocephali or juveniles have never been found at either place. Deep Lake, located in Big Cypress National Preserve, is a sinkhole 30 m deep, covering 0.81 ha. Ellis (1956: 7) wrote how "tarpon have been reported to complete their entire life cycle, including spawning, in fresh water." A few pages later (p. 10) he seemed more certain, noting that when tarpons become isolated permanently from the sea, as at Deep Lake, they "appear to complete their entire life cycle." He added, "It is also known of lakes in the island of Andros, Bahamas." None of this had been verified at the time, nor has it been since.

Simmons (1900: 329) was "inclined to think that the San Juan River and Lake Nicaragua are the principal breeding places of this fish [in Nicaraguan waters], and that it is a mere migratory visitant to our coast." Brown and Severin (2008) noted that Atlantic tarpons in Lake Nicaragua and its tributaries were large enough to have reached reproductive size and speculated whether they breed there or return instead to marine waters. They wondered whether Atlantic tarpons sometimes reproduce in freshwaters, even if spawning events might result in total failure of eggs or larvae to survive. To my knowledge the reproductive biology of Lake Nicaragua tarpons has not been assessed, and poses intriguing questions. Do gonads of tarpons living in freshwaters ever mature? If not, what could be the mechanism repressing the attainment of maturity? If these fish indeed mature, do they migrate back to the sea or shed their gametes in the lake, however futilely? Are adult tarpons in the upper Río San Juan permanent residents or transients moving back and forth haphazardly between fresh and saline waters? If migratory, are they catadromous like some Pacific tarpons are thought to be (see below)?

Limited data from a Texas reservoir indicate that growth of Atlantic tarpons in freshwaters and saline waters is comparable (Howells and Garrett 1992), hinting that size at reproduction could be similar, and the same conjecture could be applied to Pacific tarpons, the species for which the evidence against freshwater spawning is stronger. Of specimens held captive at Tamil Nadu, southeastern India, Alikunhi and Rao (1951: 100) wrote, "It is well known that while *Megalops* [*cyprinoides*] fattens quickly in fresh water ponds, it never breeds in that environment." They noted too that "adult specimens ... have generally been found to have gonads poorly developed in fresh water ponds, all the year round ...."

Thomas (1887: 124) included the Pacific tarpon among those Indian fishes having a "migratory spirit," implying it is anadromous and enters rivers to reproduce. However, his remarks in this regard were never explicit. Interpreting Thomas' musings as solid evidence of reproduction in freshwaters, Raj (1916: 252) concluded, "In support of this ... statement, I may state that I have obtained the species in various sizes from a pond about 4 miles from the sea which has no communication with river or backwater." Raj's comment is intriguing, although not proof of a reproducing population. More likely the pond had been stocked because the presence of tarpons apparently made it unique to the region. Raj admitted, "I have not seen it [the tarpon] in the Red Hills tank [reservoir] or in any of the larger lakes further [*sic*] inland."

Shen *et al.* (2009) strengthened the case for offshore spawning by Pacific tarpons. They recorded high Sr/Ca ratios in leptocephali from oceanic waters off Taiwan and suggested that deep offshore waters are the spawning sites. Leptocephali caught while recruiting into brackish waters ranged in age from 18–34 days, hinting of spawning areas nearby, although not inshore.

The strongest refutation of freshwater spawning comes from an informal, long-term study in India. Kulkarni (1983) reported having personally stocked juvenile Pacific tarpons in the freshwater lakes Valvan and Shirota at Lonavla (or Lonavala), Maharashtra during July 1939. This small population was subsequently left undisturbed and never restocked. Only adults were recovered when the lakes were seined in 1970 and again in August 1983 and 1991. Specimens from the 1983 seining averaged 670 mm TL and ranged from 2.75–3.1 kg. Gonads of these fish were "inconspicuous" (Kulkarni 1983: 232), and there was no other evidence of either reproduction or the capacity to reproduce.

The Pacific tarpon is more likely catadromous than anadromous (Bishop *et al.* 2001: 454, 460, 531–532; Merrick and Schmida 1984: 53). That is to say, they abandon freshwater rivers on reaching maturity and migrate downstream to the sea to expel their pelagic gametes. It was Mann's (2000: 66) opinion that "juveniles appear to remain in estuaries (or connected freshwater bodies) until reaching maturity." Having attained it, "Adults presumably migrate out to sea in order to spawn and may re-enter estuaries once spawning is compete ...." These comments are based on reports from northern Australia and Papua New Guinea indicating that large specimens often found in the upper reaches of tropical rivers are rarely in reproductive condition and are probably still juveniles.

For example, of 278 Pacific tarpons caught by Coates (1987) in the Sepik River system, 93 were >400 mm SL, and in all but one the gonads were small and inactive. This single mature specimen was a female of 410 mm SL with maturing ovaries composing 7% of her total body weight of 1.26 kg. She was caught in August ≈70 km upriver from the estuary. Of 155 Pacific tarpons gill-netted in the Alligator Rivers area of the Northern Territory, Australia by Bishop *et al.* (2001: 491 Table 157), just seven specimens (all females) showed evidence of maturing. They ranged from 186–360 mm TL. How Pacific tarpons are distributed in river systems by location and size is also revealing: large juveniles in the upper reaches, smaller ones prevalent in the middle and lower reaches, and very small juveniles and larvae occupying the estuaries (Bishop *et al.* 2001; Coates 1987; Merrick and Schmida 1984: 53–54; Roberts 1978). Adults of reproductive size and condition shift to floodplain billabongs (Bishop *et al.* 2001: 455–456 Table 150) and offshore waters. Pandian (1968: 570), who studied Pacific tarpons in the Cooum River, Chennai (formerly Madras), India wrote that "Larger fishes weighing 600 gm. or more, have not so far been observed in the backwaters." Small specimens were present in considerable numbers throughout the year.

3.4 When tarpons spawn

Reproduction of Atlantic tarpons occurs in the Northern Hemisphere from spring through early autumn. Florida is the northern part of the reproductive range. Cyr (1991: 18) gave spawning months for Florida in general as April–September, which brackets the ranges published by others. Smith's (1980: 136) assessment was not specific: "Previous studies have indicated that the tarpon spawns in the spring and summer, and the present material confirms this." Tucker and Hodson (1976) declared the spawning season as April or May to September. In Babcock's (1951: 50) opinion, tarpons on Florida's southwestern coast spawn in July and

August; in Breder's (1944: 226), "during May, June and July at least." Crabtree *et al.* (1997) determined that tarpons in southern Florida spawn from April–July. By August most females are spent or recovering. Another important discovery from this study is confirmation that spawning of tarpons in Costa Rican waters is not seasonal; reproductively active females were captured throughout the year, as verified by histological examination of gonads.

Storey and Perry (1933: 285) speculated that spawning off southwestern Florida took place during March, April, and May when tarpons "schooled in considerable numbers in shoal waters from one half to one mile off the southeast shore of Sanibel Island, Lee County." Anglers then – and still do – also gather in considerable numbers to pursue them, and "Milt has been seen escaping from fish in play [i.e. while hooked], and when brought to the release-hook [*sic*] quantities exceeding half a pint have been expressed from a single fish." Storey and Perry noted that females had not, to their knowledge, been taken from these aggregations. We know now that adult females are present too.

Despite the occasional presence of larvae and juveniles at inshore locations in the northern Gulf of Mexico, evidence of offshore reproduction in this region remains unconfirmed. In July 2011, three adults of breeding size (for size at breeding see Crabtree *et al.* 1997; Ferreira de Menezes and Pinto Paiva 1966, and Section 3.5) were caught by anglers ≈30 km south of Venice, Louisiana. The female was 1778 mm FL (56.8 kg), the two males 1676 mm FL (55.5 kg) and 1698 mm FL (56.6 kg). Stein *et al.* (2012) examined their gonadal tissues and deemed them capable of spawning, the female estimated to have spawned within 24 hours of capture based on late vitellogenic oocytes and 24-hour post-ovulatory follicles. Testes of the males contained spermatozoa but no evidence of active spermatogenesis, indicating the fish might have been caught at the end of their reproductive season, although while still capable of spawning.

Stein and colleagues recognized that however enticing it is to find tarpons in active or spent condition outside their expected reproductive range, this alone is not evidence of local reproduction. Atlantic tarpons can – and do – travel long distances. For example, a tarpon of 1430 mm FL (29.06 kg) in active spawning condition was removed 14 July 1974 from a pound net at Fort Pond Bay, eastern Long Island, New York. As described by Hickey *et al.* (1976: 187), "The fish was a running ripe male with a total gonad weight of 630 g (1.39 pounds) which was 2.17 per cent of the body weight."

According to Smith (1980: 140), "The small size of the larvae reported by Berrien *et al.* (1978) off North Carolina (6.5 to 21.5 mm) indicates that spawning takes place there, too." More likely it implies that spawning occurs farther south, and eggs and Stage 1 leptocephali drift north in the Gulf Stream to the latitude of North Carolina. However, big tarpons of breeding size were not strangers to the mid-Atlantic coast from June–August in years past. Shreves (1959: 1–2) wrote: "Huge schools of tarpon had been observed along the ocean frontage of Virginia's Eastern Shore, a 70-mile strip of long, narrow, outside islands with comparatively shallow bays and inlets between them and the mainland." He continued, "For many years as far back as natives of this locality could remember, local commercial fishermen had considered the 'big silver herring', as they called tarpon, to be strictly a nuisance which cruised daringly in these shallow waters and tore great gaps in their fishing nets."

The record to which Smith referred in the paragraph above is confusing. Berrien *et al.* (1978: 13, 81) seem to have listed a total of six Atlantic tarpon leptocephali caught in plankton tows during the R/V *Dolphin* cruises between December 1965 and December 1966. They were obtained at depths to 33 m and ranged from 6.5–9.9 mm NL. Wade (1962: 555) reported two much larger specimens of 145.6 and 159.8 mm SL(?) captured 3 December 1959 at a lake in the Cape Fear River system near Wilmington, North Carolina.

Crabtree (1995) added data of early larvae (range 5.1–28.1 mm SL, $n = 373$) caught off Florida's east and west coasts and Bimini, Bahamas to those of Crabtree *et al.* (1992) collected during other cruises (total $n = 702$) and falsified the null hypothesis that hatching frequency occurs uniformly within a lunar month. Hatching instead displayed clear peaks associated consistently with new or full moons. Peak hatching occurred 6.3–7.4 d after a full moon and 3.4–8.7 d after a new moon. Back-calculated hatch dates were 10 May–18 July, although a few larvae hatched from 3–14 August off Florida's west coast, which suggested additional spawning into mid-August.

Shenker *et al.* (2002) examined otoliths from 93 Atlantic tarpon leptocephali captured at Sebastian Inlet, which connects the Atlantic Ocean with Indian River Lagoon on the east coast of Florida. Assuming otolith increments are deposited daily, the larvae (13–26 mm SL) were 15–26 d old ($\bar{x}$ = 20.2 days). Back-calculation[1] yielded hatch dates ranging from mid-May to mid-August (Fig. 3.3). The data also revealed significant lunar periodicity, most larvae (71%) hatching within 10 d after summer full moons (Fig. 3.4). Peak 1994 inshore recruitment just before July's full moon showed an association with hatching around the time of the June full moon. The mid-May to mid-August hatch dates are similar to Crabtree's (1995) findings for Florida's southwestern coast.

Tucker and Hodson (1976) caught 24 Stage 2 larvae between 5 June and 7 October in the vicinity of Cape Fear, North Carolina, speculating they had hatched to the east or south and indicating a spawning season extending from April or May to September. Noting that Stage 2 Atlantic tarpons had also been collected inshore in late October near Pensacola, Florida in the northern Gulf of Mexico by Tagatz (1973), Cyr (1991: 18) wrote: "Considering that [all] these fish were probably from 40 to 55 days post-hatch, their spawn dates fall well within the range [of April-September] proposed here." (Cyr 1991: 19) suggested that the spawning season of Atlantic tarpons follows a latitudinal gradient, decreasing with increasing latitude, which seems reasonable.

[1] Back-calculation of hatch date involves estimating age of a larva from its capture date.

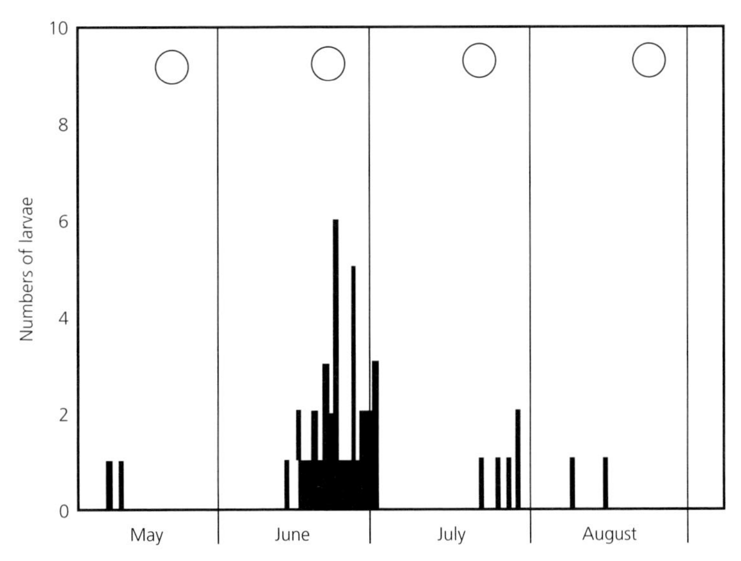

Fig. 3.3 Estimated hatching dates of Atlantic tarpon leptocephali caught at Sebastian Inlet, Indian River Lagoon Florida during summer 1994. Clear circles represent full moons. Source: Shenker *et al.* (2002: 64 Fig. 13).

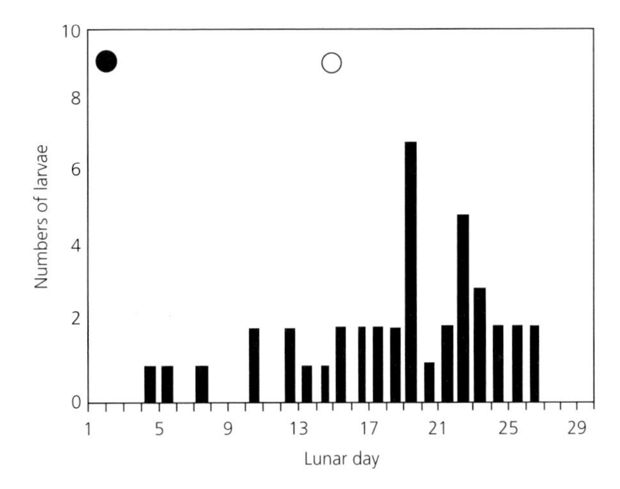

Fig. 3.4 Lunar periodicity in hatching dates for 41 Atlantic tarpon leptocephali caught at Sebastian Inlet, Indian River Lagoon, Florida during summer 1994. Rayleigh's test of circular distributions ($Z = 11.71$, $n = 41$, $p < 0.001$) indicates a non-uniform hatching distribution through the lunar months. Solid circle represents new moon, clear circle represents full moon. Source: Shenker *et al.* (2002: 64 Fig. 14).

3.5 Size and age at maturity – Atlantic tarpons

Breder (1944: 239 Table XIII) speculated that because most adult Atlantic tarpons caught in the Florida recreational fishery had seven scale rings, or checks, they had survived seven winters and were reaching the age of first spawning. His opinion took form after examining fish 1138–2040 mm TL ($\bar{x}$ = 1397 mm TL, n = 59), the size range (not coincidentally) at which the species becomes vulnerable to trophy angling. The largest immature fish measured 1422 mm TL; the smallest mature specimen was 1398 mm TL. Nonetheless, Breder (1944: 247) had reservations about length at onset of maturation, wondering whether his estimate of maturity after the sixth or seventh winter "may be a little old ...." Consistent with Breder's doubts, Babcock (1951: 49–50) stated: "I have examined many small tarpon containing eggs which I am sure were not over four years of age." He offered no evidence to back this claim.

The smallest mature female examined by Cyr (1991: 92) was 1245 mm SL and 29.09 kg; its estimated age was 13.8 y (Table 3.1). However, predicting age at first maturity was not feasible because too few fish in the smaller size classes were available to examine. The smallest mature male Cyr found was 1181 mm SL and 21.82 kg. Its age was not estimated. Again, insufficient data made age determination at first maturity impossible. The general statement has been made that Atlantic tarpons spawn at >1000 mm SL and age 10 y (Crabtree *et al.* 1997).

Atlantic tarpons in tropical waters might breed when younger and smaller, and they might reproduce throughout the year or have more than one spawning period within a calendar year. Crabtree *et al.* (1997) compared aspects of reproduction of Atlantic tarpons from Florida and Costa Rica. Specimens from Florida

Table 3.1 Weight, standard length, and age at sexual maturity of female Atlantic tarpons in southern Florida. Sample size (n) in parentheses. *Source:* Cyr (1991: 100 Table IV–2).

Weight class	% mature	SL class (mm)	% mature	Age class (years)	% mature
0–10	0 (2)	0–800	–	0–5	0 (1)
15–20	100 (1)	800–900	0 (4)	5–10	0 (2)
20–25	66 (3)	900–1000	0 (1)	10–15	50 (6)
25–30	100 (4)	1000–1100	0 (2)	15–20	93 (14)
30–35	90 (10)	1100–1200	100 (1)	20–25	91 (20)
35–40	81 (16)	1200–1300	80 (5)	25–30	100 (11)
40–45	92 (13)	1300–1400	87 (15)	30–35	100 (10)
45–50	100 (24)	1400–1500	92 (26)	35–40	100 (6)
50–55	100 (8)	1500–1600	93 (40)	40–45	100 (4)
55–60	100 (15)	1600–1700	100 (25)	45–50	100 (2)
60–65	100 (9)	1700–1800	100 (10)		
65–70	100 (1)	1800–1900	100 (3)		
70–75	–				
75–80	100 (2)				
80–85	100 (2)				

they examined (the northern extent of the range) demonstrated peak gonadal development from May–July; by August most females were spent. No seasonal pattern was event in Costa Rican fish. In this study the smallest mature Florida male measured 901 mm FL, the smallest mature female 1285 mm FL. In contrast, the smallest mature Costa Rican fish were 880 mm FL (males) and 1126 mm FL (females).

Dahl (1971: 159), however, wrote that sexual maturity in the Colombian Caribbean was attained at 1000–1200 mm SL, or about the same length as in Florida waters, and hinted at a bimodal breeding cycle without providing details: "*Su desove comienza a finales de abril o principios de mayo. Aparentemente no todos los individuos desovan en este tiempo, ya que algunos lo hacen mucho más tarde.*" [It spawns starting at the end of April or beginning of May. Not all individuals evidently spawn during this time; some wait until much later.] Chacón Chaverri (1993) examined a series of Costa Rican fish caught in the recreational fishery (range 7.0–74.0 kg, $\bar{x}$ = 29.68 kg, n = 113). A female of 19.09 kg contained vitellogenic oocytes, and a 10.0 kg male had mature sperm in its testes.

To Chacón Chaverri (1993: 17), "*Los datos preliminares sugieren que el sábalo de Costa Rica alcanza la madurez sexual a tamaños más pequeños que los de la Florida; sin embargo los datos de tamaño de la madurez sexual son insuficientes.*" [The preliminary data suggest that Costa Rican tarpons reach sexual maturity at smaller sizes than those from Florida; however, size data at sexual maturity are insufficient.] He went on to state, "*Sábalos de varios estadios de madurez han sido encontrados durante todos los meses, lo que representa una possible reproducción alrededor del año con un modelo de dos máximos acentuados entre los meses de febrero a mayo y de agosto a noviembre.*" [Tarpons in several stages of maturity have been found during every month, possibly pointing to year-around reproduction with a bimodal model having two peaks, one February–May, the other August–November.] Chacón Chaverri (1993: 16) based his tentative conclusion on monthly gonadosomatic index (GSI)[2] values, which demonstrated the bimodal pattern. Wade (1962: 589) reported a Stage 1 Atlantic tarpon leptocephalus of 12.8 mm SL "taken November 12 off French Guinea [*sic*]."

Although Crabtree *et al.* (1995) provided a much larger sample size than Chacón Chaverri (1993), their specimens were obtained from southern Florida, which probably represents a different population. I therefore used Chacón Chaverri's W vs. L equation to describe the W vs. L relationship of fish from Costa Rica, despite its likelihood of being less accurate and precise:

$$\mathrm{Log}_{10}(W) = -8413 + 3.144(\log_{10} L) \quad (3.7)$$

[2] The gonadosomatic index is the quotient of the ratio of gonad weight to total body weight. It is typically expressed as a percentage, in which case $\mathrm{GSI} = \frac{\text{gonad weight}}{\text{total body weight}} \times 100$.

Table 3.2 Frequencies of Atlantic tarpon by length class, sex, and maturation status caught in commercial fisheries at Acaraú, Ceará, Brazil during 1962 to 1964. Total males and females combined = 2469. Source: Ferreira de Menezes and Pinto Paiva (1966: 88 Table II).

	Males' Reproductive status			**Females' Reproductive status**				
Length range classes (cm FL)	**No sperm**	**Sperm**	**Total**	**Immature**	**Maturing**	**Mature**	**Spent**	**Total**
55.0–59.5	0	0	0	1	0	0	0	1
60.0–64.5	0	0	0	0	0	0	0	0
65.0–69.5	0	0	0	0	0	0	0	0
70.0–74.5	1	0	1	0	0	0	0	0
75.0–79.5	0	0	0	2	0	0	0	2
80.0–84.5	0	0	0	1	0	0	0	1
85.0–89.5	1	0	1	1	0	0	0	1
90.0–94.5	1	0	1	1	0	0	0	1
95.0–99.5	0	4	4	0	0	0	0	0
100.0–104.5	3	17	20	1	0	0	0	1
105.0–109.5	3	57	60	3	0	0	0	3
110.0–114.5	12	200	212	5	0	0	0	5
115.0–119.5	26	322	348	8	0	0	0	8
120.0–124.5	31	254	285	9	1	0	0	10
125.0–129.5	45	79	124	14	3	1	0	18
130.0–134.5	50	32	82	9	12	2	3	26
135.0–139.5	26	11	37	20	21	5	3	49
140.0–144.5	3	2	5	20	37	34	15	106
145.0–149.5	3	2	5	23	79	85	25	212
150.0–154.5	0	1	1	9	76	116	51	252
155.0–159.5	0	0	0	4	58	126	49	237
160.0–164.5	0	0	0	2	28	120	35	185
164.0–169.5	0	0	0	3	14	51	19	87
170.0–174.5	0	0	0	1	4	33	16	54
175.0–179.5	0	0	0	0	0	13	4	17
180.0–184.5	0	0	0	0	0	3	2	5
185.0–189.5	0	0	0	1	0	1	0	2
Total	205	981	1186	138	333	590	222	1283

in which W = weight (kg) and L = mm FL. Chacón Chaverri did not report values of L. Solving for L yields 986.3 mm FL for males of 10 kg and 1210.6 mm FL for females of 19 kg. These numbers conform with information of Breder's (1944: 220 Table II) for Atlantic tarpons of unknown maturity in southwestern Florida, and for mature or maturing fish in northeastern Brazil (Table 3.2).

Indications are that Costa Rican tarpons breed when smaller than tarpons from Florida and Brazil (Crabtree *et al.* 1997). Whether this is a natural phenomenon or an artifact of size selection in the local fisheries is unknown. The study of Crabtree and colleagues is important for several reasons, one being the large sample size: southern Florida (range 102–2045 mm FL, n = 1469), Costa Rica

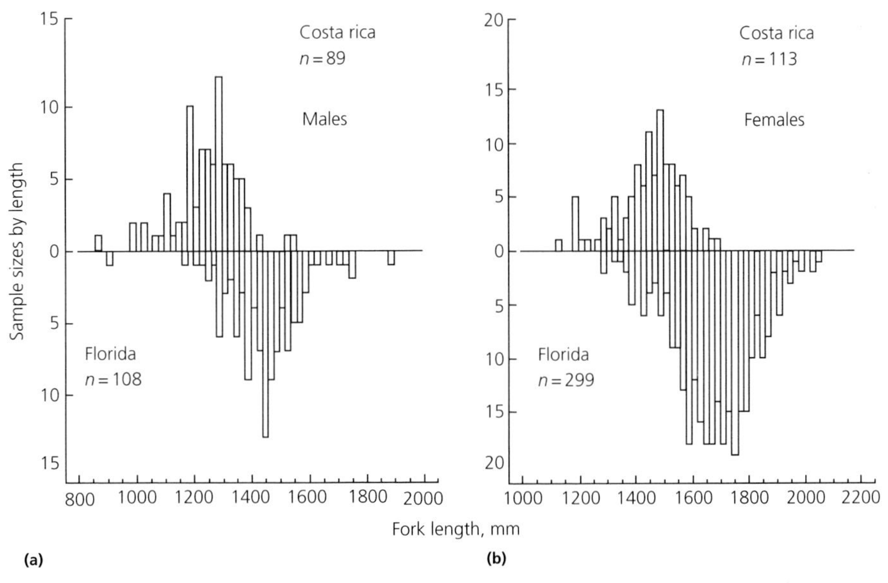

Fig. 3.5 Length-frequency distributions of mature male and female Atlantic tarpons from southern Florida and Costa Rican waters. Source: Crabtree *et al.* (1997: 275 Fig. 1).

(range 880–1860 mm FL, n = 217). Gonads of 737 Florida fish were examined histologically (n = 409 GSI values). Data for Costa Rican fish: 202 gonads examined (n = 178 GSI values).

The Florida fish were significantly longer; mature females were significantly longer than mature males from either location (Fig. 3.5). Southern Florida males had attained maturity by 1175 mm FL; just one shorter male (901 mm FL) was mature. The shortest mature female from Florida was 1285 mm FL. With a single exception all the rest longer than this were also mature, and maturity had been reached by 10 years. The exception was a mature female of 7 years. In the Costa Rican group just one immature female of 880 mm FL was examined. The rest (range 1126–1860 mm FL) were mature.

As additional tentative evidence of extended spawning in its southern range, a few Atlantic tarpon leptocephali of ≈24 mm have been caught between December and August in the Mexican Caribbean between Puerto Morelos at the eastern end of Quintana Roo and the northern border of Belize (Schmitter-Soto *et al.* 2002).

Cyr (1991: 83–116) staged and weighed gonads of 209 adult Atlantic tarpons from both Florida coasts and determined GSIs. Of these fish, 141 were obtained from taxidermists between June 1988 and August 1990, having been caught in the recreational fishery and large enough to be killed and mounted. Oocytes were of two sizes, 300–500 μm and 600–900 μm, corresponding, respectively, with early- and late-stage vitellogenesis. Female gonad weight was 0.06-6.4 kg ($\bar{x}$ = 2.47, n = 168); for males 0.03–3.34 kg ($\bar{x}$ = 1.23 kg, n = 74). In both sexes the gonad weight and fish weight demonstrated strong positive correlation.

The GSI values of males were 0.1–9.3%, peaking in May at 4.4% and decreasing from June–August. For females the GSIs ranged from 0.01–9.9%, being highest in May and lowest in August, but without a clear trend from April–July, values staying high over the entire duration of sampling. No female contained maturing oocytes. The actual time of spawning in southern Florida might not coincide exactly with gonadal development. The best estimates Cyr (1991: 93) obtained were back-calculations from the leptocephali collected, indicating spawning from late May to mid-June, reinforced to some extent by dates when leptocephali have been collected by others (e.g. Eldred 1967; Smith 1980).

Frequency of oocyte diameter was distributed bimodally through the first week of May (50–150 μm and 300–450 μm), becoming trimodal in mid-May (mode ≈600 μm), largest (700–900 μm) in the third week of May, remaining stable through July, and again becoming bimodal by August.

Fish in pre-spawning condition were 74.2% of the adult females sampled and present continuously from March–July, being most prevalent in May (79.6%) and June (81.0%). Spent females were rare during much of this time (May–July), rising to 89% in August. Developing males were most common in May and June, ripeness occurring only in May, June, and July and representing, successively, 7.4%, 28%, and 29% of the specimens examined. Spent males were also caught from May–July and composed 7.4%, 4%, and 20% of catch.

Smith (1980) collected a Stage 1 leptocephalus of 21 mm SL off Cozumel, México 11 April 1976 that had probably been hatched farther south, giving it an estimated age of 22–36 days, or a hatch date sometime in early to mid-March (Cyr 1991: 94). Chacón Chaverri (1993) reported capturing leptocephali inshore along the Caribbean coast of Costa Rica from July-October (mostly Stage 2 and early Stage 3 of 30.0–45.0 mm SL) and juveniles from December-February (45.0–60 mm SL). Mercado Silgado and Ciardelli (1972) described Atlantic tarpons in Stages 1–3 collected at the mouth of Canal del Dique, an inshore location near Cartagena, Colombia in September, October, and November, but stated that the most effective time to collect (I presume he meant when leptocephali were most numerous) was July–November.

Ferreira de Menezes and Pinto Paiva (1966) concluded that in coastal waters of Ceará State, northeastern Brazil, sexual maturity begins at approximately 950 and 1250 mm FL for males and females, respectively, most mature males being 1100–1250 mm FL and females 1500–1650 mm FL (Figs. 3.6 and 3.7). In their opinion, most males are mature after 5 y and females after 8 y, although Cyr (1991: 43) questioned the validity of these and the other age-related results because aging was based on scale annuli, which some experts consider less reliable than otolith increment counts, tending to age long-lived fishes much younger than they actually are. Crabtree *et al.* (1995: 627) concluded, "Tarpon scales do not appear to be suitable for age estimation."

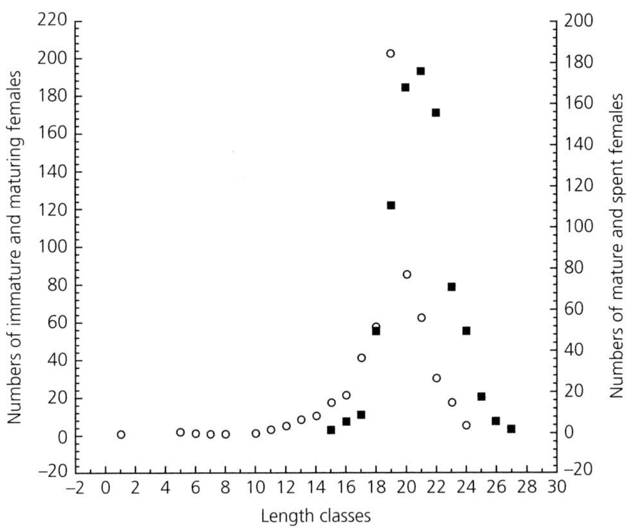

Fig. 3.6 Double-Y scatter plots of length at sexual maturity of male Atlantic tarpons caught in coastal weirs at Acaraú, Ceará State, northeastern Brazil 1962–1964 (total n = 1186): solid squares are immature fish without sperm (n = 205), clear circles mature fish with sperm (n = 981). Length categories on the abscissa represent 27 range classes of length (FL in cm): 1 (55.0–59.5), 2 (60.0–64.5), 3 (65.0–69.5), 4 (70.0–74.5), 5 (75.0–79.5), 6 (80.0–84.5), 7 (85.0–89.5), 8 (90.0–94.5), 9 (95.0–99.5), 10 (100.0–104.5), 11 (105.0–109.5), 12 (110.0–114.5), 13 (115.0–119.6), 14 (120.0–124.5), 15 (125.0–129.5), 16 (130.0–134.5), 17 (135.0–139.5), 18 (140.0–144.5), 19 (145.0–149.5), 20 (150.0–154.5), 21 (155.0–159.5), 22 (160.0–164.5), 23 (165.0–169.5), 24 (170.0–174.5), 25 (175.0–179.5), 26 (180.0–184.5), 27 (185.0–189.5). Source: Plotted from data in Ferreira de Menezes and Pinto Paiva (1966: 88 Table II).

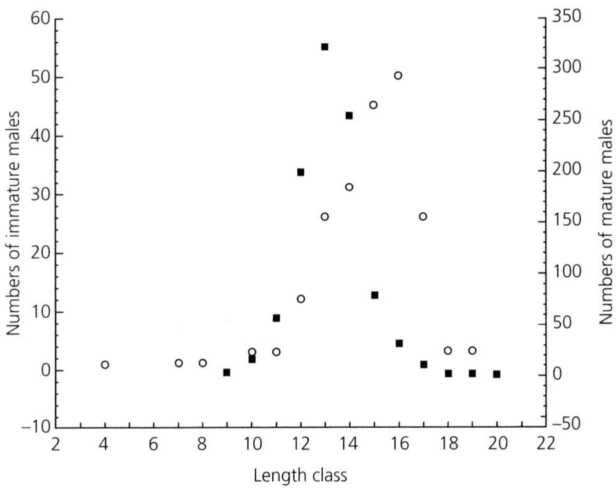

Fig. 3.7 Double-Y scatter plots of length at sexual maturity of female Atlantic tarpons caught in coastal weirs at Acaraú, Ceará State, northeastern Brazil 1962-1964 (total n = 1283): clear circles are fish with immature and maturing ovaries (n = 471), solid squares fish with mature and spent ovaries (n = 812). Length categories on the abscissa same as Fig. 3.5. Source: Plotted from data in Ferreira de Menezes and Pinto Paiva (1966: 88 Table II).

Most reproduction in the Brazilian tarpons was thought to occur from October through December, or in austral spring (Ferreira de Menezes and Pinto Paiva 1966: 90 Table III). Their findings were based on examination of a large sample comprising 1186 males and 1283 females caught in coastal fish weirs. Smith (1980) reported capturing 60 Atlantic tarpon leptocephali of 5.8–22.3 mm SL (NL in very small larvae) from surface waters ≈ 183 m above the sea bed in the Mexican and Floridian Gulf of Mexico and the northern Caribbean Sea (eastern Yucatán peninsula) during July and August 1977.

3.6 Size and age at maturity – Pacific tarpons

Pacific tarpons are much smaller than Atlantic tarpons as adults and therefore mature at smaller sizes. Specimens gill-netted by Bishop *et al.* (2001) in the Alligator Rivers area, Northern Territory, Australia ranged from 137–410 mm FL ($\bar{x}$ = 246.2 mm FL, n = 155) with a mean weight of 209.9 g; ≈ 80% of the catch fell within 170–300 mm FL. Bishop *et al.* (2001: 460 Table 153) surmised that if growth rate of wild juvenile Pacific tarpons compares with that of larvae reared in freshwater by Alikunhi and Rao (1951) of 31.4 mm/month, the largest of their Australian specimens would be less than 1 y old, indicating possible maturity in the first year.

These investigators estimated the Pacific tarpons in their samples to attain reproductive size at ≈ 300 mm FL and considered most of their longer specimens to be large juveniles (Bishop *et al.* 2001: 495 Table 159). They based this on length at first maturity being 75–100% of the longest specimen, which closely matches Wade's 1962 estimation that Pacific tarpons attain the appearance of miniature adults at ≈ 300 mm TL, or ≈ 9 months (Fig. 3.8).

More than 95% of the total catch examined by Bishop *et al.* (2001) from the Alligator Rivers area was immature, the individuals sexually indistinguishable. Fish that were clearly male showed signs of reproductive development as the dry season ended, and spent females were caught as they re-entered rivers and started upstream during the mid-wet season. The generally accepted premise is that Pacific tarpons are summer spawners (e.g. Pollard 1980: 53). However,

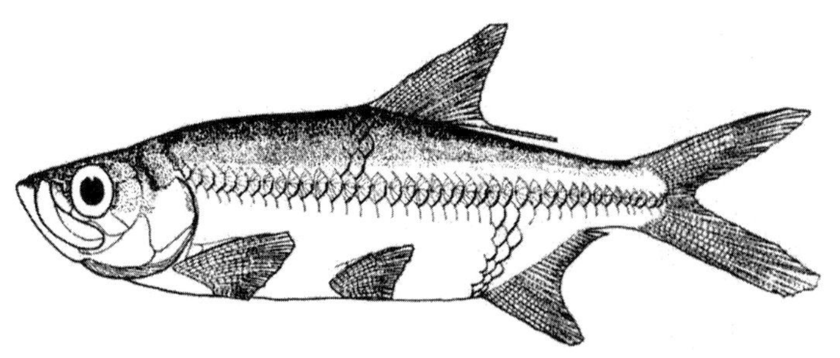

Fig. 3.8 Pacific tarpon, 300 mm SL. Source: Wade (1962: 574 Fig. 7).

some investigators have suggested they spawn all year (e.g. Wade 1962) or most of the year (Alikunhi and Rao 1951). Considering the Pacific tarpon's range spans half the globe, neither possibility seems unreasonable (see Preface). Coates (1987) reported immature stages occurring in the Sepik River system throughout the year, perhaps indicative of year-around spawning. Job and Chacko (1947: 17) wrote that in the vicinity of Madras (now Chennai), "Fingerlings of *Megalops* [*cyprinoides*] are fairly common in estuaries and backwaters throughout the year." Whether these represent a continuous influx of immigrants or residents for one or more seasons is unknown.

Water temperature affects the rate of metamorphosis, and seasonal temperatures might influence year-around reproduction, but this has not been demonstrated. Captive studies by Chen *et al.* (2008) showed Stage 2 Pacific tarpons to complete metamorphosis within the normal time of ≈ 14 d at 25°C, the process quickening at 30°C. Rearing at the winter temperature of 20°C caused a slight delay. Whether the water had practical salinity (S_P) values (S_P is explained in Chapter 6.1) of 0, 10, or 35 proved irrelevant.

Padmaja and Rao (2001) determined from gamete-staging and GSIs that Pacific tarpons sampled at two locations in the vicinity of Visakhapatnam (Andhra Pradesh), southeastern India, mature at 190 mm (males) and 180 mm (females). They did not state whether these values represent FL, SL, or TL. That females mature when smaller than males seems unusual. So is finding that fish of 100–170 mm contained immature gonads sufficiently developed to permit them to be sexed. Specimens of 180–225 mm contained spherical, translucent ova. The only fish with mature ova, however, was a single specimen of 325 mm. The highest GSIs of both sexes were recorded in November and April. Appearance of larvae in estuaries twice yearly and bimodal spiking of GSI values suggested one breeding event during June and July and another spanning December and January. Neither ripe nor spent fish could be found during their study.

CHAPTER 4
Recruitment

4.1 Introduction

Being largely metaphorical, a life-cycle requires its evaluator to step across an imaginary perimeter and look in both directions. Depending on place of entry, a fleeting image of the organism appears depicting a moment in its state of maturation at some point from egg through hatching, larval development, maturity, senility, and death. We can then follow the process, taking closer looks at our subject and measuring and describing its changing physiognomy along this continuum until every arbitrary "stage" has been assessed. But if the subject is an oceanic creature, two questions usually remain incompletely answered: where did it come from, and how did it get *here*?

The corollaries are many. In the case of tarpons, why do adults travel to distant locations to spawn? At what depths are the fertilized ova abandoned to drift and develop, and why might depth be important? What factors affect growth, dispersal, and survivorship of larvae? The specific answers are unknown, although some predictions can be made based on what we understand about physical oceanography and plankton ecology generally. The dispersion of planktonic larvae, in other words, depends on a melding of biology and behavior with proximate chemical and physical abiotic forces impinging on everything.

4.2 Life in the plankton

Atlantic tarpons from fertilized egg through most of Stage 1 remain in the offshore plankton ≈15–25 d (Shenker *et al.* 2002). Pacific tarpons have a planktonic phase of ≈25–30 d (Tzeng *et al.* 1998). The eggs of both species hatch at sea, and larvae eventually migrate inshore where they metamorphose into fry (Chapter 1). Afterward they stay in brackish habitats or move into inshore freshwaters.

Tarpons: Biology, Ecology, Fisheries, First Edition. Stephen Spotte.

A larval fish adrift in the ocean is a single infinitesimal component of a worldwide planktonic community that outnumbers the stars in the Universe (Armbrust and Palumbi 2015). We can start with the obvious: a newly hatched larval tarpon either lives and grows or it dies. The faster it grows, the stronger its chances of surviving. Assuming adequate availability of food, it drifts and swims along a trajectory carried by currents, its small size rendering it far more constricted by the sea's physical characteristics compared with an adult, notably light, currents, and the viscosity of seawater (Section 4.6). Despite these handicaps, made worse by a fish larva's tiny size, increasingly flexible and responsive behaviors eventually exert a certain control.

Locomotion is a prerequisite to acquiring the food needed for growth (Kaufmann 1990). Learning and reinforcement allow prey and predators to become recognizable, and continuously developing structural, neurological, and sensory systems gradually loosen the bonds of viscosity, one result being increased burst swimming speed (Chapter 5) for attack and escape, increased *endurance* (the length of time a larva can swim at a fixed speed without rest or food) for prolonged swimming to locate and stay in favorable stratified currents, and enhanced awareness of the spatial relationship linking feeding and risk; that is, the trade-off between the need to eat and self-preservation. This last association, as stated by Fiksen *et al.* (2007: 196), "is driven by the exponentially decreasing vertical profile of underwater light – the key determinant for encounter rates with both predators and prey ...."

A fish larva's vertical location in the water column, in combination with its state of development and time of day, influence how well it sees, which in turn affects rates of prey encounters and adeptness at avoiding predators lurking beyond the background spacelight. The situation then becomes one of higher risk during prey-seeking. That the capacity to cover larger distances seeking prey becomes commensurate with more extended exposure to predators is apodictic. In this configuration, growth must be offset by risk. However, growing larger also reduces the degree of danger from classes of predators that are small, abundant, and of fixed size, and larvae leave behind a certain quantity of risk by outgrowing their smallest and most vulnerable stages. Thus fast-growing larvae often have better overall survival (Nielsen and Munk 2004; Takasuka *et al.* 2007).

High growth rate, despite the attendant risk, has measurable utility from a fitness perspective. But predators never go away, and getting bigger means becoming more visible and vulnerable to still another class of predators. Many planktonic creatures make maximum use of whatever size they are by capitalizing on diel changes in illumination, rising vertically to the surface at night and descending into twilight or deeper into darkness during the day.

Fiksen *et al.* (2007: 196) stated correctly that "It is often tempting to simply impose behaviour on individuals or to implement caricatures of observed behaviours in models, making them more descriptive than explanatory."

Interpreted in a slightly different context, a population comprises not simply a sum of individuals but of individuals each contributing to that sum. Vertical positioning in daylight sharpens the orientation of fish larvae in pelagic space, permitting diel periodicity to act in concert with tidal rhythms and aiding eventual recruitment into estuaries (Steffe and Westoby 1991). At locations of strong tidal currents, for example, some larvae use the reduced shoreward flow just above the bottom to assist their recruitment. The orientation behavior model predicts that accuracy of orientation becomes compromised at night when vision is reduced, and movements then attenuate to passivity. This aspect is less well explained by observation.

Fish larvae are not mere riders of currents. They participate actively in their dispersal and migrations, in part by positioning themselves strategically in the water column to take advantage of favorable shear forces and currents, in the process influencing their own drift trajectories (Fiksen *et al.* 2007 and references; Paris and Cowen 2004). Many fish larvae have considerable control over their dispersals when only a few millimeters long. They are able to swim long distances at speeds sufficient to alter drift trajectories, and they can detect cues in the pelagic environment (e.g. light, sound, odors) and orient accordingly (Leis 2007). Swimming performance increases with development and, in most cases, with size (Clark *et al.* 2005). That many late-stage larvae can swim faster than local currents – usually considerably faster – is well documented. For example, Fisher *et al.* (2005) evaluated data from late-stage larvae of 89 species of coral fishes from Australia and the Caribbean of which 95% could swim faster than the average current speeds measured in the vicinity of Lizard Island, Great Barrier Reef. Some larval fishes are able to swim continuously for long distances (>20 km upon reaching ≈10 mm SL), evidence not just of the capacity to overcome current speeds, but also of remarkable endurance (Clark *et al.* 2005).

A few meters of vertical separation can radically alter a drift trajectory, resulting in larvae ending up in different – and potentially less auspicious – geographic areas. Stated differently, *the vertical positioning of fish larvae in pelagic zones strongly influences their horizontal displacement.* Vikebø *et al.* (2005) dropped particles designed to drift passively at depths of 10 m and 20 m above Atlantic cod spawning areas in the Norwegian Barents Sea, tracked them for 100 d, and discovered they ended up hundreds of kilometers apart. I find it useful to think of vertical positioning by fish larvae in pelagic ecosystems as habitat selection, no different than locating a suitable inshore area later in life.

Behavioral processes are critical to surviving the early stages of planktonic life (Vikebø *et al.* 2007). The ecology of a species in its larval state is driven and shaped by patterns of behavior that in turn are limited by physical oceanographic factors, some already mentioned (e.g. illumination, temperature, viscosity, turbulence, turbidity). A pelagic fish larva's behavior helps sustain its life-cycle, a loop powered by steady feedback and reinforcement, even exerting evolutionary influence over where and when spawning occurs at maturity.

Before a fish can recruit it must outlive the larval stage. In tarpons this means persisting in the oceanic plankton for the requisite duration while somehow factoring in the timing of metamorphosis to coincide with arrival inshore, or perhaps vice versa. At this point other difficulties arise, such as the possible need to travel alongshore until encountering the mouth of an estuary or inlet, then moving through the opening on a suitable tide and selecting a habitat once there. This last often involves additional upstream migration into tributaries or backwaters. The sum of these passages requires periods of drift interrupted by active swimming, the former modulated by the latter, and everything accommodated by flood tides, favorable currents, wind-driven surface disturbances, and local upwellings. If some elements of recruitment involve drift and random movements (*kineses*), others are directed taxes relying on timely execution of movements. Interactions of physical factors, along with a larva's age (or ontogenetic stage), influence the depth it selects in the water column. Because few physical forces are vectorial, taxes are restricted mostly to those directed by light (phototaxis) and currents (rheotaxis) and modulated by such scalar factors as salinity, temperature, turbidity, turbulence, and olfactory cues (Boehlert and Mundy 1988).

4.3 Inshore migration

Shenker *et al.* (2002) cited Ferrell (1999) as having caught 100 Atlantic tarpon larvae during nocturnal flood tides at Sebastian Inlet, a passage between the Atlantic Ocean and Indian River Lagoon on Florida's east coast, during September–October 1998.[1] In the work of Shenker and colleagues, recruitment occurred as late as 22 October, and larvae ranged from 11–24 mm SL ($\bar{x}$ = 17.8 mm SL). Hatch dates extending into autumn, perhaps in combination with storms, could also explain the late recruitment of tarpon larvae into Georgia estuaries and salt marshes (Rickards 1968).

Rickards (1968: 231) postulated that Atlantic tarpon leptocephali remain offshore (maybe as far east as the Gulf Stream) throughout much of the year and thought that larvae he caught in summer and autumn "must have been in the Sapelo Island [Georgia] area in May and June." He cited Richard A. Wade in a personal communication as having found juveniles in winter along Florida's east coast north to Fort Pierce (≈27°26′20″N, 80°20′8″W). According to their sizes, Wade believed they had moved shoreward in November and December, perhaps in the company of larvae that made it to Rickards' sampling sites in Georgia during the same months.

[1] The citation was omitted from their bibliography. I was unable to find Ferrell's report and therefore have not seen it.

Gehringer (1959) described a tarpon leptocephalus caught in a Gulf Stream plankton tow directly east of Brunswick, Georgia (≈31.1589° N, 81.4892° W) in August. Harrington and Harrington (1960) noted that the earliest larval stages were present during autumn in salt marshes and mangroves of Indian River Lagoon, eastern Florida. In another personal communication to Rickards, Thomas Linton reported catching a tarpon leptocephalus farther north at Sapelo Island in April 1964. Rickards (1968: 232) concluded: "Therefore, it would seem that tarpon larvae are present in the Gulf Stream at least from November until August." This statement remains unverified.

Dahl (1965) pointed out that many of the streams flowing into Colombia's Golfo de Morrosquillo are dry, or nearly so, their mouths closed to the sea until start of the rainy season when flooding opens them. At Arroyo de Pechilín (Pechilín Creek), for example, recruitment of leptocephali occurred in late June and early July, after which only a few specimens were found. Spawning had apparently occurred in April. After entering the stream and undergoing metamorphosis, fry migrated into bordering mangrove swamps.

At locations where tidal amplitude is consistently low, recruitment depends heavily on seasonal rains to flood streams and rivers, permitting leptocephali access to inshore brackish and freshwaters. Chacón Chaverri (1994) documented highest recruitment during or just after rainy periods in Costa Rica's Glandoca-Manzanillo National Wildlife Refuge.

The body length at which Atlantic tarpons recruit into estuaries varies. Cyr (1991: 35) reported an average of 206 mm SL in October, Harrington (1958) 16.0–17.9 mm SL at Indian River Lagoon from 10–12 September. The fish at that time were growing at ≈20 mm SL/month, which extrapolates to 46.0–57.9 mm SL in early October. Farther north, recruits of 29.7–111 mm SL were invading the salt marshes of Sapelo Island by 20 July, and specimens 20–30 mm SL were still present in November. Cyr (1991: 36) stated, "The four month span during which these small tarpon were collected argues for protracted recruitment."

Boehlert and Mundy (1988) wrote, "In general, species that recruit to estuaries from offshore do so at advanced developmental stages, usually near or after metamorphosis ...." This is true of members of the order Elopiformes, which are typically undergoing metamorphosis as they enter inland waters. Advanced development goes along with improved orientation, swimming ability, and endurance, and probably helps keep new recruits from being swept back out to sea on ebb tides; that is, aids in their *retention*.

Tarpon larvae can recruit rapidly. Occasionally the timing of inshore movement coincides so closely with a stochastic weather event as to appear deterministic. Kuthalingam (1958: 9) wrote about Pacific tarpon larvae entering India's Cooum River: "As soon as the bar was opened on night of November 23rd 1956 plankton collection made on the next morning in Coovum [*sic*] revealed the larvae of *Megalops*." Whether the bar had opened as the result of a storm or unusually high tide was not mentioned, but evidently larval tarpons were poised to enter the estuary from the Bay of Bengal, and, from Kuthalingam's observations, had already begun metamorphosis. Although still ribbon-shaped and transparent, these specimens possessed villiform teeth, well-formed fins, and noticeable air bladders and anal openings. Moreover, they had commenced to feed on copepods, crustacean larvae, and even larval fishes, indicating their status was late Stage 2 development. These were doubtfully new recruits. For example, metamorphosing Atlantic tarpons do not feed until Stage 2 phase VIII (Chapters 1.6, 7.7, Appendix B).

The smallest Pacific tarpons examined from the Alligator Rivers area, Northern Territory, Australia by Bishop and colleagues were obtained during the last half of the wet season and first half of the dry season, the largest in the mid-wet season. Bishop *et al.* (2001: 20) concluded: "The juveniles therefore recruited to the catchments during the mid-wet season when connection was made with the estuaries." Having entered the river and moved upstream, they recruited into floodplain billabongs and shallow areas of backflow billabongs (Bishop *et al.* 2001: 22). This pattern was revealed by length-frequency distributions across seasons. Mean lengths were greater in the late-dry and early-wet seasons than during the mid-wet time, when juveniles recruited in greater numbers. Russell and Garrett (1983: 808 Table 1) recorded small juveniles (22–25 mm, whether FL, SL, or TL is unstated) from the Norman River estuary, northern Queensland in December, or during the early-wet season. Davis (1988) reported thousands of recruits entering tidal creeks in Leanyer Swamp north of Darwin, Northern Territory in December and January, at the rainy season's peak.

Shen *et al.* (2009) presented length and age data for Pacific tarpon leptocephali recruiting into the Tadu Creek estuary, western Taiwan, along with lengths, weights, and ages of juveniles from the stream's freshwater upper reaches, the mouth of its estuary, and adjacent offshore waters (Table 4.1).

4.4 Offshore migration

Griswold (1913: 97) believed that juvenile Atlantic tarpons "remain for several years in brackish water before going to sea." Such observations might be true in some cases, but emigration out of and immigration into isolated bodies of water depends when access to the sea is restored, usually after weather-driven breaches through narrow barriers, submersion of barriers by overwash during storms or spring tides, or rainfall sufficient to cause flooding (e.g. Dahl 1965). Marshall B. Bishop, a museum collector, witnessed such an event shortly after two storms had passed over Sanibel Island, Florida. According to Breder (1944: 218): "In one case Mr. Bishop actually saw them [juvenile Atlantic tarpons] entering a newly storm-made cut which was again closed in a short time, securely land-locking these fishes." How long they stay when offered an opportunity to escape has not

Table 4.1 Lengths and weights of Pacific tarpon leptocephali and juveniles from Tadu Creek (estuary and freshwater upper reaches) and adjacent offshore waters, western Taiwan. L = length, W = weight, $\overline{X}$ = mean, ± SD = standard deviation, n = sample size. Source: Shen *et al.* (2009: 258 Table 1).

Location	Life stage	*n*	*L*, mm FL Range	$\overline{X}$ (± SD)	*W*, g Range	$\overline{X}$ (± SD)	Age Range	$\overline{X}$ (± SD)
Estuary	Leptocephalus	88	21.9–30.2	25.4 (±1.7)	—	—	18–34 days	25.2 (±4.7) d
Upper reaches	Juvenile	62	174.4–343.7	251.7 (±28.4)	69.4–558.8	227.9 (±79.3)	1–3 years	2.2 (±0.6) y
Offshore	Juvenile	19	276.6–435.1	368 (±41.1)	286.8–1145.2	780.0 (±244.0)	2–5 years	4.1 (±0.9) y

been determined. William Beebe, having discovered juvenile Atlantic tarpons trapped in an isolated lagoon in Haiti, returned one day to find the connection with the sea re-established and the tarpons gone (Beebe 1927; Beebe 1928: 67–74; Beebe and Tee-Van 1928). In such instances immigration and emigration seem entirely opportunistic.

That adults also frequently remain in Florida rivers, lagoons, and estuaries throughout the year is well documented, Storey and Gudger (1936) and others having reported them in winter fish kills, often in large numbers (Chapter 7.4). Griswold (1913: 99–101), an avid tarpon angler, noted that "some fish remain in the deep rivers of the east coast of Florida all winter." He knew this because he caught them regularly. Gunter (1941) did not include the tarpon among species killed during a cold spell on the Texas coast in January 1941, indicating that tarpons of all sizes are largely absent from the northern Gulf of Mexico in winter.

Low seasonal water temperatures might also drive emigration at locations in the northern part of the range. Rickards (1968) stated that juvenile Atlantic tarpons left the marshes at Georgia's Sapelo Island by the end of October after reaching ≈150 mm SL, presumably before the onset of cold weather. This could limit their time in residence to <3 months. Rickards speculated that these emigrants then became part of the oceanic tarpon population. He offered no evidence of this, nor to my knowledge has any been found in the years since his report. Where they go is unknown.

Allen *et al.* (1982: 7–22) wrote: "A 22 mm tarpon (*Megalops atlantica*) larva (leptocephalus) was collected with the [epibenthic] sled in August at [South Jones Creek]. Large tarpon are summer visitors to the ocean adjacent to Winyah Bay [South Carolina]." He noted that "larval and juvenile tarpon (to about 300 mm) are commonly caught in upper marsh waterways during the summer." This could mean that at least some depart during their recruitment year. Winyah Bay is ≈345 km north of Sapelo Island, Georgia. Zerbi *et al.* (2005) reported Atlantic tarpons in lagoon backwaters of Puerto Rico staying no longer than 56 d when they attained 180 mm SL. Cyr (1991: 37) noticed that most first-year recruits at his southern Florida sites abandoned their inshore habitats when ≈400 mm SL and 500 g, or about a year old.

Irregular weather-dependent movements into and out of estuaries and lagoons are characteristic of both species. Alikunhi and Rao (1951: 100) described recruitment of Pacific tarpon leptocephali to inshore areas of Tamil Nadu, southeastern India: "During the north-east monsoon months, October to December, when connection between the sea and the backwater is established by breaches in the sand bars across the mouths of the rivers Adyar and the Cooum in Madras [now Chennai], the ribbon-shaped post-larvae of … *M. cyprinoides* migrate in shoals into brackish water and gradually and naturally get acclimatized to less saline conditions." Years earlier Raj (1916: 253) had stated that Pacific tarpons reproduce in May and June, and "larvae are common in the river Cooum in October and November in fresh water." October and November were also the months when Kuthalingam (1958) captured 24 larvae in the Cooum River, all 24.6 mm long (whether FL, SL, or TL is unstated) and starting to metamorphose.

Pacific tarpons at some locations might recruit into brackish and freshwater habitats and not emigrate seaward until mature. Coates (1987) believed that Pacific tarpon larvae in the Sepik River system of northern Papua New Guinea migrate from coastal areas or estuaries into freshwaters at ≈100 mm SL, entering the river during April near the end of the wet season. Inshore migration then continues until July, about midway into the dry season. Kowarsky and Ross (1981) captured specimens of 139–370 mm TL (n = 6) in the Fitzroy River as they migrated upstream through the barrage at Rockhampton, Queensland during March and April 1979. The barrage is located 59.6 km from the river mouth.

Coates (1987) strung nets in the Sepik River and divided his catch of Pacific tarpons into four arbitrary size classes. Fish of 150–250 mm SL were most numerous in April (austral autumn). Those <150 mm SL appeared from May through August and disappeared by September, suggesting either migration or growth into a larger size category. Coates (1987: 532–533) speculated, "The general increase in catches of smaller fish (size categories <150 and 150–250 mm [SL]) between April and July, inclusive, probably represents recruits of the current year class entering the river from the estuary or sea." He noted that the abundance of specimens >250 mm SL displayed no seasonal pattern, and fish in this size category evidently were present all year.

4.5 Mechanisms of recruitment

The means by which larval tarpons locate, orient, and recruit into lagoons and estuaries is likely based on one or more chemosensory mechanisms. Chemical attractants are known to drive the recruitment and subsequent metamorphosis of many coral fishes and invertebrates, including corals themselves (Dixson *et al.* 2014). In some situations, odorants in plumes of water flowing outward from lagoons are putative attractants (Atema *et al.* 2002). That low salinity *per se* exerts a positive tropism and serves as an influential factor in recruitment of larval tarpon seems problematical. As Boehlert and Mundy (1988: 57) emphasized, "One must use caution in interpreting cause and effect in relationships between recruitment and single physical factors."

The low-salinity hypothesis, while superficially interesting, fails to account for recruitment into estuaries like the Sine-Saloum and Casamance of Sénégal and saltpans and tidal pools along the Norman River estuary in Queensland (Russell and Garrett 1983), all of which are hypersaline, or the Rio Grande de Buba in Guinea Bissau (western Africa) with its nearly constant ionic strength

(Diouf 1996: 16–17). Observations of this sort say little about a tarpon's motivation to enter waters of varying ionic strength but lots about its tolerance.

4.6 Factors affecting recruitment

Point-source stimuli affect recruitment. These are biotic or abiotic factors that elicit a fish larva's short-term response. Many are scalar. Among the abiotic factors easily listed are temperature and salinity and their gradients, current speed, water depth, composition of the substratum (e.g. muddy, sandy, silty), bottom slope, turbulence, and turbidity. Lunar phase influences nighttime illumination, affecting visual acuity and consequently nocturnal movements. Most recruitment of larval fishes occurs at night (Boehlert and Mundy 1988 and references), and tarpons of both species apparently recruit most heavily on nocturnal flood tides (Shenker *et al.* 2002; Tzeng *et al.* 2002).

Biotic factors based on audition and olfaction are the most poorly understood. Extensive work with salmon revealed that imprinting of olfactory cues (i.e. specific inland water odors *sensu lato* Hasler *et al.* 1978) from the natal stream are crucial to homing (e.g. Dittman and Quinn 1996; Hasler *et al.* 1978; Lema and Nevitt 2004; Nevitt and Dittman 1999). Food odors are possibly olfactory stimulants; the sounds generated by turbulence could trigger ancestral auditory "memories." Little of this has been tested at the mechanistic level, and we have no evidence that tarpon leptocephali are even drawn to locations they occupied as early juveniles.

The arrival of fish larvae at predictable places is obviously not just a result of random drift, although drift is an important component; instead, it represents the culmination of sustained oriented behavior. This has been demonstrated in tests using a passive particle-drift null model against an oriented-behavior model in which larval fishes position themselves vertically in the water column to take advantage of water moving shoreward (Steffe and Westoby 1991).

Buoyancy control by many planktonic fish larvae must be accomplished without aid of a completely developed swim bladder, and the gelatinous matrix perhaps serves this function in Stage 1 tarpons (Chapter 1.6). Certainly the capacity to choose and maintain a particular depth in the water column is crucial. The Atlantic menhaden (*Brevoortia tyrannus*), like the tarpon, spawns offshore. Its larvae, pushed by wind-driven surface currents, are transported inshore and enter estuaries (Power and Walsh 1992 and references). Larval menhaden tend to sink. In growing from 10–15 mm SL at 19–30 d post-hatch (dph) a menhaden larva's weight in water – the force applied by the fish to maintain its vertical position – increases 18-fold and coincides with development of the swim bladder.

Recent evidence indicates that recruitment of benthic species to specific natal sites (e.g. a particular coral reef) is not universal. Leis *et al.* (2009: 221) wrote that "low precision [of orientation] at an individual level, and a general lack of directionality among individuals, means that horizontal swimming is less likely to have a clear, direct influence on dispersal than in larvae of reef species that have more precise orientation." The tarpon is perhaps among those fishes lacking finely honed philopatry, unlike larvae of many coral fishes that recruit to a natal reef and mature salmonids that home into a natal stream to spawn. Tarpon larvae might be more flexible about which estuary they enter, having been hatched in the open sea (i.e. there *is* no natal inshore location). Any capacity to orient horizontally with high accuracy and precision might be an unnecessary adaptation, but this is conjecture.

More probable is that stratified currents are selected by larval tarpons according to ontogenetic stage, as they are by other fish larvae, allowing them to drift and swim eventually to a geographic region within the latitudinal and longitudinal boundaries of their genetic population. The process, we could assume, was initiated by the adults that spawned them having chosen a location and depth in the open sea along which a drift trajectory was ancestral and thus predictable within broad (i.e. populational) limits. If true, this suggests that the ecology of tarpon larvae strongly affects the ecology of their progenitors (and vice versa), requiring adults to migrate considerable distances offshore so that eggs can be shed at locations with favorable currents, illumination, and other factors influencing survivorship in the planktonic stage and the timing of later metamorphosis.

The transport of fish larvae inshore is facilitated by seasonal winds, estuarine fronts, wave motion and direction, convergence of currents, and other abiotic factors acting alone and in combination (Steffe and Westoby 1991 and references). The viscosity of water is a major impediment to the free movement of larval fishes. At 1027 kg/m^3, surface seawater is roughly a thousand times denser than dry air at sea level (1.275 kg/m^3) and ≈850 times more viscous. *Viscosity* is a measure of the resistance of a fluid undergoing deformation by shear or tensile stress. In practical terms, viscosity measures the friction, or "thickness," within a fluid. For example, molasses has a higher viscosity than water, and a newly hatched fish trying to swim in the sea is perhaps analogous to a human attempting the Australian crawl in a swimming pool filled with molasses. An early fish larva is handicapped by its small size and lack of a working swim bladder. In addition, its appendages and supporting skeletal structures needed for propulsion, turning, braking, and steering have not yet emerged or are incompletely developed.

Viscosity is affected by temperature and density (a substance's mass per volume), decreasing with rising temperature much as butter flows more fluidly when hot than warm. Density, however, increases as the temperature drops; cold seawater is more dense than warm seawater and more viscous. In other words, its mass per volume increases with falling temperature and in the process it becomes "thicker," or gains viscosity.

To a swimming animal the greater the viscosity of the water the more effort required to move through it. As a result, metabolic costs of such activities as locomotion and breathing become greater than experienced by a terrestrial animal breathing and moving about in air. Ions in solution worsen matters by increasing the density. These combined effects are insignificant on the swimming performance of adult tarpons, although not their larvae, and the entry of larvae into brackish estuaries and lagoons where the temperature is often higher and the salinity generally lower than in the open sea might make swimming easier and less energy-intensive. Whether this offers an evolutionary advantage is unknown.

Brackish lagoons and estuaries have long been considered "nurseries" for many species of coastal fishes, places ostensibly offering havens from predation and foods of the right sizes and concentrations to promote survival and growth (e.g. Beebe 1928), but any such mechanistic benefits are poorly documented. Even the safety aspect is not guaranteed, unless the area in question is isolated from the sea and predators are denied access (e.g. Beebe 1927). Small fishes sheltering in mangroves fronting seagrass beds often experience heavy predation that actually diminishes seaward (Hammerschlag *et al.* 2010).

That the tarpon's ability to survive – even thrive – in backwaters of poor environmental quality, thus indirectly excluding other water-breathing predators, was proposed years ago (e.g. Harrington 1966; Wade 1962). Cyr (1991: 38) was circumspect, and rightly so. He wrote that "once the high-predation size threshold is exceeded, tarpon would find advantage in exiting the confining ... habitat in search of more salubrious growth conditions." He added that additional motivation for emigration could be shifting food requirements that confined habitats are unable to meet.

Tidal amplitude affects recruitment of tarpons into lagoons and marshes (Davis 1988; Russell and Garrett 1983; Zerbi *et al.* 1999). Post-larval tarpons apparently enter Georgia salt marshes in late spring and summer on spring tides associated with the new and full moons, and in autumn when swept into these habitats during storms (Rickards 1968). Both have been termed *overwash events* (e.g. Cowley *et al.* 2001; Reddy *et al.* 2011) because in many locations and at certain times of year a sandbar or some other natural structure is breached temporarily, after which the remnant of a pool or lagoon is once again isolated from the sea along with any fishes left behind. Heavy rainfall at other locations or unusually high tides simply flood the backwaters, bringing new recruits, and the receding water leaves them trapped until the next overwash event (e.g. Breder 1933; Zerbi *et al.* 2005). Sometimes a bar remains as a result of low fluvial flow during dry periods. Babcock (1951: 50) noted that because young tarpons were developing during Florida's hurricane season starting in mid-summer, "It is not strange they are carried into the interior pools by the high water, which usually accompany [*sic*] these violent storms." Franks (1970: 35) suspected hurricanes to be factors in the recruitment of tarpons into landlocked ponds at Horn Island, Mississippi.

Several investigators (e.g. Beebe 1927, 1928; Breder 1944; Harrington 1958; Kuthalingam 1958; Russell and Garrett 1983) discovered young tarpons in isolated backwaters opening to lagoons or the sea during storms, unusually high tides, or heavy rains, allowing fishes trapped in them to enter or escape. Sometimes these locations are vernal, containing water only during the rainy season (Babcock 1951: 49; Breder 1944: 218; Ferreira de Menezes and Pinto Paiva 1966; Storey and Perry 1933). The fishes retained had arrived earlier during similar periods of high water. When such events happened rarely, early observers wondered how tarpons found their way into these isolated bodies of water, which led to interesting speculation. The angler Gray Griswold sent a letter to Beebe (1928: 72) in which he described a small lake on Den Island, San Carlos Bay, southwestern Florida containing juvenile Atlantic tarpons, noting that it "has no inlet to the sea nor has the sea invaded it in the memory of man. It is supposed the [tarpon] spawn was dropped there by birds, or carried to the lake on the back [*sic*] of alligators."

Beebe (1927) captured several juvenile Atlantic tarpons at two linked Haitian lagoons called Source Matelas, separated from the sea by a narrow strip of land. Later, in mid-January, he collected 36 healthy young tarpons at the same site (Beebe 1928). The water was 46 cm deep and lay stagnant above a sulfurous muddy substratum. The fish ranged in length from ≈50–200 mm.[2] A week or so later (23 January) he seined the area again, capturing 154 tarpons, all ≈76–180 mm with the exception of a single specimen measuring ≈330 mm. The only other fish were three small snooks (*Centropomus undecimalis*). Beebe went back again on 21 March and found the lagoon had re-opened to the sea. He managed to catch only six tarpons of ≈114–180 mm, a snook of ≈25 mm, and four small mojarras (family Gerreidae). The snook and mojarras he (Beebe 1928: 70) called "recent emigrants from the open water outside."

Ferreira de Menezes and Pinto Paiva (1966) suggested that tarpons caught at Ceará in northeastern Brazil hatch at sea during the year's last quarter, reach coastal lagoons at the beginning of the next year's first quarter, and stay there about a year. These authors claimed that young Atlantic tarpons seek access to the sea on attaining 400–500 mm FL, citing as evidence the fish weirs of Ceará State, which to their knowledge had never yielded a tarpon <500 mm FL. Ferreira de Menezes and Pinto Paiva (1966: 88 Table II) listed 2469 specimens examined from the weirs, just one within the range 550–595 mm FL. Having seen grainy black and white photographs of such structures, I wonder if the openings in their barriers were sufficiently spaced to prevent fishes this small from escaping. The weirs pictured by Seraine (1958: Figs. 2–10), for example, consisted of irregularly shaped sticks (tree branches) driven into the mud in a picket fence arrangement and anchored by thick pilings (Fig. 4.1).

[2] Beebe measured these specimens in inches, which I converted to millimeters. Whether he used FL, SL, or TL is unknown.

Fig. 4.1 Part of a coastal fish weir set at Acaraú, Ceará State, northeastern Brazil. Note the spaces between the vertical sticks, which appear wide enough to allow small Atlantic tarpons to escape. Source: Seraine (1958: Fig. 7).

Vertical positioning in the plankton has its inshore equivalent in the tidal stream during recruitment. Shenker *et al.* (2002) used channel-nets moored to fish the upper 1 m of Sebastian Inlet, which links Indian River Lagoon in eastern Florida and the Atlantic Ocean. During two summers of sampling, 253 Atlantic tarpon leptocephali were caught from 25 May–12 September 1994 and 723 from 8 June–2 September 1995, with 74% of total recruits (976 leptocephali) caught in 1995. Leptocephali ranged from 13–26 mm SL. Peak recruitment was 2–4 d preceding summer periods of full moon, and larvae moved through the inlet only on nocturnal flood tides and nearly always nearer the surface. This behavior is consistent with selective tidal stream transport observed in other fish larvae that actively position themselves higher and lower, respectively, in the water column depending on whether the tide is at flood or ebb (e.g. Barletta and Barletta-Bergan 2009; Hare *et al.* 2005).

Variability of recruitment during lunar cycles was large at Sebastian Inlet, occurring in pulses. In 1994, 73% of recruits were captured over 13 nights in mid-July when the moon was waxing, the catch peaking on nights 11–13 (full moon on night 15). Most of the rest were netted during smaller pulses in June and August. June 1995, in contrast, yielded two short pulses of 172 leptocephali, but little recruitment through most of July until the last days when >50 larvae were captured (Fig. 4.2).

A Category 1 hurricane passed over the study site during 1 August 1995, followed by 4 d when samples could not be obtained. Then came a 5-d period

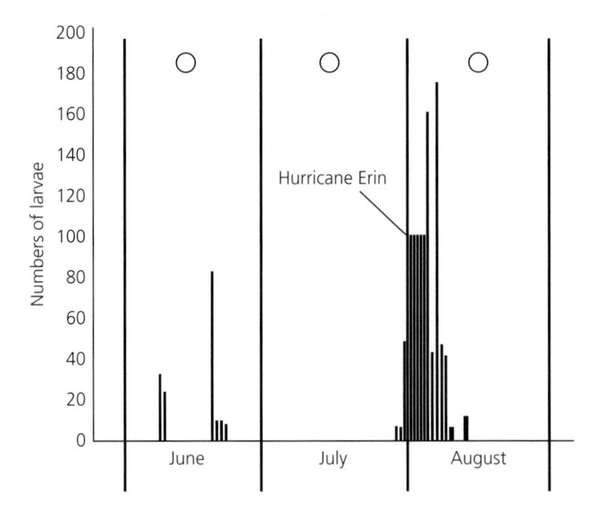

Fig. 4.2 Daily catches of larval Atlantic tarpons at channel-net stations set at Sebastian Inlet, Indian River Lagoon, Florida in summer 1995 (total n = 723). Clear circles represent full moons. Source: Shenker *et al.* (2002: 62 Fig. 7).

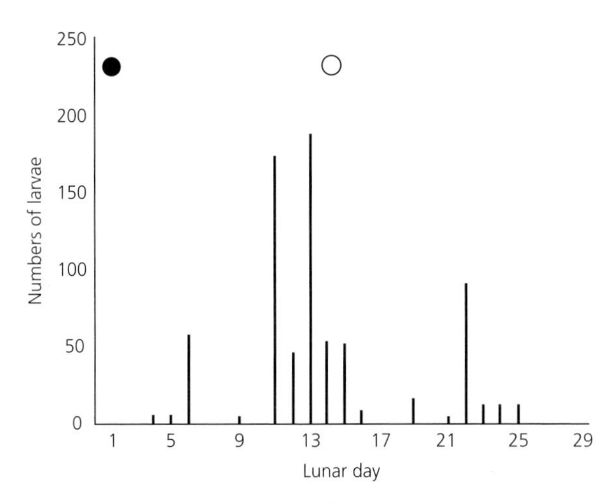

Fig. 4.3 Lunar cycle of Atlantic tarpon leptocephali caught at Sebastian Inlet, Indian River Lagoon, Florida during summer 1995. Rayleigh's test of circular distributions (Z = 8983.0, n = 723, $p < 0.001$) indicates non-uniform hatching through the lunar months. Solid circle represents a new moon, clear circle a full moon. Source: Shenker *et al.* (2002: 63 Fig. 9).

when 491 leptocephali (70% of the total 1995 catch, or an average of 175 larvae/d) were obtained (Fig. 4.2), reinforcing earlier observations that storms can push larval tarpons toward salt marshes, lagoons, and other inshore areas (Breder 1933; Rickards 1968). Consistent with 1994, recruitment within the lunar month was not distributed uniformly (Fig. 4.3).

As the figure shows, the mean mass of larvae moved onshore during night 13, two nights before full moon. The uptick in recruitment, however, actually started 2–3 days prior to the 1995 hurricane, with 50 specimens caught in the last sample before the storm temporarily shut down collecting. Whether there was any correlation with the advancing front is unknown. That autumn numerous juveniles were seen in northern salt marshes of Indian River Lagoon >50 km from the nearest inlet from the Atlantic, their dispersal perhaps aided by the hurricane (Shenker *et al.* 2002). Such situations are not unusual. Breder (1944: 218) wrote presciently that juvenile Atlantic tarpons "may get into the odd places in which they are sometimes found by being driven in on the wings of a hurricane or late summer storm."

CHAPTER 5
Breathing and respiration

5.1 Introduction

Oxygen, which is necessary to sustain aerobic metabolism, is simultaneously a deadly toxin, and biological evolution is partly a history of the compromises organisms have devised to maintain the delicate balance between assimilating sufficient amounts to sustain life, but not enough to cause lethal cell damage.

Life arose in anaerobic environments. Mitochondria are thought to have once been facultative anaerobic bacteria capable of detoxifying oxygen (Hsia *et al.* 2013). In an aerobic environment this would require a "stepping down" of partial pressure in oxygen taken up to balance homeostasis while simultaneously staving off cell death (Fig. 5.1). The struggle to sustain this equilibrium projects well beyond organs and species; it forms the very foundation of organic existence. As Hsia *et al.* (2013: 849) wrote, "The respiratory organ is the 'gatekeeper' that determines the amount of oxygen available for distribution."

All major teleost groups except sturgeons (Chrondrostei) and herrings (Clupeomorpha) include air-breathing species. Whether lungs originated from the swim bladder, the reverse, or neither remains unresolved (Hsia *et al.* 2013). The overwhelming majority of today's air-breathing fishes live in tropical freshwaters, and the traditional view has been of evolution in warm swamps of the Late Silurian and Early Devonian, where conditions of low oxygen often prevailed (Barrell 1916). To Packard (1974), more likely environments were hypersaline seas or estuaries in which the low solubility of oxygen at high temperatures restrained aerobic metabolism even when these locations contained sufficient oxygen for water-breathing.

The tarpon is a *bimodal breather*, capable of extracting oxygen from both water and air. It has two "gatekeepers" where gases are exchanged, the gills and the *swim bladder* (also called *gas bladder* and *air bladder*). Among elopomorph fishes, only tarpons (family Megalopidae) and eels (family Anguillidae) can make use of air.

Tarpons: Biology, Ecology, Fisheries, First Edition. Stephen Spotte.

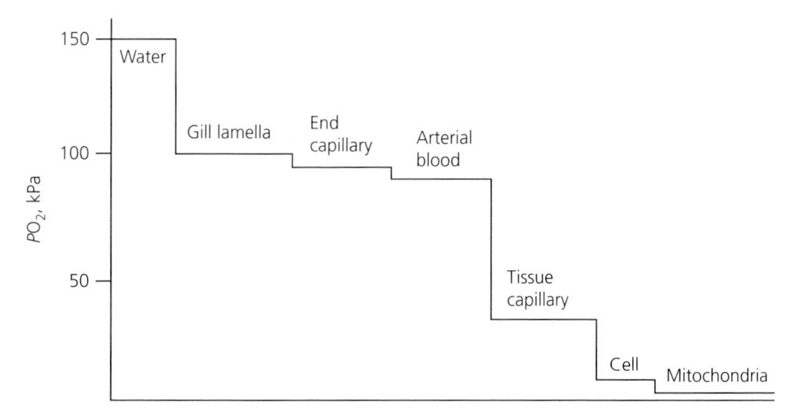

Fig. 5.1 Oxygen cascade: the series of convective, diffusive, and biochemical barriers that progressively lower the partial pressure of oxygen until it reaches near-anoxic levels necessary for optimal mitochrondrial intracellular function. Modified for water-breathers. Source: After Hsia *et al.* (2013: 855 Fig. 6).

Ventilation, or *breathing*, refers to the mechanical process by which water or air is moved into contact with gas-exchange tissues. To clearly separate the mechanics of breathing from the physiological aspects, I restrict the term *respiration* to physiological processes involving gas exchange between the external environment (water or air) and blood.

5.2 Water-breathing

Gills are the dominant structures for breathing water and, as discussed elsewhere (Chapter 7), regulating internal ion composition, acid-base balance, and excretion of nitrogenous wastes (Brauner and Rombough 2012). The *gills* comprise *gill*, or *branchial*, *arches* (usually four pairs) with attached pairs of *gill filaments* (also called *primary lamellae*), each supporting numerous epithelium-covered, plate-shaped *lamellae* (also called *secondary lamellae* when the filaments are referred to as primary lamellae) on top and bottom (Fig. 5.2). Four major types of cells occur in the epithelium: pavement cells; mitochondria-rich cells (MRCs), or ionocytes (Chapter 6.3); mucous cells; and cells that are undifferentiated (Evans 1999; Huang *et al.* 2008; Monteiro *et al.* 2010). Of all these, pavement cells take up > 90% of the gill surface (most of them in the lamellae) and compose the loci of gas exchange (Evans 1999).

The lamellae are therefore principal sites of gas transfer (this chapter) and exchange of ions (Chapter 6) with the external environment. Oxygen dissolved in water diffuses across thin layers of lamellar tissue into capillary blood. Diffusion rate is driven by steepness of the partial pressure gradient between blood and water, which flow in counter-current, increasing contact time and

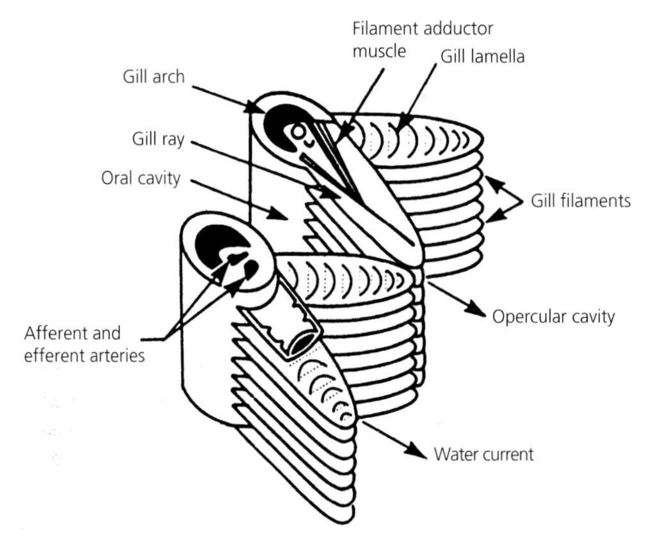

Fig. 5.2 Diagrammatic structure of a teleost gill. Arrows indicate direction of water flow. Source: Spotte (1992: 202 Fig. 4-1) after Hughes (1961: 347 Fig. 1b). ©1961 Reed Business Information – UK. All rights reserved. Distributed by Tribune Content Agency, LLC.

augmenting transfer down a concentration gradient between the oxygen-rich water outside the lamellae and comparatively oxygen-poor blood inside. Contact time between the water and lamellar surfaces is ≈2500 ms in a resting fish, but <30 ms at the highest flow rates (Randall 1982).

On being taken up by the bloodstream, oxygen rapidly assimilates into the physiological milieu by bonding with hemoglobin. This process, which occurs continuously, keeps the partial pressure of blood oxygen low relative to that of the water, the last symbolized by P_wO_2. For this to happen efficiently, fish gills must: (1) maximize the surface area available for oxygen diffusion; (2) minimize diffusion distances between water and blood; and (3) maximize perfusion of the lamellar tissues to expose sufficient surface area to the water and thus guarantee adequate oxygen uptake (Evans 1999).

Fishes have accomplished all three remarkably well (Park *et al.* 2014). The gill filaments are densely packed with lamellae, offering a large cumulative surface area for gas exchange while simultaneously generating viscous resistance (Chapter 4.6), which can become a liability in certain circumstances but ordinarily serves as a braking mechanism to increase the contact time of water flowing past the lamellae and enhancing diffusion of oxygen into the blood. Oxygen uptake is thus diffusion-limited, dependent on the surface area of available lamellae and the rate at which water moves past them. Diffusion of oxygen from water across the lamellar tissues and into the bloodstream is dynamic too, controlled by mechanical and cardiovascular elements (Section 5.4).

Flow rate also depends on interlamellar gap width. Park *et al.* (2014) collected published interlamellar distance data for 75 species of teleosts ranging in body weight from 0.1 g to 100 kg and found gap width to vary within a small range (≈20–100 μm). In seeking an explanation they modeled the data assuming geometric similarity across species and applied the cube law (Chapter 2.2), finding that body weight and lamellar length scale isometrically (i.e. in proportion). They also built a microfluidic chip that mimics gas-transfer characteristics of a gill lamella. The model is necessarily crude because scaling in nature is unlikely to be perfectly isometric, and an artificial lamella by definition is not a real one. However, the results offer insight into why the variation in interlamellar distance over six orders of body weight is so small. Park *et al.* (2014: 8069) wrote, "The optimum is reached at an interlamellar [gap] distance that increases the surface area for oxygen diffusion but does not markedly impede water flow." The conclusion that evolution has produced gills adapted for "maximization of the oxygen transfer rate for a given pumping pressure" is perhaps not unexpected.

During water-breathing, oxygen uptake occurs at the gill filaments. Diffusion is slower because water is denser than air. With oxygen's low solubility and diffusivity in water,[1] augmentation of gas transfer is promoted by *ventilation*, or pumping action, of the opercula, which serves to flood and refresh the gill filaments continuously with oxygenated water. The pumping of water is hard work, consuming 10–20% of the metabolic activity of a resting fish and even more during exercise (Packard 1974 and references).

As described by Coolidge *et al.* (2007), ventilation is accomplished by the dual pumping action of the buccal and opercular cavities, which expand at the same moment. This results in suction that closes the opercula and draws water into the mouth. Both cavities are then compressed when the mouth closes, forcing the inspired water over the gill filaments and out the opercula, minus some of its dissolved oxygen. Most teleosts using active gill ventilation generate pressures of 5–50 Pa (Hughes 1960). However, many fast-swimming pelagic species (e.g. mackerels, tunas, lamnid sharks) use *ram ventilation*, during which the gills are ventilated passively as the fish swims open-mouthed, providing a constant flow of aerated water (e.g. Farrell and Steffensen 1987; Hughes 1960). *Aquatic ventilation rate* (or simply *ventilation rate*, V_R) is then controlled by varying the gape size. In fishes using ram ventilation and cruising at 0.7–2 m/s, pressures rise to considerably higher values of ≈0.2–2 kPa (Park *et al.* 2014).

During active ventilation the nearly continuous flow of water needed for gill irrigation is sustained by alternating positive and negative pressure differentials between the buccal and opercular cavities (Fig. 5.3). Backflow is prevented in the buccal cavity by the *velar fold*, or *oral valve*, which seals the mouth, and by closing of the opercula (i.e. the *opercular valves*), which seals the opercular cavity.

---

[1] The diffusion coefficient of O_2 in water is 2×10^{-9} m²/s.

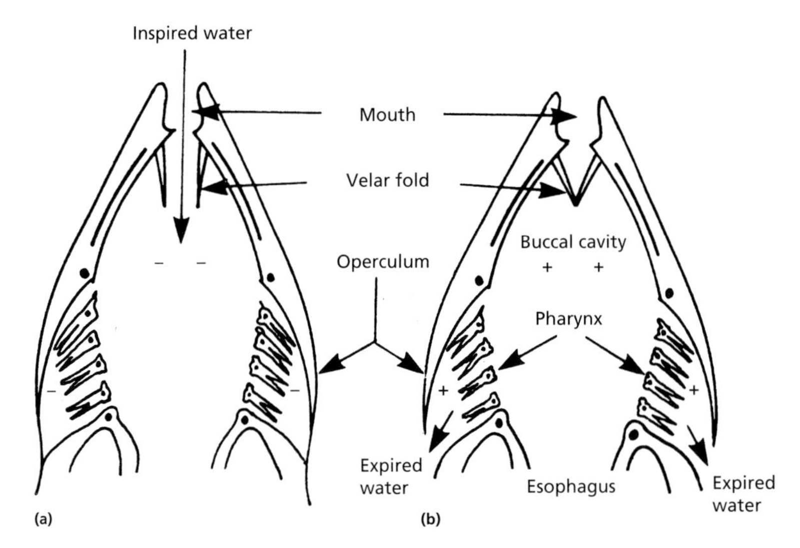

Fig. 5.3 Water-breathing (aquatic ventilation) by bony fishes. **(a)** Inspiration: the opercula close, the velar fold (oral valve) opens, the buccal cavity dilates with expansion of the gill arches, and water enters. **(b)** Expiration: the velar fold closes, the gill arches contract, the opercula open, and water is forced over the gill filaments and into the pharynx. Plus and minus signs indicate positive and negative pressure. Source: Spotte (1992: 202 Fig. 4-2).

5.3 Air-breathing

Air-breathing broadens a fish's *metabolic scope* – that is, the difference between its standard (i.e. resting) and maximum metabolic rates – compared with water-breathing alone. The need to expand metabolic scope becomes especially timely during heightened activity, when the additional oxygen requirement can be augmented by imbibing air. Water is not only many times denser than air and much more viscous, its oxygen concentration is only 3% that of an equal volume of air. These combined properties make extraction of oxygen from water more metabolically expensive.

As outlined by Graham (1997: 65–133), Hsia *et al.* (2013), Junk (1984: 224–225), and Seymour *et al.* (2004), *air-breathing organs* (ABOs) that evolved modifications to use atmospheric oxygen have included lungs and *physostomous swim bladders* (i.e. swim bladders connected by a duct to part of the alimentary canal, allowing the uptake of air)[2]; alternatively, skin, tissues of the mouth, pharynx, stomach, and intestines became specialized to absorb oxygen from the

[2]The connection between the swim bladder and esophagus has been lost in the more derived (i.e. "advanced") fishes. The *physoclistous swim bladder* is used almost exclusively for buoyancy control, with gases removed or added as necessary from the bloodstream. Some physoclistous species take air into suprabranchial cavities where gases are exchanged.

atmosphere (e.g. Florindo *et al.* 2006, Jucá-Chagas and Boccardo 2006, Nelson and Dehn 2011, Rantin and Kalinin 1996, Richards 2011, Saint-Paul 1984, Scarabotti *et al.* 2011, Tripathi *et al.* 2013, Winemiller 1989).

Nearly all fishes with ABOs occupy freshwaters (Sloman *et al.* 2009). Seawater species typically rely on epithelial tissues of the skin, gills, and buccal cavities for gas exchange. The tarpon is exceptional in this respect, being among the few fishes living in seawater to possess an ABO. Anatomical features of the tarpon physostomous swim bladder have been described and illustrated by Babcock (1936: 48–53), de Beaufort (1909: 530 Fig. 2), Graham (1997: 82 Fig. 3.14 and Fig. 3.15), and Seymour *et al.* 2008: 283 Fig. 1 and Fig. 2). The last reference is the most detailed and informative, and figures from it are reproduced here in grayscale (Figs. 5.4 and 5.5).

Of the four major teleostean lineages, auxiliary air-breathing bypassed the Clupeomorpha (herring-like fishes), becoming most varied in the Osteoglossomorpha and Euteleostei. Of today's Elopomorpha just the tarpons are equipped with ABOs, and excepting *Megalops atlanticus* and *M. cyprinoides* every air-breathing species of fish with a physostomous swim bladder lives in freshwater throughout its life. Juvenile tarpons often occupy freshwater habitats before migrating to the sea and maturing there (Chapter 4.4); conversely, adults of both species often live permanently in brackish and freshwaters (Chapter 7.2).

The top of the tarpon swim bladder is attached firmly to the dorsal body wall, from which it lies suspended within the mesentery, connected to the gut by the *pneumatic duct*. Air taken in by mouth enters through this passage, which is short, broad, and thick-walled. Seymour *et al.* (2008: 284) described the ABO's general appearance as "a thin-walled, elongate sac positioned dorsal to the digestive tube." The anterior part – the section extending forward of the juncture with the pneumatic duct – projects under the skull in contact with the otic capsules, an arrangement believed to promote sound transmission, allowing it to serve as a hearing aid.

Parts of the tarpon swim bladder not involved in respiration make up ≈25% of the total weight. These sections are evenly-surfaced, translucent, and histologically unremarkable, featuring nondirectional arrays of smooth muscle strands. The tarpon swim bladder's salient internal feature comprises four thin longitudinal and nearly parallel bands – one dorsal, one ventral, and two lateral – that arise from the basal surface in *respiratory ridges*, each continuing almost the entire length of the organ (Fig. 5.4a-c). Oxygen uptake occurs within them. As shown, the ridges become visible when the swim bladder is sliced lengthwise and its edges laid back. The dorsal ridge extends anteriorly to underneath the skull. The shapes of the ridges and their proximity to one another vary, perhaps by age or size or even by individual, but in sum they account for the remaining ≈75% of the swim bladder's total weight. The respiratory ridges are distinguished by folds and convolutions that increase their surface areas for gas exchange.

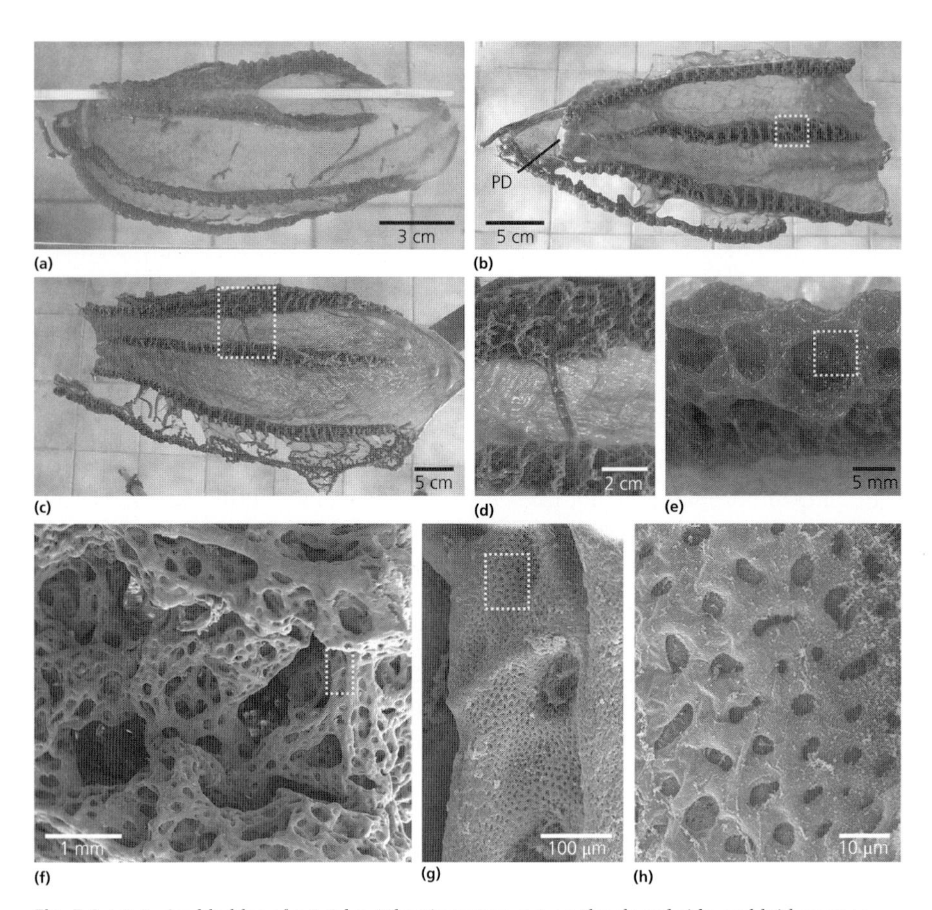

Fig. 5.4 (**a**) Swim bladder of a 1.3 kg Atlantic tarpon cut on the dorsal side and laid open to reveal its internal surface. The four ridges of respiratory tissue extend along the length of the swim bladder, and the ventral ridge and one lateral ridge (top) are interconnected in the organ's anterior region. (**b**) Swim bladder of a 13.7 kg specimen laid open as in (**a**) showing little connection between respiratory ridges. The dorsal ridge (bottom) extends anterior of the pneumatic duct (PD) into the narrow tubular projection of the swim bladder under the skull. (**c**) Swim bladder of a 38.6 kg specimen laid open as in (**a**) and (**b**). Some vascular connections between respiratory ridges are present. (**d**) Magnified view of dashed box in (**c**) showing a large blood vessel connecting two ridges of respiratory tissue. (**e**) Enlarged image of the dashed box in (**b**) showing the faveolata surface of the ventral respiratory ridge. (**f**) SEM image of the dashed box in (**e**) revealing the interconnected septa forming air chambers of different sizes. (**g**) Magnified view of the dashed box in (**f**) showing the respiratory surface of the interconnected septa. (**h**) Enlarged view of box in (**g**) showing the respiratory epithelium. Source: Seymour *et al.* (2008: 283 Fig. 1).

Seymour *et al.* (2008: 282) described the composition of the respiratory ridges as "cellular, spongy aveolar-like tissue …." and having the obvious function of trapping, retaining, and absorbing gases. The ridges are fused to respiratory epithelium on the swim bladder's inner surface where capillaries of 5.60 (± 0.94) μm perfuse the contact points, their erythrocytes assimilating oxygen from the

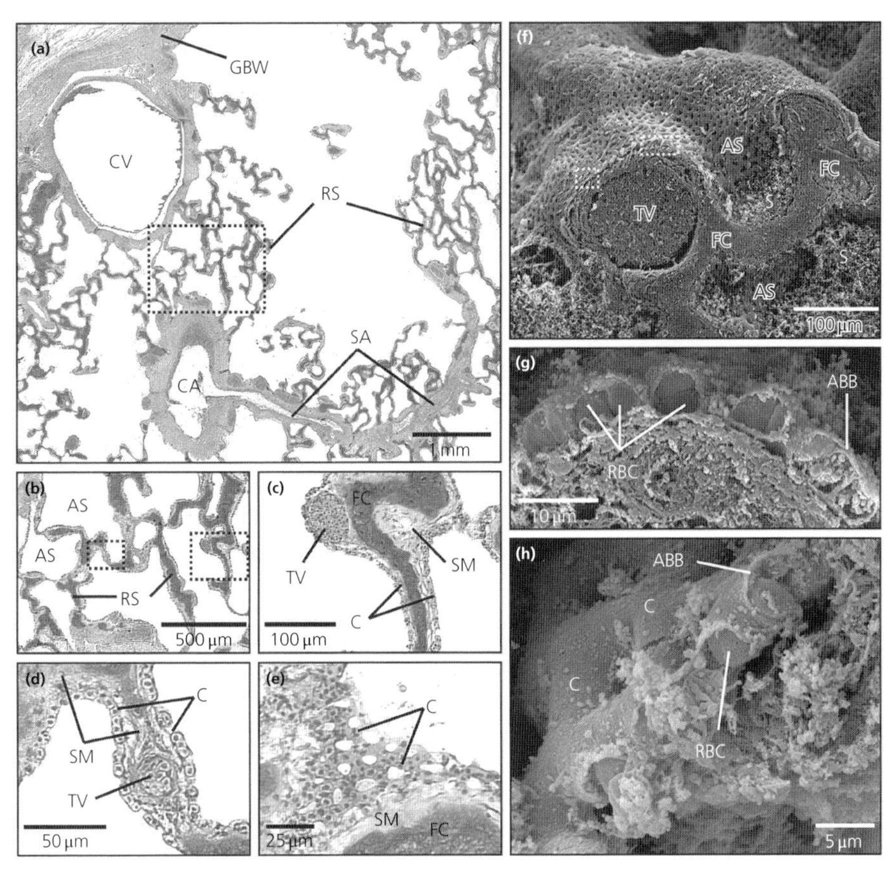

Fig. 5.5 Light and scanning electron microscope (SEM) images of transverse sections through the swim bladder respiratory tissue of a 13.7 kg Atlantic tarpon. (**a**) The main conduit artery and vein extend along the base of each ridge of respiratory tissue and appear to distribute and collect blood along its length. A segmental artery leaves the main conduit artery to deliver blood to the respiratory septa. (**b**) Magnified image of dashed box in (**a**), showing the interconnected respiratory septa and associated air spaces. (**c**) Enlarged view of dashed box in (b,right), showing a cross section through a respiratory septum composed of a fibrocartilage base surrounded by smooth muscle, a large tributary vessel, and respiratory epithelial capillaries. (**d**) Magnified image of dashed box in (**b**,left), depicting a small tributary vessel and epithelial capillaries filled with red blood cells. (**e**) Blood-filled capillaries on the surface of the respiratory septum. (**f**) SEM section through the respiratory septum, revealing a large tributary blood vessel. A layer of surfactant lines the respiratory epithelium in the air-filled chambers. (**g**) Magnified image of dashed box in (**f**,right), showing cross sections through the epithelial capillaries containing red blood cells. (**h**) Enlarged view of dashed box in (**f**,left), depicting epithelial capillaries and a cross-section through the air-blood barrier. Abbreviations: AAB, air-blood barrier; AS, air space; C, capillary; CA, main conduit artery; CV main conduit vein; FC, fibrocartilage; GBW, gas-bladder wall; RBC, red blood cell; RS, respiratory septa; S, surfactant; SA, segmental artery; SM, smooth muscle; TV, tributary vessel. Source: Seymour *et al.* (2008: 284 Fig. 2).

inspired air. There the thickness of the air-blood barrier is reduced to a miniscule 0.287 (± 0.072) μm (Fig. 5.4h). What apparently is surfactant on the respiratory epithelium (Seymour *et al.* 2008) aids the process (Daniels *et al.* 2004).

Water-breathing requires two phases (Liem 1989): intake (buccal expansion with mouth open) and outflow (opercular compression with mouth closed). Air-breathing by fishes with physostomous swim bladders makes use of an additional four aerial phases (Brainerd 1994; Liem 1989):

Phase 1: Water-breathing momentarily pauses as an air-breathing fish rises to the surface. The mouth stays closed while the buccal cavity expands, lowering pressure at the mouth to below that of the ABO. Air then moves from the ABO into the buccal cavity either passively from reduced hydrostatic pressure or by active transfer as a result of buccal muscular activity, maybe aided by muscles surrounding the swim bladder.

Phase 2: The buccal cavity compresses, forcing exhaled air out the mouth (as in the bowfin, *Amia calva*), or more commonly from the opercula.

Phase 3: The buccal cavity expands again, imbibing fresh air from above the surface.

Phase 4: The buccal cavity compresses (presumably with the opercula closed), pumping the new air through the pneumatic tube and into the ABO. Excess inspired air in the form of bubbles is expelled from underneath the opercula after the swim bladder has filled.

Thus the buccal cavity expands and compresses twice in rapid succession, one expansion-compression cycle for expiration followed by another for inspiration.

Brainerd (1994 and references) termed this pattern *four-stroke buccal pump ventilation* and labeled it the "Type I" pattern of expiration before inspiration and typical of most primitive fishes, including bichirs (*Polypterus* spp.), gars (*Lepisosteus* spp.), the bowfin, and the osteoglossomorphs *Arapaima gigas*, *Gymnarchus niloticus*, and *Notopterus chitala*. During "Type II" ventilation, some species (e.g. the bowfin) only partially inspire on occasion and bypass expiration (Brainerd 1994; Hedrick and Jones 1999), probably for buoyancy control.

Two highly derived fishes, the ostariophysans *Gymnotus carapo* and *Hoplerythrinus unitaeniatus*, employ a variation of four-stroke buccal pumping, reversing the order of expiration and inspiration by taking in fresh air upon surfacing and pumping it directly to the ABO (Brainerd 1994 and references; Liem 1989). Air from the ABO is then transferred into the buccal cavity and released from the mouth (*G. carapo*) or opercula (*H. unitaeniatus*). Call it "Type III."

Many observers have reported seeing tarpons expel air bubbles either after surfacing for air, after taking in air and submerging, and both, but usually in superficial terms. A century ago Raj (1916: 252) briefly described the air-breathing sequence of Pacific tarpons in aquariums:

> *At frequent intervals they rise slowly till they are quite near the surface, when they make a sudden dash to the surface and descend down in a moment having swallowed air, which escapes through the opercular cleft. This habit is natural to the fish, as it often rises to the surface in ponds and the act is accompanied by a characteristic splash of water and the escape of air bubbles.*

We can speculate that because tarpons are elopomorphs, and thus primitive fishes too, they use four-stroke ABO ventilation. Further bolstering this notion, the tarpon ABO closely resembles that of the bowfin and pirarucu (*Arapaima gigas*), the latter among the most primitive of extant species of the three major teleostean lineages in which ABOs evolved. As Liem (1989: 339) explained: "Its [the tarpon's ABO] configuration and connection to the esophagus are identical to those in *Amia* ... [and] when using *Amia* as the outgroup for teleosts, the gas bladders in *Megalops* and *Arapaima* must be considered primitive among teleosts."

Assuming the above reasoning is true (i.e. that the air-breathing cycle of tarpons is the same as other conventional four-stroke ventilators), a tarpon surfacing to inspire air has its mouth shut while spent air is being transferred from ABO to buccal cavity, making oral exhalation unlikely. I presume too that during this first phase the opercula are closed. The four-phase sequence next calls for compression of the buccal cavity, forcing exhaled air out the mouth or opercula, the latter locations being common in most primitive fishes except the bowfin. The buccal cavity expands again, taking in fresh air and compressing afterward, forcing new air into the ABO via the pneumatic duct. During these last two phases the opercula are presumably closed to prevent air from escaping.

A bowfin slows its forward movement when about to take a breath, even stopping momentarily and sticking its snout out of the water to expire and then inhale (Liem 1989), but a tarpon surfaces to breathe while in motion. When near the surface both Atlantic and Pacific tarpons often swim using a dolphin-like rolling motion, exposing their backs to the air momentarily (e.g. Bishop *et al.* 2001: 23, Breder 1925a: 140, Dailey *et al.* 2008, Edwards 1998, Ellis 1956, Grey 1919: 3, Griswold 1913: 91, Herbert and Peeters 1995: 21, Hildebrand 1963: 116, Ley 2008: 9, Merrick and Schmida 1984: 53, Moffett and Randall 1957, Randall and Moffett 1958, Schultz 1952: 33, Swanson 1946, Wells *et al.* 1997). Kaplan (1937: 101) gave the unlikely explanation that rolling aided tarpons in "locating their whereabouts ...." This behavior is evident in Atlantic tarpons as small as 18.1 mm SL, indicating to Harrington (1966: 869) that "they are already dependent on atmospheric air for survival as they are at larger sizes ...." Bishop *et al.* (2001: 23) commented that rolling behavior of Pacific tarpons was more noticeable "in the more anoxic billabongs ...."

Spent air exhaled by a tarpon at the surface could likely not be seen by an observer looking down at the water. Any visible bubbles released during Phase 1 could represent either spent air or newly inspired air in excess of the volume required to inflate the swim bladder. Whether from the mouth or opercula, they would likely be released just beneath the surface. Exhalation through the mouth has been reported, but whether this is an important pathway is doubtful. For example, Griswold (1913: 92, 1922: 9) once caught a juvenile tarpon of 5 in. (≈ 12.7 cm) in a gill net in Cuba and kept it alive in a tub. "From time to time he would rise to the surface as large fish do in the rivers, then go to the bottom of the tub again, and in a moment the bubbles would slowly issue from his mouth." Besides representing a sample of one, Griswold's fish had been netted and was undergoing capture and captivity stress and perhaps was injured. Its behavior might not have been typical.

A tarpon having just taken a breath and in the act of diving often releases air from the opercula (Geiger *et al.* 2000), but is it spent air or new air in excess of that required to re-charge the swim bladder? In either case the opercula would be slightly more elevated than the mouth during diving and therefore under less hydrostatic pressure, even if the angle is shallow. Assuming hydrostatic pressure influences location of the release site, we should then expect the reverse; that is, passive release through the mouth in a surfacing fish instead of through the opercula and driven by the transfer of buccal pressure posteriorly. No bubbles would be visible from air released orally above the surface.

Geiger *et al.* (2000: 184), after watching captive Atlantic tarpons through aquarium glass, wrote: "An air bubble is expelled from beneath each operculum as the fish approaches the surface [at an angle of ≈ 30°], suggesting that the gas bladder is emptied *during* ascent [emphasis added]." In other words, *before* breaking the surface following a repertoire different than described so far. For this to work, hydrostatic pressure must exceed pressure in the opercular cavity and gas transfer is passive; alternatively, air is forced out the opercula via compression of the buccal cavity, which, if expanding at that instant, would mean the opercula are likely closed. Geiger and colleagues did not mention the status of the opercula, but the question could be easily answered by filming the sequence. Perhaps the opercular openings are slight and nearly imperceptible, just enough to let loose a little air. Regardless of circumstances, the matter is largely irrelevant: *any air released before reaching the surface must originate in the swim bladder because the fish has not yet inhaled.*

Geiger and coauthors implied that both buccal and opercular spaces expand when the mouth breaks the surface, and that this maneuver is accompanied by "an abduction of the operculi [*sic*] and gular plate." However, whether inspiration or expiration occurs first in most air-breathing fishes remains to some extent controversial (Graham 1997: 89–90, Liem 1989), no doubt because it varies.

As discussed above, most species exhale first upon surfacing to rid the swim bladder of deoxygenated gas. In others the process is reversed. In still others, retained gases are mixed with newly inspired air prior to expiration. Graham speculated that this last method might lessen time at the surface and help in avoiding predators: "Alternatively, it may be that muscles in the chamber wall cannot compress the organ [swim bladder] sufficiently to cause ventilation or that expiration briefly compromises the organ's buoyancy and sound detection functions." However, this action, as it typically happens, occurs in rapid stages measured in milliseconds, the opercula closing as the buccal cavity expands to retain its momentary pressure. It seems to me that the swim bladder could not be filled without simultaneously pressurizing the pneumatic duct by closing the opercular valves. If so, presuming adduction of the opercula rather than abduction after expansion of the buccal cavity during inspiration appears more likely.

Geiger and colleagues continued: "As the fish descends [also at an angle of ≈ 30°] the buccal and opercular cavities are compressed ...." Again, these events probably occur in tandem, not simultaneously. Geiger and colleagues then noted that following inspiration, "a small bubble is expired from beneath the most dorsal end of each operculum." They contended that tarpons therefore exhale air from the swim bladder before inhaling fresh air, which is probably true. Air bubbles released from the opercula during descent are more likely to represent excess inspired air, which is the case with the pirarucu (Liem 1989). Seymour *et al.* (2007) confirmed that ventilation by Pacific tarpons consists of a single exhalation before subsequent inhalation and speculated that air bubbles released after inhalation probably represent excess air retained in the buccal cavity.

It seems reasonable that tarpons control the volume of air in the swim bladder by regulating the amount expired, inspired, or both. Seymour *et al.* (2007) pointed out that inconsistent expansion of the ABO would affect buoyancy, which seems to have been confirmed in extreme situations by Geiger *et al.* (2000), when some of their juvenile Atlantic tarpons held in captivity were prevented from surfacing to breathe air (Section 5.5).

Coolidge *et al.* (2007: 194) speculated that the first phase of air-breathing in fishes with ABOs is perhaps "assisted by elastic recoil of the ABO as well as compression of muscles in the wall of the ABO." In the Cladistia, which comprises the air-breathing bichirs (*Polypterus* spp.) and closely related reedfish (*Erpetoichthys calabaricus*), recoil and subsequent ABO re-inflation is enhanced by rigid skin and body walls latticed with ganoid scales. The tarpon has ctenoid scales, and whether they serve a similar function is untested. Fish scales of many kinds are well known for their high tensile strength and flexibility (e.g. Ikoma *et al.* 2003).

The evolution of air-breathing is remarkable in its protean variation, but also for how form and function have converged at times across such diverse lineages. This is seen in three that are unrelated. *Gymnotus* (a gymnotiform) and *Hoplerythrinus* (a characiform) have evolved unusual and almost identical swim bladders, along with highly specialized esophageal pumps (Liem 1989). Ventilation in both commences with inhalation, not exhalation. Air drawn in through the mouth is forced into the esophagus by buccal pressure and then under esophageal peristalsis into the swim bladder. Next comes exhalation, when air in the swim bladder flows back into the esophagus which, when filled, undergoes anti-peristalsis to expel the gas via the buccal cavity. *Gymnarchus* (an osteoglossiform) also uses peristalsis and anti-peristalsis to push along volumes of air, but in reverse sequence from the other two forms (Liem 1989).

Specialized esophageal pumping seems restricted to physostomous fishes having pneumatic ducts near the posterior end of an elongated esophagus, which establishes what Liem (1989: 347) described as "a functional problem of moving air to and from the buccal cavity." The problem is one of distance.

Esophageal pumping, in other words, is analogous to the pumping stations used by drinking water plants to push water through their pipes to far away locations, the distances being too great for a single pump to sustain line pressure and be effective. The tarpon has no such problem. Its esophagus is comparatively short, the pneumatic duct located closer to the front of the swim bladder.

Atlantic tarpons commonly jump, and the behavior appears to be spontaneous. During his stay in Panamá, Breder (1925a: 140) reported them "constantly leaping near the foot of the [Gatun] dam." In the early 1970s I lived at Key Largo, Florida directly on Florida Bay, where at dusk the tarpons began to jump, and their jumping continued through the night into dawn. I offer no explanation of why this occurs, but it seems unnecessary as a mode of air-breathing. Simply rising to the surface would be easier. Other functions have been attributed to jumping, including attempting to escape from predators, shaking off remoras or parasites, or "play" (Hildebrand 1963: 116–117), although none of these hypotheses has been tested.

5.4 Cardiovascular function

Juvenile tarpons live in waters containing variable quantities of oxygen. A habitat in which the partial pressure of oxygen is within the normal range at a given temperature is *normoxic*. One that is below normal is *hypoxic*. Habitats where oxygen is almost nonexistent are *anoxic*, and where water is poorly mixed and photosynthesis intense, oxygen can exceed expected concentrations at saturation, becoming *hyperoxic* (Spotte 1992: 221–223). Adults inhabiting well-mixed seawater seldom encounter conditions other than normoxia.

Authorities once considered the hearts of bony fishes to be two-chambered, consisting of an atrium and ventricle. Farrell (1993), who provided a detailed description of the teleost heart, discussed four sequential chambers: sinus venosus, atrium, ventricle, and bulbus arteriosus. Icardo (2012 and references) later included two more: the conus arteriosus (Senior 1907a, 1907b) and a distinguishable atrioventricular segment. Primitive "fishes" (e.g. hagfishes, lampreys) retain both a conus and bulbous arteriosus, as do all extant elopomorph teleosts (genera *Albula, Elops, Megalops,* and *Pterothrissus*).

Venous blood returning to the heart collects in the sinus venosus and is pumped in sequence by the atrium and ventricle to the bulbus. Heartbeat is maintained by the sinus venosus, which serves as a "pacemaker"; atrial contractions keep the ventricle filled. The ventricle is the largest and most varied of these chambers, its shape and size species-dependent. It controls ventral aortic blood pressure (and *stroke volume*, V_S, to a lesser extent), which among teleosts is highest in active species. Its elasticity allows it to dampen the pulse of ejected blood to create a smooth, continuous stream into the major arteries. The fish ventricle has two kinds of *myocardium*: the *spongiosa*, making up a structural, sponge-like network of *trabeculae*, and the *compacta*, an outer layer encapsulating the spongiosa (a *trabecula* is a unit of small, sturdy, rod-like, collagenous tissue that serves a mechanical support function). In almost all fishes the mass of the spongiosa exceeds that of the compacta (Farrell 1993).

Cardiac output ($\dot{Q}$) is channeled through the bulbus and ventral aorta and distributed to the gills by means of afferent branchial arteries. In water-breathing fishes, this encompasses the entire volume of $\dot{Q}$ (Evans 1999; Farrell 1993). Pressure inside the lamellae oscillate with each heartbeat. Mean pressure is ≈3.5 kPa, with oscillations of ≈± 0.5 kPa (Randall 1982). Afferent branchial arteries are arranged in 4–7 bilateral pairs. Instead of returning to the heart after passing through the respiratory system as in other vertebrates, blood is shunted straight into systemic circulation. Respiratory and systemic circulation, in other words, are arranged in series, a result being that pressure in the ventral aorta is approximately two-thirds greater than in the dorsal aorta.

The systemic circulatory system is composed of primary and secondary systems. The aortas (dorsal and ventral) are the main conduits of primary circulation. Secondary circulation is carried out by arteries branching off the primary system and services the exposed surface structures (e.g. gills, scales, skin) where gases are exchanged, although the primary system remains the principal provider of blood to the gills. Both primary and secondary returning blood drains into common veins of the primary system. Primary veins send blood from the head and gills directly to the heart; returning blood from the trunk muscles and gastrointestinal tract is first shunted sequentially through the kidney and liver before returning to the heart.

Cardiac output is the product of heart rate (f_H) and stroke volume (V_S), although it also correlates positively with temperature. Overall, values of $\dot{Q}$ are species-dependent and based on routine level of activity. Active species have higher resting values of $\dot{Q}$ than quiescent ones. The tarpon belongs in the active group, although not so active as tunas and some other purely pelagic fishes.

Bursts of activity are fueled anaerobically and thus limited to short duration. During *burst* (*anaerobic*) *exercise*, $\dot{Q}$, f_H, and arterial blood pressure (P_aO_2) rise, and these responses are often accompanied by less oxygen delivered to the heart; that is, the partial pressure of oxygen in the venous blood (P_vO_2) declines. Blood pH drops too, exposing the heart to acidosis that extends through most of the recovery time to homeostasis. These critical events affect the survival of tarpons and other fishes captured and then released during angling (Chapter 8.4).

In conditions of moderate aquatic hypoxia (P_wO_2 of ≈9.3 kPa), $\dot{Q}$ can ordinarily be maintained. At ≈5.3 kPa *bradycardia* (reduced heart action) is apparent in many species, which seems a peculiar response under the circumstances. Mammals, in direct contrast, increase their rates of lung ventilation, and $\dot{Q}$ increases, mainly as a result of *tachycardia* (increased heart rate) as the available oxygen diminishes. Farrell (2007) suggested at least two potential benefits of bradycardia by fishes. First, oxygen supplied to the spongiosa is controlled mainly by P_vO_2, which is not consistently reliable. Second, because the fish heart is capable of large increases in V_S, diastolic residence time is extended, enhancing oxygen diffusion and sustaining $\dot{Q}$. Under severe hypoxia and anoxia, $\dot{Q}$ decreases in most species studied to date except hagfishes, air-breathers, and certain water-breathers adapted over evolutionary time to hypoxic conditions (Farrell 1993 and references, Farrell 2007). Some species can also respond by increasing the blood's *oxygen carrying capacity* (determined by the types and concentrations of hemoglobin present), and by depressing aerobic metabolism (Tripathi *et al.* 2013).

Based on structural properties of the ventricle, Tota *et al.* (1983) partitioned elasmobranch and teleost hearts into four categories, or "types," depending on the presence or absence of a compact myocardial layer, and, if one is present, the extent of its development (Fig. 5.6). As Icardo (2012) emphasized, the distinctions are not trivial. For example, it has been assumed that completely trabeculated ventricles fail under increasing *afterload* (i.e. tension that accumulates during ejection of blood), although this appears to be untrue of teleost hearts, which capably maintain high levels of performance in the absence of a compacta; that is, ventricles of Type I. A *Type I heart* has no compacta. Because

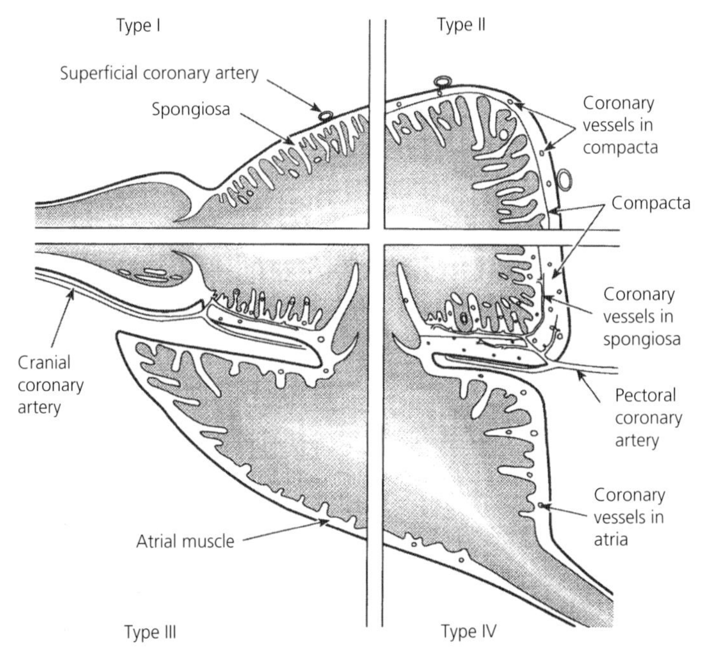

Fig. 5.6 Diagrammatic illustration of the main features distinguishing the four types of fish ventricles. *Type I*: one myocardial layer (spongiosa) only, no capillaries. *Type II*: inner spongiosa and outer compacta, coronary circulation, capillaries only in compacta. *Type III*: similar to type II except capillaries in both myocardial layers. *Type IV*: larger percentage of ventricle as compacta (>30%), more extensive capillary system of atrium. The tarpon heart is Type II. Source: Farrell and Jones (1992: 7 Fig. 1).

the ventricle is avascular and operates in the absence of coronary circulation the Type I heart is also referred to as a venous, or avascular, heart and is entirely trabeculated, consisting only of spongiosa. Venous blood in Type I hearts pools in the lumens within trabecular spaces of the spongiosa. Most teleosts (≈80%) have type I hearts (Farrell 1993 and references).

Coronary circulation occurs in hearts of the other three categories, but in the *Type II heart* only in the compacta. In hearts of *Type III* and *Type IV*, coronary circulation extends into the spongiosa as well. Type IV hearts differ from those of Type III in having a greater proportion of compacta forming the ventricular mass. In Types II, III, and IV, the spongiosa augments the supply of venous oxygen to the myocardium and probably enhances cardiac function, although this might contribute only 1–2% of $\dot{Q}$ (Farrell 1993).

In general, active species have large ventricles in proportion to body mass, which allows them to quicken heart rate, generate larger stroke volumes, and sustain higher blood pressures, all beneficial during prolonged (aerobic) swimming, burst (anaerobic) exercise, and when confronted with low dissolved oxygen (Farrell 1993).

If, as suggested by Farmer (1997), air-breathing evolved mainly to supply oxygen to the myocardium and not as a mechanism of coping with hypoxia, then whether the heart has coronary circulation under either condition becomes relevant (Farrell 2007). In strictly water-breathing fishes, aquatic hypoxia induces a decline in the partial pressure of oxygen of both arterial (P_aO_2) and venous (P_vO_2) blood, resulting in *functional* (*internal*) *hypoxia* when oxygen demand of the tissues exceeds the available concentration of circulating oxygen. In species with an ABO, air-breathing causes P_vO_2 to spike. The newly oxygenated blood departing the ABO is obviously advantageous to a Type I heart, provided it goes to the heart directly. Any such benefit is lessened in a fish with coronary circulation because oxygenated blood is then shunted to the heart from the gills. In addition, P_aO_2 is sustained during swimming, but not during aquatic (i.e. environmental) hypoxia (Farrell 2007).

If a fish is an air-breather with a Type II heart, oxygen originating in the ABO supplies the spongiosa but not the compacta. This configuration applies to tarpons, which have coronary circulation and a Type II heart (Farrell 2007).

More than half the myocardial mass of adult Pacific tarpons is composed of compacta, placing them among other species having high activity levels (Santer and Greer Walker 1980). Thus air-breathing by adults boosts oxygen concentrations supplied to the spongiosa but not necessarily to the compacta. This means that >50% of myocardial oxygen demand in adult tarpons is supplied by coronary circulation (Clark *et al.* 2007).

Juvenile tarpons seldom breathe air when in normoxic waters, either while resting (Shlaifer and Breder 1940) or swimming slowly (Seymour *et al.* 2004). Although oxygen concentrations in the swim bladder are often high, retained ABO oxygen is apparently not used by Pacific tarpons in normoxic situations even while swimming at moderate speeds (Seymour *et al.* 2007). It contributes little or nothing to the myocardium, and myocardial oxygen demand is compensated by increased coronary blood flow (Clark *et al.* 2007, but see Wells *et al.* 2005).

Juvenile Pacific tarpons breathe air in times of aquatic hypoxia. They also increase air-breathing frequency and consequently oxygen uptake by the swim bladder when swimming in hypoxic waters. According to Farrell (2007: 395), "This leads to the conclusion that the primary drive for air breathing in juvenile Pacific tarpon is aquatic hypoxia and exercise is a secondary drive." The high percentage of myocardial compacta of adults implies they perhaps gain less from air-breathing during hypoxia from the standpoint of supplying oxygen to the heart and maybe have a diminished reliance on breathing air after migrating more or less permanently into normoxic seawaters.

Juvenile Atlantic tarpons breathe air regardless of the dissolved oxygen concentration, but air-breathing becomes substantial when values of P_wO_2 fall to ≤5 kPa (Geiger *et al.* 2000). Seymour *et al.* (2007) established that transition from water- to air-breathing by juvenile Pacific tarpons commences when P_wO_2 reaches ≤8.3 kPa, and that V_R becomes nearly imperceptible at P_wO_2 values ≤6 kPa (Fig. 5.7). In Pacific tarpons, blood oxygen in equilibrium with P_wO_2 at ≈ 6 kPa would be 60–85% saturated, or within their physiologically relevant pH range of 7.4–7.8 (Wells *et al.* 2005), and water-breathing alone would be incapable of saturating the blood with oxygen. Continued gill ventilation in hypoxic water could actually reverse the diffusion gradient, causing oxygen to be lost from the ABO to the water. If oxygen in swim bladder air and in blood transporting it away have equilibrated, PO_2 can climb to 13 kPa as a result of air-breathing.

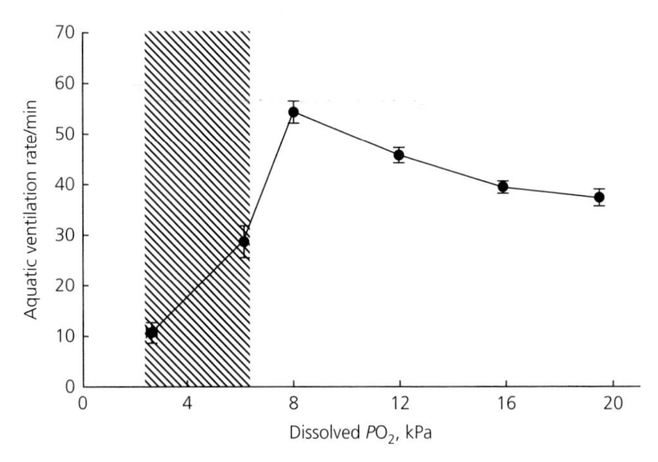

Fig. 5.7 Relationship between dissolved oxygen (P_wO_2) and aquatic ventilation rate (V_R/min) in juvenile Pacific tarpons ($n = 10$). Data are $\bar{x} \pm 95\%$ CI. The value of V_R declines at low P_wO_2. The shaded region highlights where opercular stroke amplitude was very low and often visibly imperceptible. Source: Seymour *et al.* (2007: 582 Fig. 1).

However, as Seymour *et al.* (2007) pointed out, all oxygenated blood exiting the ABO flows directly to the heart, and from there to the gills. A steep diffusion gradient could result in oxygen shifting rapidly from lamellar blood to the water outside.

5.5 Hypoxia

As mentioned, the overwhelming majority of air-breathing fishes alive today are tropical freshwater species, and Roberts (1972) considered hypoxia and drought the two defining abiotic factors in their evolution. A prominent hypothesis has been that adaptations to breathe air are evolutionary mechanisms for coping in unstable habitats where dissolved oxygen fluctuates and often declines to dangerously low levels (e.g. Clack 2007; Graham 1997: 6–7). Extant species survive in swamps, tidal pools, oxbows, billabongs, *várzeas*, and lagoons, habitats often subjected to daily or seasonal shifts in water quality severe enough to render conditions temporarily unsuited for aquatic aerobic respiration (Graham 1997: 6–7, Seymour *et al.* 2004, Sloman *et al.* 2009, Wells *et al.* 1997). One cause can be stratification, during which some layers become hypoxic or anoxic (Kramer *et al.* 1978, Wells *et al.* 1997) as a result of weak mixing. Stagnation induces eutrophication through the microbial decomposition of vegetative matter, which uses up dissolved oxygen (Roberts 1972). Abetting the process are persistently high water temperatures that lower oxygen solubility and raise overall aquatic respiration, increasing biological oxygen demand (Wells *et al.* 2005).

Fishes breathing water respond to aquatic hypoxia by raising the ventilation rate. In other words, they increase V_R as measured by pumping of the opercula. While a habitat becomes hypoxic, water-breathing fishes living there experience bradycardia, although blood output from the heart is maintained by the simultaneous increase in stroke volume as more blood is pumped with each heartbeat.

The available lamellar surface area probably approaches total use only during maximum aerobic demand (Duthie and Hughes 1987). Blood flow in the gills is altered in several ways. For a resting or slowly swimming fish, one way is to "recruit," or bring into service, some of its underused lamellae. In combination, bradycardia and increased V_S cause blood pressure to rise, which in turn perfuses gill lamellae not being used at the moment to augment gas transfer between water and blood (Perry and McDonald 1993). Most of the lamellar surface area (> 80%) is occupied by blood vessels (Farrell 1993 and references). Gas exchange then becomes more efficient as additional lamellae get actively involved (i.e. are "recruited") and the total lamellar surface area expands (Randall 1982).

Other evolutionary adaptations have been described. Some water-breathing teleosts cope with hypoxia by temporarily "remodeling" the structure of their gills, sometimes substantially reducing epithelial thickness to promote more effective gas transfer, elongating the lamellae or expanding them, and increasing enzymatic activity (Matey *et al.* 2008). Transfer to hypoxic water induced crucian carps (*Carassius carassius*) to increase the respiratory surface area of their lamellae ≈7.5-fold (Sollid *et al.* 2003); raising the water temperature had a similar effect (Sollid *et al.* 2005). Gill remodeling can sometimes occur over just a few hours, and is reversible following return of normoxic conditions.

Assimilation of oxygen by the gills of a quiescent fish is *perfusion-limited;* that is, limited by the rate of blood flow during normoxic conditions. At such times the uptake of oxygen and rate of blood flow are proportional and reminiscent of what occurs in mammalian lungs. Under hypoxic conditions the gills instead become diffusion-limited, resulting in diminished oxygen uptake from the water into the blood. At the same time, levels of organophosphates (ATP and related compounds) in the erythrocytes diminish too, leading to a sharp increase in the affinity of the hemoglobin for oxygen. Put simply (Spotte 1992: 222): "In effect, the hemoglobin becomes miserly with what little oxygen is available and subsequently binds it more tightly." This works to make distribution to tissues downstream from the gills more efficient.

Oxygen in the blood exists as dissolved O_2 in circulating plasma and bound with the respiratory pigment hemoglobin (Hb). Plasma O_2 accounts for $<5\%$ of the total (Perry and McDonald 1993). The reversible binding of O_2 to iron in hemoglobin molecules is the most crucial of the blood's oxygen transport properties, and the simplified reaction can be written:

$$Hb + O_2 \leftrightarrow HbO_2 \tag{5.1}$$

When blood oxygen is high, hemoglobin molecules combine with O_2 to form *oxyhemoglobin*, HbO_2, and the reaction is driven to the right. At low blood oxygen levels O_2 is released as the reaction reverses and shifts left. In the first situation all available sites on the hemoglobin molecule are occupied, and the hemoglobin is oxygen-saturated. At a given PO_2 in the blood the relationship between blood oxygen and oxyhemoglobin (equation 5.1) can be demonstrated by plotting an oxyhemoglobin dissociation curve. Its shape will be species-dependent (Figs. 5.8 and 5.9).

Temperature and the combined effect of carbon dioxide and pH influence the binding of oxygen to hemoglobin, causing equation 5.1 to shift left and the oxyhemoglobin dissociation curve to shift right. If the shift is caused by increased temperature it has the seldom-mentioned benefit to the heart of increasing and sustaining P_vO_2 (Farrell 2007). An increase in the partial pressure of carbon dioxide in the blood lowers blood pH, subsequently weakening the oxygen-hemoglobin bond and resulting in a leftward shift, a phenomenon called the *Bohr effect*. A reduction in blood pH (i.e. *acidosis*), induced by the sudden appearance of metabolic acids (e.g. lactate) or carbon dioxide during strenuous exercise can loosen the Hb-O_2 oxyhemoglobin bond. In some fishes the Bohr effect is exaggerated and complete saturation of the hemoglobin is impossible after

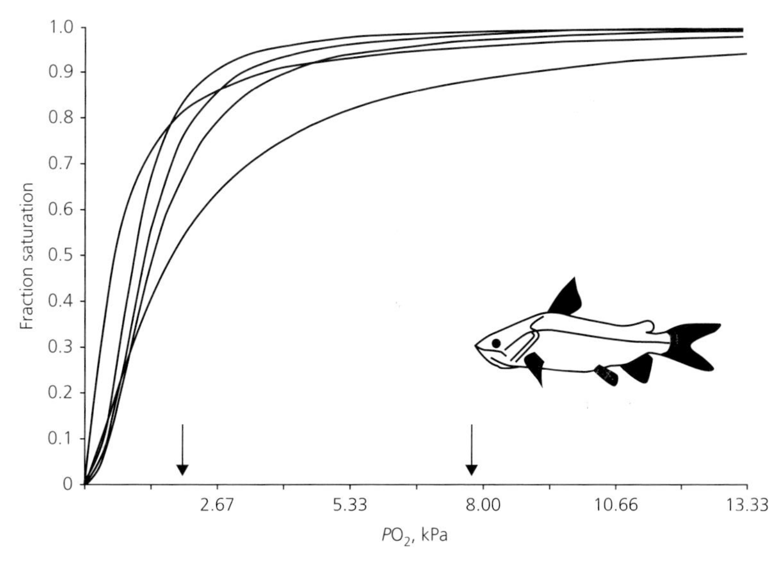

Fig. 5.8 Oxyhemoglobin dissociation curves (means of $n = 8$ fish) at 25°C for sagor venous blood, *Hexanematichthys sagor* (= *Arius leptaspis*). The sagor is a catfish sympatric with Pacific tarpons in some Australian billabongs. Curves are displayed in order of decreasing affinity for O_2 from left to right: pH = 8.2, 7.8, 7.4, 6.6. Vertical arrows indicate bottom (left) and surface (right) PO_2. Source: Wells *et al.* (2005: 89 Fig. 1).

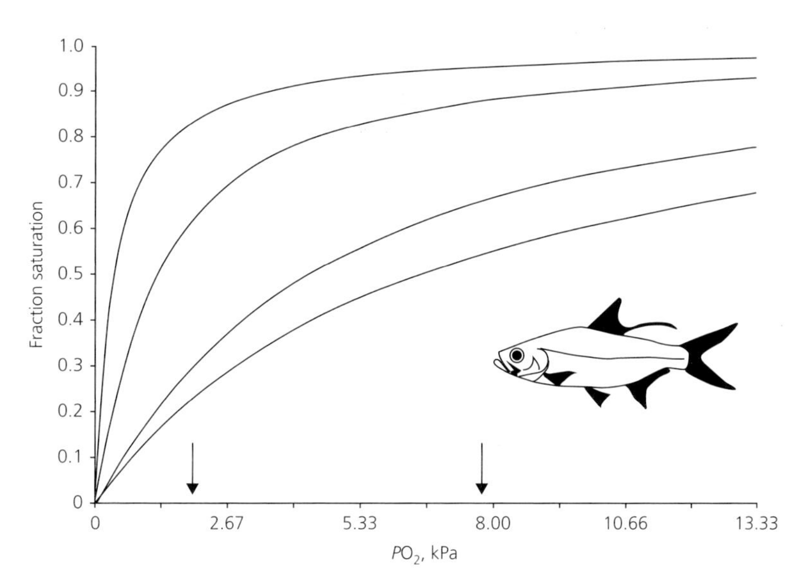

Fig. 5.9 Oxyhemoglobin dissociation curves (means of $n = 7$ fish) at 25°C for Pacific tarpon venous blood in order of decreasing affinity with declining pH. Curves are displayed in order of decreasing affinity for O_2 from left to right: pH = 8.2, 7.8, 7.4, 7.0. Vertical arrows indicate bottom (left) and surface (right) PO_2. Source: Wells *et al.* (2005: 89 Fig. 2).

blood pH drops, no matter how high the oxygen concentration. This is the *Root effect* (Brittain 1985, 1987). The distinction, according to Alexander (1993), is this. The Bohr effect is a reduction of the extent to which hemoglobin becomes oxygen-saturated at low blood pH and *any* PO_2. The Root effect is a reduction of the blood's oxygen carrying capacity at low pH. Tarpons normally exhibit a high Bohr effect and a low Root effect (Wells *et al.* 1997).

In sum, the amount of oxygen assimilated by the gills and shunted to the tissues of a fish is controlled by several factors, among them the existing partial pressure of dissolved oxygen in the water, the shape of its species-specific oxyhemoglobin dissociation curve, its blood carrying capacity (capacity of the blood to transport oxygen), and its cardiac output ($\dot{Q}$).

Two inhabitants of northern Australian billabongs – the sagor, or salmon catfish (*Hexanematichthys leptaspis*), and Pacific tarpon – respond differently to hypoxia (Wells *et al.* 2005). In experiments by these authors, mean temperature of the water during the dry season was 24.3 (± 1.9)°C. At pH 5.0, water at 10 cm under the surface was acidic. Near the surface P_wO_2 measured ≈7.88 kPa but was ≈1.95 kPa just above the bottom, a value approaching anoxia. Note from Table 5.1 that values of hemoglobin and hematocrit (Hct) were lower for the water-breathing catfish than the bimodal-breathing tarpon. The tarpon's red blood cells were more numerous (compare the RBC counts). Cell size was smaller, as a comparison of MCV values shows. This partly explains the tarpon's higher (actually, unusually high) oxygen carrying capacity, although the higher hemoglobin was probably not a factor because mean corpuscular hemoglobin concentrations (MCHCs) in the species were not statistically different.

The oxyhemoglobin dissociation curves reveal different patterns (Figs. 5.8 and 5.9). The curves show the catfish's venous blood as having the higher oxygen affinity. At an aquatic pH of 7.4 its P_{50} (the PO_2 at which 50% of the Hb is bound to O_2 as HbO_2) measured ≈1.20 kPa. When affinity of the hemoglobin for oxygen decreases (i.e. when P_{50} increases), an oxyhemoglobin dissociation curve is shifted right and unloading of oxygen to the tissues needing it for aerobic metabolism is enhanced. The tarpon, with a P_{50} of ≈4.27 kPa, seems incapable of saturating its blood with oxygen during functional (internal) acidic conditions, as seen by the right-shifted curves. This finding has important implications during strenuous anaerobic exercise, such as when a tarpon is caught on a hook (Chapter 8.3, 8.4).

As already mentioned, water's high viscosity compared with air extracts a metabolic cost. Simply ventilating the gills can require 10–20% of even a quietly resting fish's aerobic metabolism, and this rises with activity level (Packard 1974). Living in saline water adds an additional estimated burden of 25–30% (Packard 1974). Although the lower viscosity of warmer seas reduces the expense somewhat, any advantage is offset by needing to raise the ventilation rate (thus reducing the efficiency of oxygen extraction), and by oxygen's lower solubility in saline waters. To use Packard's (1974) example, well-aerated seawater

Table 5.1 Hematological values of sagors (*Hexanematichthys leptaspis* 280–570 mm FL) and Pacific tarpons (258–303 mm FL from a northern Australian billabong. Source: Wells *et al.* (2005: 89 Table 1).

	Sagors (*n* = 8)	Pacific tarpons (*n* = 7)
Hb (g/l)	90.9 (±12.0)	117.0 (±10.8)
Hct (%)	39.3 (±2.5)	47.7 (±5.2)
MCHC (g/l)	233.1 (±26.8)	246.5 (±18.8)
RBC (10^6/μl)	1.17 (±0.12)	2.11 (±0.44)
MCV (fl)	342.0 (±21.1)	219.1 (±18.2)
MCH (pg)	74.3 (±3.1)	53.4 (±1.8)
O_2 capacity (vol%; ml O_2 %)	12.2 (±1.6)	15.6 (±1.2)

of 30°C with a P_wO_2 of 20.3 kPa is equivalent to a well-mixed freshwater environment of the same temperature having a P_wO_2 of 18.9 kPa, a considerable difference of ≈7%.

Clark *et al.* (2007) measured changes in heart function, including cardiac output ($\dot{Q}$), of Pacific tarpons under hypoxic conditions (Fig. 5.10). At normoxia both juveniles ($\bar{x}$ = 0.49 ± 0.08 SEM kg, $\bar{x}$ = 318 ± 11 SEM mm, *n* = 4) and adults ($\bar{x}$ = 1.21 ± 0.07 SEM kg, $\bar{x}$ = 435 ± 5 SEM mm, *n* = 3) displayed routine air-breathing frequencies (f_{ab}) of ≈0.03 breaths/min and $\dot{Q}$ values of ≈15 mL/min/kg of body weight. Swimming in a flume at ≈1.1 *L*/s under normoxia raised f_{ab} by eight times to ≈0.23 breaths/min, increasing $\dot{Q}$ threefold in juveniles and two-fold in adults. Aquatic hypoxia (2 kPa O_2) during resting raised f_{ab} 19-fold to ≈0.53 breaths/min in both juveniles and adults, while $\dot{Q}$ increased threefold in resting juveniles and remained unchanged in resting adults. Exercise (as described above) during aquatic hypoxia raised f_{ab} by 35 times to ≈0.95/min compared with fish held in normoxia. Although juveniles experienced a spike in $\dot{Q}$ of nearly double during hypoxic exercise, values of adults stayed the same during resting and exercising. In place of a $\dot{Q}$ response they became agitated, tired quickly in the flume, and had to be held upright by investigators while the water was returned to normoxia.

Values of f_{ab} rise in juvenile and adult tarpons experiencing severe hypoxia and exercise simultaneously. Air-breathing under these conditions is advantageous to both juvenile and adults, and juveniles seem especially well adapted to survival in severely hypoxic water, superior to many exclusively water-breathing species. Juveniles exposed to progressive hypoxia during exercise demonstrated increased $\dot{Q}$, f_H, and f_{ab} without a rise in V_S. The results demonstrate that hypoxia – and perhaps exercise to a lesser extent – initiate air-breathing and cardiac output, responses that enhance oxygen transfer to the myocardium and other tissues.

Although f_{ab} by adults matched that of juveniles, adults failed to show a positive $\dot{Q}$ response with increasing hypoxia, either while resting or exercising.

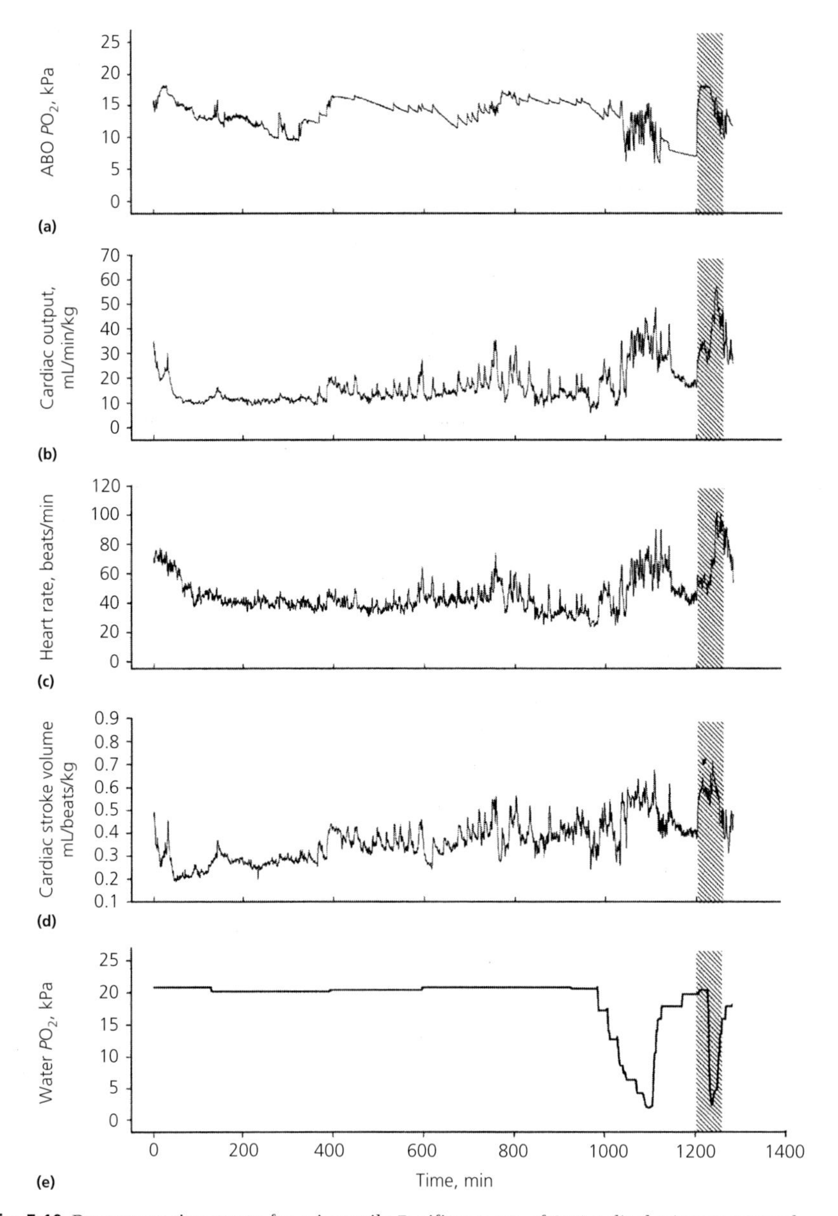

Fig. 5.10 Representative traces for a juvenile Pacific tarpon of 364 g displaying measured variables for the time it was in a flume at 27°C. Elevated values for the first 3 h are associated with recovery from surgery. The traces are for swimming speed of 0.27 *L*/s, except for the shaded region during which it was increased to 1.3 *L*/s. Source: Clark *et al.* (2007: 565 Fig. 2).

Such a result might be expected of an exclusively water-breathing fish, which experiences bradycardia during the extreme hypoxia (2 kPa) reproduced by Clark *et al.* (2007). The unusual response of adults is maybe an attribute of their high percentage of myocardial compacta, as shown by Farrell (2007).

As discussed earlier (Section 5.4), the compacta is more oxygen-limited than the spongiosa during hypoxia unless the higher venous concentration produced during aerial breathing reaches the coronary circulation before being lost down a concentration gradient and into the low-oxygen inspired water.

Coronary circulation in tarpons is restricted to the myocardium by a layer of connective tissue separating it from the spongiosa (Farrell *et al.* 2007). As Farrell (2007: 1719) wrote: "Therefore, the only way the spongy myocardium of the ventricle could benefit from the coronary circulation is through oxygen left in the coronary veins, which drain into the heart near the atrio-ventricular region." As pointed out by Clark *et al.* (2007), when confronted by decreased coronary oxygen and restricted $\dot{Q}$ in tandem, adult Pacific tarpons perhaps shift some blood flow from the systemic circulation to the ABO, and this might explain why they tired so easily in the flume with progressive hypoxia. Also, the ventricle's *compliance* (its capacity to expand) is possibly reduced by the greater mass of compacta, which could be expressed by the lowered V_S and the increased importance of f_H. Whatever the reason, aquatic hypoxia is more a liability to adult tarpons than to juveniles, a condition seldom encountered by adults in well-mixed ocean waters.

Less work has been done on physiology of the Atlantic tarpon, but what there is reveals a similar pattern of how the environment affects both water- and air-breathing. Geiger *et al.* (2000) measured the effect on f_{ab} and V_R of P_wO_2, water temperature, pH, and sulfide concentration (see Chapter 7.2 for results of sulfide on tarpons). In addition, some specimens were evaluated to find if air-breathing is obligate or facultative. The test fish (n = 76) were juveniles of 165–285 mm TL ($\bar{x}$ = 218 mm) and 50–332 g (mean not stated) acclimated to captivity at S_P 22 and specific temperatures between 19° and 33°C.

For the test, 10 fish were subjected to forced submergence during normoxia at 22–26°C. Seven survived until experiments ended at 14 d, although all lived at least 8 d. The three that died did so at 8, 9, and 10 d, but in the end all were negatively buoyant. Loss of buoyancy occurred within 2 d, and the suspected cause was deflation of the swim bladder. All 10 fish had anterior abrasions on the lower jaw, putatively from trying to breathe air against the closed cover (undescribed). Abrasions on the ventral surfaces and ragged fins were considered evidence of increased density and the subsequent inability to sustain neutral buoyancy. Whether these experiments revealed obligate or facultative air-breathing depends on how the terms are defined. Geiger *et al.* (2000 and references) categorized juvenile tarpons as continuous facultative air-breathers, in part because they survived submergence for 1–2 weeks. Still, what distinguishes obligate from facultative air-breathers remains vague, open-ended, and presently indeterminate.

Holstvoogd (1936: 5) briefly mentioned the captive rearing of young juvenile Pacific tarpons captured along the coast at Batavia (now Jakarta), Dutch East Indies (now Indonesia) at the beginning of Stage 2 metamorphosis. "When we put 5 fishes in an aquarium in which a tulle screen prevented their rising to the surface, only one of them survived after a day." Dahl (1965: 7), in observing his captive Atlantic tarpons shrink to 16 mm SL and then resume growth (i.e. start of Stage 3), wrote: "*En este estado, la respiración de aire atmosférico ya era una necesidad fisiológica; ejemplares en los cuales se obstaculizó la subida a la superficie murieron después de un interval entre una y tres horas.*" ["At this stage, breathing atmospheric air was already a physiological necessity; specimens hindered from rising to the surface died within 1–3 hours."] He noted that surfacing behavior coincided with the swim bladder becoming fully formed, an event occurring early in metamorphosis (Fig. 5.11).

Geiger *et al.* (2000) monitored air-breathing frequency at temperatures of 19, 22, 26, 29, and 33°C and P_wO_2 values ranging continuously from 0.133–21.2 kPa (0.5–100% air saturation). Both variables and their interactive effects influenced f_{ab} as measured in breaths/h. At ≤26°C, f_{ab} increased at low values of P_wO_2, being highest below ≈5.3 kPa. At 29–33°C, f_{ab} varied considerably but the rate was high at 33°C when P_wO_2 was <5.3 kPa. Values of f_{ab} also varied with temperature within intervals of P_wO_2. At normoxia (P_wO_2 > 10.7 kPa), f_{ab} was highest at 33°C, lower at 29°C, and lowest at 19–26°C. At P_wO_2 of 5.3–10.7 kPa, f_{ab} increased gradually with rising temperature; below 5.3 kPa, f_{ab} varied but without a consistent trend. At 19–22°C, f_{ab} increased with decreasing P_wO_2 from 10.7–0 kPa, with P_wO_2 accounting for 21 and 24% of the variability in f_{ab} (Geiger *et al.* 2000: 186 Table 1). Air-breathing frequency was independent of P_wO_2 at high oxygen levels. At 26–29°C, f_{ab} fell over the entire range of P_wO_2, becoming nearly independent, and at 33°C, f_{ab} was indeed independent, regardless of P_wO_2.

The partial pressure of dissolved oxygen and temperature independently affected aquatic breathing rate as measured by V_R, but their interactive effects were not significant. Direct correlation was recorded between P_wO_2 and V_R when P_wO_2 was < 5.3 kPa (Geiger *et al.* 2000: 186, Table 1), and P_wO_2 explained 18–72%

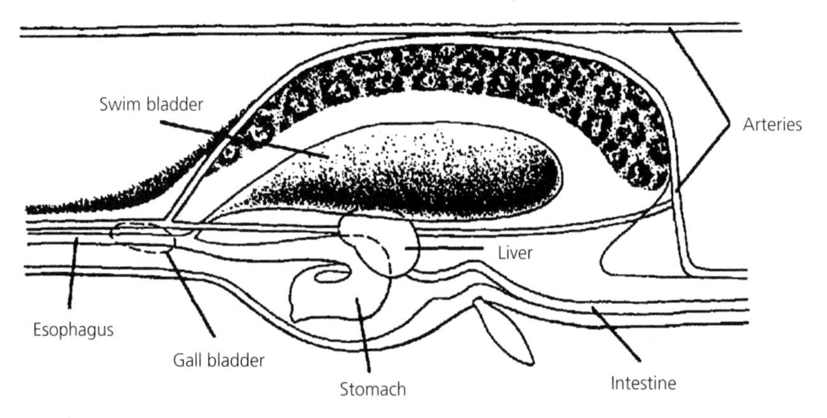

Fig. 5.11 Abdominal cavity of a Stage 2 Atlantic tarpon larva of 21.1 mm SL illustrating the well-developed swim bladder, which becomes functional at even shorter lengths. Also see Chapter 1. Source: Mercado Silgado and Ciardelli (1972: 181 Fig. 10C).

of the variability of V_R in hypoxic water. At ≥ 5.3 kPa, V_R varied little with changes in P_wO_2. In the dilute saline water used in all the experiments, mean V_R rose during normoxia from 31.5/min at 19°C to 42.2/min at 33°C. Low P_wO_2 throughout would seem to be a potential confounding variable; the value deemed normoxic (10.7 kPa) could just as well be considered borderline hypoxic. Resting juvenile Pacific tarpons, for example, show a hypoxic response in freshwater by commencing to breathe air at P_wO_2 values <11.6 kPa (Seymour *et al.* 2004).

As mentioned, hypoxia in water-breathing fishes forces a decline in PO_2 of both arterial and venous blood, which in tarpons continues when they exercise in hypoxic water unless compensated completely by air-breathing (Clark *et al.* 2007). Both f_{ab} and oxygen uptake via the ABO are stimulated during exercise (Seymour *et al.* 2007), but whether compensation is total and $\dot{Q}$ can be sustained by breathing air is unknown.

Seymour *et al.* (2004) evaluated the partitioning of water- and air-breathing and its effect on V_R and respiration using juvenile Pacific tarpons (150–330 mm TL, 59–620 g, n = 10). The fish were forced to swim in a freshwater flume for 1 h at 30°C in combinations of low flow speed (0.11 m/s), high flow speed (0.22 m/s), normoxia (21 ± 1.0 kPa), and hypoxia (8.4 ±0.4 kPa). Aerial oxygen uptake was measured and its assimilation (in practical terms, consumption) calculated. Dissolved oxygen uptake at the gills was also calculated. Aquatic ventilation rate increased significantly at both flow speeds during hypoxia, but not at high flow speed in normoxic conditions. At high flow speed, f_{ab} was significantly higher in hypoxic water than during normoxia. There was a tendency for greater oxygen removal during hypoxia at high speed, but this parameter was independent of fish weight and not significant overall (Fig. 5.12).

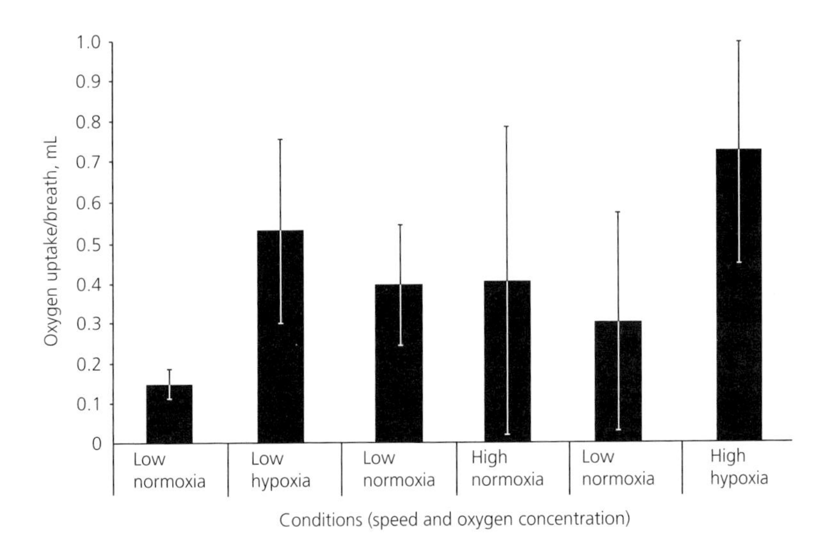

Fig. 5.12 Oxygen extracted with each breath by juvenile Pacific tarpons; treatments and statistics as in Fig. 5.13 (see text and source publication). Source: Seymour *et al.* (2004: 764 Fig. 4).

Water-breathing dominated at low flow speed in normoxia. Air-breathing was nearly nonexistent, and the rate of oxygen uptake from the water ($\dot{M}O_{2w}$) was 1.53 mL/kg$^{0.67}$/min. Fish at low flow speed during hypoxia responded with significantly more aerial breathing (f_{ab} = 0.6 breaths/min at 8 kPa), with the gills accounting for $\dot{M}O_{2w}$ rates of 0.71 mL/kg$^{0.67}$/min and the ABO ($\dot{M}O_{2air}$) for 0.57 mL/kg$^{0.67}$/min.

At high flow speed during normoxia, f_{ab} was only 0.1 breaths/min; rates of oxygen uptake at the gills and ABO were, respectively, 2.08 and 0.08 mL/kg$^{0.67}$/min, demonstrating low use of air when dissolved oxygen is adequate. Oxygen uptake by the gills during hypoxia rose to 1.39 mL/kg$^{0.67}$/min and to 1.28 mL/kg$^{0.67}$/min by the ABO, showing enhanced gill performance even during exercise.

Normoxia at high flow speed affected neither aquatic nor aerial uptake. Air-breathing occurred most often at high flow speed and hypoxia. Total oxygen uptake fell 27% at low flow speed in hypoxic water compared with normoxia, and at high speed increased 23% during normoxia. Oxygen uptake reached a remarkable 52% at high speed during hypoxia, presumably from the switch to air-breathing, evidence of considerable oxygen transfer capability during exercise and proof of the tarpon's moderately high activity level but potential athleticism when stimulated.

The tarpon's life-style requires sustaining high oxygen uptake. Seymour *et al.* (2004: 766), citing results of Wells *et al.* (1997) in the context of theirs, stated: "This level of aerobic activity is reflected in higher blood hematocrit and hemoglobin concentrations, a higher Bohr effect, a lower oxygen affinity, and a higher concentration of red muscle ... than two other similarly sized but less active obligatory water-breathing fish that live in the same environment ...." These other species were the saratoga (*Scleropages jardinii*, family Osteoglossidae) and barramundi (*Lates calcarifer*, family Centropomidae). Ultimately, air-breathing by tarpons is marked by three post-exercise characteristics: (1) elevated hematocrit; (2) low affinity of the hemoglobin for oxygen; and (3) a high Bohr effect (Wells *et al.* 2003, 2005, 2007).

An important finding by Seymour *et al.* (2004) was the threshold at which air-breathing commenced in resting Pacific tarpons (11.6 ± 1.6 kPa, n = 7) subjected to gradually increasing hypoxia. They reasoned that buoyancy is the primary purpose of intermittent air-breathing in normoxic waters, respiratory (and thus metabolic) functions being sustained by aquatic ventilation. The swim bladder supplies only a small part of the oxygen taken up in well-mixed waters even following a period of hypoxia or exercise. The differing physical and physiological requirements of buoyancy control and aerobic metabolism could account for the substantial variation seen in the quantity of oxygen assimilation per breath (Fig. 5.12). The importance of air-breathing by tarpons exposed to hypoxia is undeniable, as shown by the sharp rise in f_{ab} (Fig. 5.13).

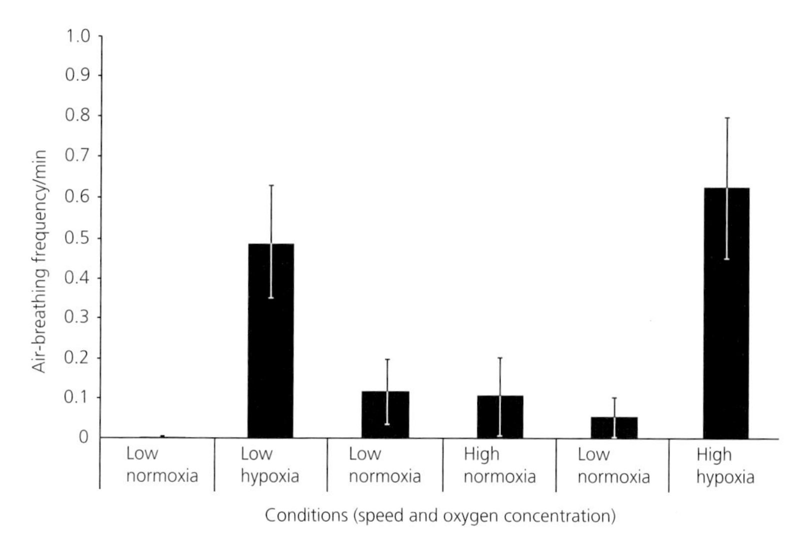

Fig. 5.13 Air-breathing frequency (f_{ab}) of juvenile Pacific tarpons during six successive treatment periods in a respirometer. Vertical bars represent $\bar{x}$ ± 95% CIs. Source: Seymour *et al.* (2004: 763 Fig. 3).

What signal initiates air-breathing in tarpons? Seymour *et al.* (2007) evaluated oxygen exchange by the Pacific tarpon swim bladder, analyzing expired gas breath by breath and concluding that hypoxia was the principal stimulus. The uptake of atmospheric oxygen increases with activity (Seymour *et al.* 2004), as discussed above. Ten juvenile tarpons of 171–362 g ($\bar{x}$ = 238 ± 38 g, 95% CI) and mean FL of 262 (± 18) mm were held at 26°C (presumably in freshwater) with oxygen sensors implanted in their swim bladders. Mean swim bladder volume (V_{ABO}) was measured using other fish that had been killed and dissected ($\bar{x}$ = 176 ± 22 g, n = 6). The value obtained ($\bar{x}$ = 55.6 ± 8.1 mL/kg) was applied when estimating breath-by-breath volumes of the test fish. Ventilation rate at first rose with declining P_wO_2 until ≈8 kPa, when it began to decrease. The rate of oxygen uptake from the swim bladder ($\dot{M}O_{2air}$) was $\bar{x}$ = 1.19 (± 0.16) mL/kg/min (n = 7). Oxygen taken up with each inhalation (V_{ab}) was $\bar{x}$ = 0.49 (± 0.15) mL (n = 7), equivalent to 2.32 (± 0.79) mL/kg (n = 7).

Accompanying the decline in V_R was a drop in its amplitude. Opercular movements became arrhythmic until stopping, in most cases at < 6 kPa. In this experiment, onset of air-breathing during hypoxia started at $\bar{x}$ = 8.3 (± 2.3) kPa (n = 8). In a previous test (Seymour *et al.* 2004, see above) the mean value had been 11.6 kPa. Conditions in the two experiments were obviously different, as was the method of measuring swimming speed.

Air-breathing frequency (f_{ab}) and P_wO_2 did not correlate well over the range 4–8 kPa, and f_{ab} was $\bar{x}$ = 0.73 (± 0.27)/min (n = 8). The volume of a single inhalation (V_{ab}) was $\bar{x}$ = 23.8 (±3.5) mL/kg (n = 7), or a tidal volume of 43% of V_{ABO} (55.6 mL/kg). Swimming speed did not affect $\dot{M}O_{2air}$ from the ABO, V_{ab}, or f_{ab}.

However, $\dot{M}O_{2air}$ showed strong positive correlation with swimming speed, attaining ≈ 1.90 mL/kg/min, with some fish taking up more oxygen per breath, others increasing air-breathing frequency. Air-breathing stopped when normoxic conditions were established, and swim-bladder PO_2 stabilized. Afterward $\dot{M}O_{2air}$ and PO_2 in the swim bladder appeared to function independently.

Fishes inhabiting incompletely mixed brackish and freshwaters are sometimes subjected to hyperoxia, especially if the habitat is heavily vegetated. Hyperoxic situations are usually transient, occurring most often during daylight when photosynthesis overrides respiration and the oxygen concentration rises above saturation at ambient temperature. At such locations the lowest oxygen concentrations are observed in early morning (Chapman *et al.* 2002; Kramer *et al.* 1978; Wells *et al.* 2005, 2007), often after intense overnight respiration by plants and microorganisms in the absence of photosynthesis. Thickly vegetated locations can thus become hypoxic in darkness, and plant decay can further reduce water quality (Perna *et al.* 2011). Hyperoxia is fleeting at some places, seasonal at others. *Várzea* lakes formed in the Brazilian Amazon during flood season can experience dissolved oxygen levels falling to <1.5 kPa at night and shifting to hyperoxic conditions in daylight (Sloman *et al.* 2009). The salient point is that susceptible waters can range from hyperoxic through normoxic to hypoxic over a diel cycle, and the organisms living there must adapt (Kramer *et al.* 1978).

Persistently high oxygen concentrations eventually can induce bradycardia and respiratory acidosis. During environmental hyperoxia, V_R values of water-breathing fishes decrease, inducing a rise in blood H^+ with subsequent hypercapnia (see below). Long-term exposure to hyperoxic conditions can cause permanent effects that include morphological changes in the gill lamellae, such as reduced surface area for gas transfer and increased diffusion distance for oxygen exchange with the blood (Saroglia *et al.* 2010). As I wrote elsewhere (Spotte 1992: 221):

> *The most remarkable lesson to be learned from how fishes respond to environmental hyperoxia is that oxygenation of the tissues overrides any requirement to rectify simultaneous disturbances in acid-base balance. In other words, a fish placed in a hyperoxic environment "forgets" to breathe because its tissue oxygen requirements are satisfied, even though the subsequent reduction in gill ventilation rate causes a rise in P_aCO_2. The latter effect could be offset to some extent by hyperventilation, which would clear the opercular cavities and increase the P_aCO_2/P_wCO_2 diffusion gradient, but hyperventilation is an atypical response to hyperoxia. Ultimate restoration of the acid-base balance is left to adjustments in [arterial blood bicarbonate], a prolonged, extra-ventilatory mechanism.*

The respiratory physiology of fishes, and subsequently the acid-base balance of fish blood, can be disrupted occasionally if the aquatic pH drops (Spotte 1992: 217–218), although this situation is more likely in poorly buffered brackish and freshwaters than ocean waters. I found no specific mention of the tarpon in published studies in this area of research, but the response again appears to be widespread among elasmobranchs and teleosts.

5.6 Hypercapnia

Environmental hypercapnia defines an abnormally high partial pressure of carbon dioxide in a body of water (P_wCO_2), which if poorly buffered lowers the pH. *Physiological hypercapnia* describes the same situation in blood (for purposes here, arterial blood, P_aCO_2). The first can induce the second, with two consequences: (1) a leveling of the diffusion gradient of carbon dioxide from blood to water, thus slowing its transfer; and (2) reduced blood pH. The high relative rate of V_R generates only ≈ 5 mL/μmol of O_2 consumption compared to pulmonary ventilation in a human, producing 0.6 mL/μmol O_2 (Heisler 1993). The result is a shallow PCO_2 gradient between blood and water, ordinarily 0.13–0.53 kPa in a fish but comparable to 4.0–6.0 kPa in a terrestrial vertebrate (Heisler 1993). As a consequence, water-breathers are forced to function within a narrow range when attempting to maintain respiratory homeostasis, and hyperventilation – the means exploited to good effect by mammals – offers little relief or compensation to a fish. As Heisler (1993) pointed out, an increase in PCO_2 of the inspired water induces an internal rise in P_aCO_2 of 3–6 times with deflections of arterial pH of ≈ 0.4–0.5 units. In a mammal, increasing breathing 30%, which Heisler called a slight increase in ventilation, reduces the PCO_2 in atmospheric air from 5.3 to 4.0 kPa and results in a near-instantaneous restoration of blood acid-base balance.

The second effect results from increased inspired CO_2. Excess H^+ ions produced as blood pH falls cause *respiratory acidosis*, a condition that persists until they can be either buffered by a subsequent rise in plasma bicarbonate ions (HCO_3^-) or transferred to the external environment. However, with V_R now lowered, CO_2 in the blood accumulates instead of being excreted into the environment, as it would during normoxia. Recovery in *normcapnic* waters (those within the expected ranges of carbon dioxide concentration and pH) ordinarily is measured in hours but sometimes days, depending on species and specific circumstances (Spotte 1992: 217–218).

Meanwhile, attaining normal acid-base balance in the blood can be delayed by interactive effects of hypercapnia and either hypoxia or hyperoxia. Here too the response is general among elasmobranchs and teleosts (Spotte 1992: 218–221). One event is a rise in P_aCO_2 if the water becomes both hypercapnic and hyperoxic, as it could in daylight if poorly mixed, low in solutes, and thickly vegetated. The carbon dioxide of both inspired and expired water would also increase, the value of the latter being greater because of CO_2 released at the gills. During conditions of normoxia-hypercapnia the partial pressure of inspired water, which is controlled by the external environment, remains unchanged; values in the blood and expired water decline. All three factors fall on return to normoxia-normcapnia.

Carbon dioxide is excreted at the gills, resulting in a 10–20% loss from the venous blood and accompanied by a reduction in plasma HCO_3^- of ≈ 20%. The latter occurs as HCO_3^- entering erythrocytes from the plasma is dehydrated to

CO_2 (the process catalyzed by carbonic anhydrase), which then diffuses into the water across lamellar epithelia (Randall 1982 and references). Because fish gills are highly permeability to protons such as H^+, blood acidosis results in their passive exit into water. Low environmental pH, however, reverses the direction, causing a net uptake of acid across the lamellar surfaces.

Kaufmann (1990) surmised that because energy for burst swimming is high compared with the activity of glycolytic enzymes, breakdown of phosphocreatine might provide the most likely source of energy. Whatever the driving mechanism, burst exercise in teleosts reduces blood pH as lactic acid accumulates, raising excretion rates of both CO_2 and H^+. The coho salmon (*Oncorhynchus kisutch*) is an athletic species like the tarpon with a physostomous swim bladder, although it strictly breathes water. When exposed to low seawater pH of 7.10, test specimens excreted HCO_3^- at a high rate during the first 30 min of swimming in a closed respirometer, and this was associated with a slow rate of H^+ transfer (van den Thillart 1983). The reverse occurred after burst activity: the rate of H^+ transfer increased, but HCO_3^- excretion into the water was depressed. Such responses are typical.

5.7 Air-breathing as social behavior

When confronting aquatic hypoxia some tropical freshwater fishes in places like the Amazon and Congo river systems rely on *aquatic surface-breathing* (*ASB*), also abbreviated as *ASR* for *aquatic surface respiration* (Kramer and McClure 1982). The top few millimeters of even anoxic waters are usually near saturation (Lewis 1970), and several adaptations have evolved to make use of this resource. When dissolved oxygen falls to ≈6.5% saturation (≈1.14 kPa at 30°C), the tambaqui (*Colossoma macropomum*) grows an extension on its lower jaw within a few hours, allowing it to skim the oxygen-rich surface layer more efficiently (Saint-Paul 1984). Africa's upside-down catfish (*Synodontis nigriventris*), its counter-shading reversed, spends most of its life on its back skimming the surface, both to feed and make use of ASB (Chapman *et al.* 1994).

Spending too much time exploiting the surface is an obvious disadvantage if threatened by aerial predators (Kramer *et al.* 1978) or aquatic predators that have also moved higher in the water column (Sloman *et al.* 2009). The evolution of air-breathing supposedly allows a fish to inhabit deeper waters and make only intermittent trips to the surface, reducing its exposure. Thus air-breathing permits a fish to survive even in anoxic environments, avoid the surface most of the time, and take advantage of the atmosphere, which is even richer in oxygen than the surface layer (Kramer *et al.* 1978).

Many social aquatic animals from fishes to whales practice synchronous breathing with conspecifics (e.g. Gee and Graham 1978, Hastie *et al.* 2003, Kramer and Graham 1976, Senigaglia and Whitehead 2012, Sloman *et al.* 2009), behavior believed by some to provide anti-predator benefits through a "dilution effect" analogous to schooling in aquatic environments (Kramer and Graham 1976), and no doubt herding and flocking by terrestrial species. For example, isolating individuals of two species of air-breathing social fishes from their conspecifics delays trips to the surface (e.g. Sloman *et al.* 2009). However, by not addressing what drives the behavior, such studies have descriptive value only, leaving any causal factors unexplained.

Harrington (1966) reported synchronous breathing in captive Atlantic tarpons as small as 18.1 mm SL. Shlaifer and Breder (1940) observed it too in captive juveniles, labeling it "imitation," although a more accurate term (presuming it actually is a form of mimicking) is *allelomimesis* (behavioral matching), similar to the self-organized swimming patterns of schooling fishes described by Parrish *et al.* (2002). I could find no evidence of actual mimicking during synchronous breathing, which at least within the context of learning is exceedingly rare even in mammals (Spotte 2014: 116–136).

Also, this behavior is doubtfully synchronous. Shlaifer and Breder (1940: 496) wrote, "One tarpon will rise to the surface and its rise may be followed immediately by that of one or more other tarpon." And, "An imitative rise … occurs within a second after the initial rise." Steps of the sequence take place in tandem, in other words, not in synchrony. However we might term it, such behavior is stimulated visually, as demonstrated by experiments using transparent glass partitions in aquariums through which two tarpons, one on each side, could see the other, and by Shlaifer (1941) using blinded fish that failed to respond to other tarpons rising to breathe. Shlaifer and Breder (1940) devised a 10-cm wooden model that was eyeless, mouthless, finless, painted silver, and vaguely tarpon-shaped, that proved moderately successful at stimulating juvenile tarpons to surface when manipulated by strings. Spacing also influenced imitative breathing, with more fish tending to rise in tandem when closely confined.

CHAPTER 6
Osmo- and ionoregulation

6.1 Introduction

Stenohaline organisms are restricted to freshwaters or seawater, solutions of limited range in terms of ionic strength and composition. *Euryhaline* species can survive in aquatic habitats that are fresh, brackish, full-strength seawater, and often hypersaline, thus maintaining osmotic and ionic homeostasis over a broader range. However, this alone does not make them euryhaline. A truly euryhaline species spends at least a segment of its life-cycle in freshwater or brackish water (Movahedinia *et al.* 2009). So defined, only about 5% of teleosts qualify (McCormick 2001 and references). Many such species inhabit intertidal freshwater or brackish habitats during certain life-history stages and later adapt to stenohaline conditions as adults – either freshwater or seawater – while retaining the capacity to live permanently in either. The tarpon is among them.

Marine biologists are typically cavalier when using the term *salinity*, often treating it as simply a number and applicable to any solution approximating the "saltiness" of water, regardless of its composition. Consequently, a concoction of dissolved salts or even brine might yield a "salinity" of 35 (considered the value typical of surface ocean waters) while differing substantially in composition from seawater. Wright *et al.* (2011) listed seven measurements of salinity now in use, reiterating that based on the principle of constant relative proportions chlorinity remains the standard. Chlorinity provides a measure of the total amount of dissolved matter in seawater in terms of the concentration of halides. Chlorinity has inherent problems of measurement and interpretation, and the analytical procedures involved are inconvenient.

In 2010 oceanographers adopted the *absolute salinity scale*, S_A, values of which are true SI units expressed as mg/kg *of seawater* (i.e. of solution) and predicated on seawater's thermodynamic properties.[1] Being a mass-fraction, S_A values are

---

[1] http://www.teos-10.org. Downloaded 12 July 2015. Also see http://www.argo.ucsd.edu/Adopting_Absolute_Salinity.pdf. Downloaded 13 July 2015.

Tarpons: Biology, Ecology, Fisheries, First Edition. Stephen Spotte.

temperature- and pressure-independent, equivalent in this respect to the mokal scale expressed in mol/kg *of solution* (Bidwell and Spotte 1985: 3; MacIntyre 1976). However, "absolute salinity" was not defined specifically for seawaters having anomalous compositions until Wright *et al.* (2011). In practice, despite chlorinity's assumption of constant relative proportions this principle does not hold true, and the actual composition of seawaters varies slightly by location.

Electrical conductivity is easy to measure, and the *practical salinity scale*, S_P, has been used since the 1970s (Anonymous 1978; Lewis and Perkin 1981; Wright *et al.* 2011), although it too has several theoretical deficiencies (e.g. Le Menn 2011). These are seldom troublesome for biologists, and the S_P scale is acceptable provided users recognize its shortcomings, the most important being that conductivity measurements from estuaries or salt ponds do not imply compositional similarity to ocean waters even when the numeric values match those of seawaters. *Practical salinity* is formally defined as the conductivity of the sample solution (e.g. seawater) at specific temperature and pressure divided by that of a standard potassium chloride solution at the same temperature and pressure (32.4356 g of KCl at 15°C in a 1 kg solution). For example, a sample of seawater at 15°C having a conductivity equal to that of the KCl standard solution has a salinity of exactly 35. *The quotient, being the ratio of two conductivities, has no units.*[2]

The *ionic strength*, *I*, of a solution is the sum of the concentrations of all ions present. *The term is conceptually generic and applies to solutions of any composition.* Seawater, however, is considered to be a specific solution in having a constant relative composition. As such, S_A defines seawater alone, not just any saline solution having an identical S_P value (e.g. hard freshwaters, brackish waters, hypersaline waters, brines, artificial seawaters). Any factor that alters the ratios of ions in seawater renders it "nonstandard," in which case "salinity" in its truest, narrowest, and most useful theoretical sense no longer applies. Ionic strength is defined by

$$I = \frac{1}{2} \Sigma z \frac{2}{i} m_i \qquad (6.1)$$

[2] According to Millero (1993), this includes "psu" (practical salinity units) sometimes given after the numeric value (e.g. "the sample measured 35 psu."). The designation *parts per thousand* (ppt) following a numeric value is ambiguous and should be avoided too, and so is its symbol, ‰. Used properly, ppt designates a mass/mass relationship, but in practice is often mass/volume because the solutions used were not prepared volumetrically. The reader is then left wondering whether ppt means mg/kg or mg/L. In practice, salinity is determined from empirical relationships between temperature and the conductivity ratio of the sample solution to International Association for the Physical Sciences of the Ocean (IAPSO) Standard Seawater. Comparison of results with other laboratories necessitates using IAPSO Standard Seawater for calibration (http://www.toptotop.org/climate/psu.php).

where z_i represents the ionic charge of the ion *i* and m_i its molality. Ionic strength is related to S_A by:

$$I = \frac{19.92 S_A}{1000 - 1.006 S_A} \tag{6.2}$$

where S_A is expressed in g/kg *of seawater* and *I* in mol/kg. The ionic strength of typical offshore seawater is 0.72 mol/kg.

The artificial seawaters used by investigators when the real thing is unavailable are seldom devised and mixed volumetrically or prepared using an appropriate concentration scale (e.g. g/L *of water* is a common error), rendering their ionic species compositions unknown from the start and ultimately unknowable (Bidwell and Spotte 1985: 3). In any case the "salinity" of a solution of artificial seawater can only refer to S_P. In the interest of correctness and consistency, I consider all such data to be practical salinity determinations and express the values without units. Often it seems more appropriate to simply refer to the solutions used in terms of their ionic strengths if making general comparisons about the "saltiness" of solutions. Here and elsewhere in the text this approach will apply even if the experimental solution described in a published report was ocean water, simply because measurements of "salinity" obtained indirectly (e.g. by conductivity alone) are theoretically unsuited to the S_A scale. Therefore, where I mention "salinity" and "ionic strength," take the meaning implied with a grain of salt.

6.2 Osmo- and ionoregulation

I found no reports documenting the mechanisms of osmo- and ionoregulation in tarpons specifically, but those by which other teleosts adapt are general enough to assume tarpons differ little, especially when occupying brackish waters and seawater. *Osmoregulation* is the control of internal water balance, a physiological characteristic of all bony fishes, and the same is true of *ionoregulation* (or *ionic regulation*). In this regard I prefer to think of fishes as solutions within solutions and treat them as such here.

Water and ion balance (osmo- and ionoregulation) are controlled by activities occurring in the integuments, gills, alimentary system, and kidneys, of which those in the gills predominate. Integration among participating tissues is modulated by drinking and urination rates, by hormones, and by "remodeling" the affected surface tissues (primarily of the gills) as needed to accommodate ionic changes in the external environment.

The discussion in the next few paragraphs is based mainly on Spotte (1992: 205–210) unless other citations are given. Two solutions are *isosmotic* if their total solute concentrations are the same. A *hyposmotic* (or *hypo-osmotic*) solution is more dilute, and one that is *hyperosmotic* more concentrated, than the comparison solution. If you continue thinking of a fish as a solution of water and ions and the external environment a different solution through which it swims, then its task is to maintain an internal milieu either hyperosmotic or hyposmotic, depending on composition of the external environment (freshwater, brackish water, or seawater). The most prevalent elements in seawater (after hydrogen and oxygen) are sodium and chloride at respective concentrations of 4.63×10^{-1} and 5.43×10^{-1} mol/kg (Bidwell and Spotte 1985: 7, Table 1). *All teleosts – regardless of habitat – maintain internal concentrations of sodium and chloride at about one-third seawater values* (Grosell 2006; McCormick 2001), or ≈ 100 times higher than most freshwaters (Wendelaar Bonga 1997), although as I shall explain, other ions of lesser concentration are regulated too. This means that most bony fishes, no matter where they live, maintain a blood osmolality within the approximate range 280–350 mOsm/kg (e.g. Lorin-Nebel *et al.* 2006; Varsamos *et al.* 2005).

When a solute causes water to move differentially through a semipermeable membrane it exerts *osmotic pressure*, the intensity of which is directly proportional to the numbers of molecules or ions, not to their concentrations. The fact of this distinction permits fishes to control internal water and solute balance. Ions cross tissue barriers passively, actively, or both. Diffusion, a form of *passive transport*, does not require energy. However, "pumping" ions through membranes – often "uphill" against concentration gradients (*active transport*) – consumes cellular energy generated by mitochondria.

The regulation of water and ions by freshwater fishes is often more variable and more likely to be species-specific than in seawater (Hwang *et al.* 2011), perhaps because freshwaters vary widely in composition and ionic strength, forcing the fishes living in them to be more physiologically facile. Seawater is remarkably invariant except near shore where diluted by fluvial flow, rain and melting ice, and periodic terrestrial runoff. Freshwater teleosts are hyperosmotic: their internal fluids contain greater concentrations of ions than the water outside (i.e. are "saltier"). They face continuous net diffusional losses of ions to the external environment and internal *osmotic loading* (i.e. a net gain of water). Although they drink very little, water enters anyway through the permeable gills and skin, and water and ions are also ingested with food. In the absence of suitable "physiological pumps" to counteract osmotic loading, internal fluids of freshwater fishes would quickly become isosmotic, matching the external environment and leading to loss of homeostasis. To survive in freshwaters, a fish negates the passive gain of water and loss of ions by producing and discharging large volumes of dilute urine, while actively taking up ions from the external environment across its gills and in food.

Seawater teleosts, in contrast, are hyposmotic, having internal fluids that are less "salty" than the surrounding ocean water. They face a net diffusional gain of ions (mainly sodium and chloride) and a net osmotic loss. Their main problems are therefore *ionic loading* (accumulation of excess ions) and dehydration (e.g. Lorin-Nebel *et al.* 2006). Although it seems paradoxical, seawater fishes must drink continuously to gain physiological water osmotically after the ions it contains have been taken up by the pertinent tissues. The seawater imbibed goes directly to the gut, where subsequent adsorption of monovalent ions like sodium (Na^+), chloride (Cl^-), and potassium (K^+) is accompanied by absorption of water (Evans *et al.* 2005). Initial removal of ions, however, starts in the esophagus (Grosell 2006). Most of the ion load is ultimately transported by blood to the gills, where it adds to a pool of still more monovalent ions diffusing inward. In the end, balance is achieved by the gill epithelium, the excess ions excreted into the environment.

Urine is not useful for removing monovalent ions because it is isosmotic with the blood and excreted only in small amounts by seawater fishes. The ions it contains are mostly multivalent species such as calcium (Ca^{2+}), magnesium (Mg^{2+}), and sulfate (SO_4^{2-}), and these are excreted via the kidney. After adsorption of monovalent ions in the gut, Mg^{2+} and SO_4^{2-} become concentrated. Along with these divalent ions, HCO_3^- also attains high levels, raising the pH of the gut fluid (Grosell 2006).

6.3 Ionocytes

Gas exchange occurs mainly through the double-layered epithelium of the gill lamellae (Chapter 5). Ion exchange between blood and the water is conducted by specialized *ionocytes* (also called *mitochondria-rich cells*, MRCs, and *chloride cells*, CCs), their function first confirmed by Foskett and Scheffey (1982). All three terms are accepted, but the last has fallen out of favor by referring implicitly to just Cl^-. Labeling a cell as "mitochondria-rich," although descriptive, omits any mention of its function. Ionocyte therefore seems preferable.

The ionocytes of most teleosts are round or oval and located in the multi-layered epithelium of gill filaments and the inter- and basal-lamellar regions (Hwang *et al.* 2011, Kang *et al.* 2013, Li *et al.* 1995, Saadatfar and Shahsavani 2011). They span the gill epithelium and are in contact with both the water and the blood, the latter via the basolateral membrane (van der Heijden *et al.* 1997), making them major contributors in the active transport of ions between these environments. Besides osmo- and ionoregulation, their duties include participating in internal acid-base regulation and nitrogen (mainly as ammonia) excretion (Sections 6.4 and 6.5).

In euryhaline species, ionocytes dampen swings in homeostasis during a fish's movements into and out of waters of varying ionic strength, increasing or decreasing their numbers and dimensions as necessary and changing morphology to accommodate the uptake or excretion of ions for the regulation of plasma osmolality, acid-base balance, and nitrogen excretion. Ionocytes of freshwater-adapted Mozambique tilapias (*Oreochromis mossambicus*), for example, had an ionocyte density of 6233/mm^2 of gill tissue, of which a subpopulation (3458/mm^2) was in direct contact with the water (van der Heijden *et al.* 1997). Ionocyte density decreased to 3061/mm^2 following seawater-adaptation after which 2445/mm^2 remained in contact with the water. There was also a cross-sectional size change from 87 μm^2 while in freshwater to 217 μm^2 in seawater. The gills of euryhaline species, having evolved to face such challenges, can often make these and other required changes within hours while withstanding the physiological trauma of adaptation and eventually regaining homeostasis.

In seawater, ionocytes excrete ions into the environment; in freshwaters they perform a reverse function and take them up. As the synonym mitochondria-rich cells makes clear, ionocytes contain numerous mitochondria, evidence of their high capacity to supply energy for active ion transport in either direction (Evans 1999, Evans *et al.* 2005, Hwang *et al.* 2011, van der Heijden *et al.* 1997). *Apical crypts* (bite-shaped cavities, or pits) are another feature of ionocytes (e.g. Alderdice 1988: 226–227, Evans 1999, Evans *et al.* 2005, Ouattara *et al.* 2014, van der Heijden *et al.* 1997), their openings facing the water where exchange occurs (Fig. 6.1).

The pattern of the openings and general phenotypic appearance of ionocytes indicate a fish's osmotic and ionic requirements (Movahedinia *et al.* 2009) and indirectly predict the direction of ion transfer – mainly Na^+ and Cl^- – into or out of the cell. In freshwater-adapted fishes the apical surfaces are typically flat, their openings filled with projections resembling microvilli, thought to increase surface area and aid in the uptake of ions from dilute environments (Hirose *et al.* 2003). Ionocytes of fishes adapted to seawater contain deep invaginations appearing as crypts in histological sections, and they lack microvilli-like projections (Kang *et al.* 2013 and references) that function to excrete ions into the environment. Apical cells from fishes adapted to brackish waters are intermediate (Kang *et al.* 2013, Tang and Lee 2011). Kang and Lee (2014) identified a homolog of the protein villin 1 in ionocytes of the euryhaline medaka (*Oryzias latipes*) adapted to freshwater, which is not activated in specimens kept in saline water.

These three different cellular subtypes have been labeled subtype I ("wavy-convex"), subtype II ("shallow-basin"), and subtype III ("deep-hole"), based on phenotypic features of the apical surfaces, as seen by scanning electron microscopy (Hwang *et al.* 2011, Kang *et al.* 2013 and references, Lee *et al.* 1996, 2003 and references, Movahedinia *et al.* 2009, van der Heijden *et al.* 1997, Wilson and Laurent 2002) and staining characteristics (Hwang *et al.* 2011). Some authors add a "seawater" subtype IV (Hwang *et al.* 2011 and references, Wilson and Laurent 2002).

Ionocytes are also distinguished by abundant *tubular systems* formed as an expansion of their basolateral membranes (e.g. Li *et al.*, 1995 and references, Mir *et al.* 2011, Movahedinia *et al.* 2009, Ouattara *et al.* 2014, Saadatfar and Shahsavani 2008, van der Heijden *et al.* 1997, Wilson and Laurent 2002) where proteins assisting with ion transport are concentrated. Examples include the

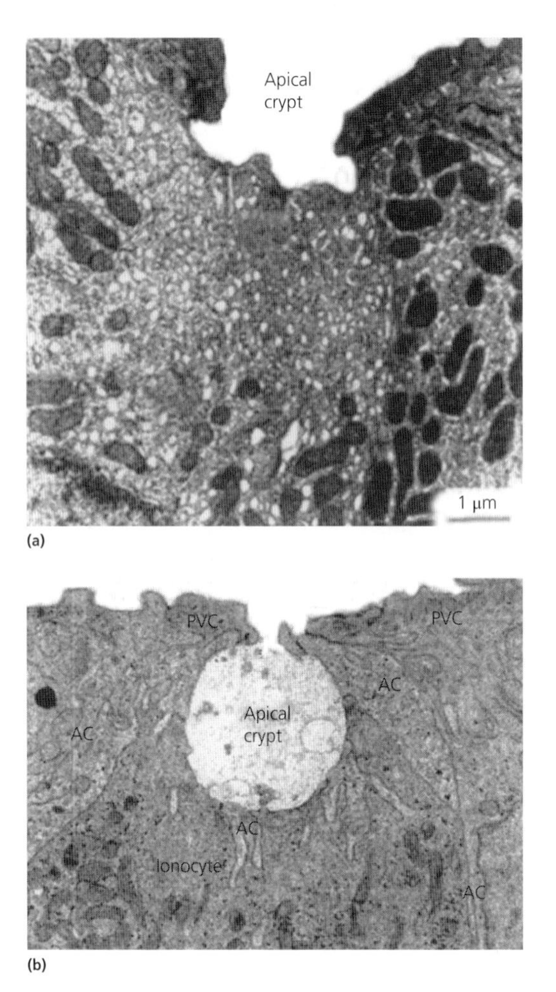

Fig. 6.1 Transmission electron micrographs of ionocytes from gills of seawater teleosts. (**a**) Ionocyte from a common sole (*Solea solea*) containing numerous mitochondria (m), a tubular system (ts), a subapical tubulovesicular system (tvs; not discussed in the text), and an apical crypt, or pit. (**b**) apical region of an ionocyte from a mummichog (*Fundulus heteroclitus*). This cell forms deep tight junctions with surrounding pavement cells (PVCs) and shallow tight junctions with surrounding accessory cells (ACs), which share an apical crypt with the ionocyte. Source: Evans *et al.* (2005: 104 Fig. 7).

enzyme Na^+/K^+-ATPase (NKA, the basolateral "sodium pump" that functions to establish electrochemical gradients), Ca^{2+}-ATPase, and Ca^{2+}-ATPase exchangers (Sucré *et al.* 2011 and references). Present too is the transmembranous protein $Na^+/K^+/2Cl^-$ (NKCC) co-transporter 1 symbolized as NKCC1 (the secretory isoform of NKCC, NKCC2 being the absorptive isoform), also in the basolateral membrane (Lorin-Nebel *et al.* 2006), and an apical anion channel homologous with the cystic fibrosis transmembrane conductance regulator (CFTR). The most abundant isoform of NKCC is NKCC1, and it increases with salinity in ionocytes,

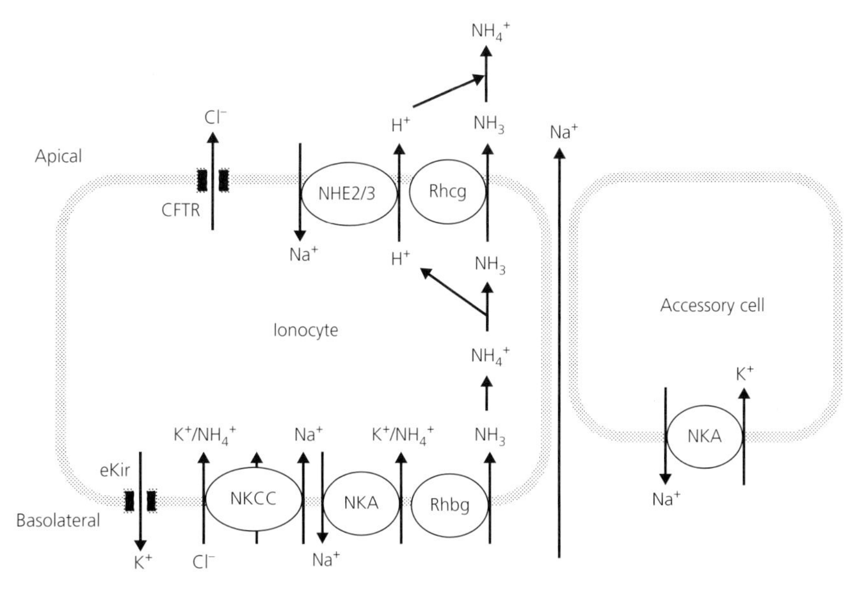

Fig. 6.2 Model of ionocytes of seawater-adapted teleosts. See text for details. CFTR = cystic fibrosis transmembrane conductance regulator; eKir = inwardly rectifying K^+ channel; NHE2/3 = Na^+/H^+ exchangers, isoforms 2 and 3; NKA = Na^+-K^+-ATPase; NKCC1 = Na^+-K^+-CC cotransporter; Rhcg and Rhbg = Rhesus glycoproteins. Source: Hwang *et al.* (2011: R29 Fig. 1).

a demonstration of its importance in ion excretion. Note that apical membranes are primary sites of ion transport between ionocytes and the water outside (Kang *et al.* 2013). Chloride is exchanged apically for HCO_3^- and discharged down an electrical gradient based on ionocytes having a stronger negative charge than the blood (McCormick 2001).

The mechanistic functions are these (Fig. 6.2). Sustaining tight junctions between epithelial gill cells are the proteins claudins and occludins. Claudin-10d and -10e increase in ionocytes of the euryhaline green spotted puffer (*Tetraodon nigroviridis*) adapted to seawater, contributing to the "leaky" epithelium characteristic of salt-secreting gills of seawater fishes (Bui and Kelly 2014). Basolaterally, NKCC1 mediates entry of Na^+ and Cl^- down a concentration gradient established by Na^+/K^+-ATPase (NKA), which is followed next by passive diffusion outward of Cl^- and ultimately Na^+ in sequence. This occurs apically via CFTR and between-cell junctions. Expression of the branchial NKCC gene decreases significantly in the European seabass (*Dicentrarchus labrax*) during long-term freshwater adaptation (Lorin-Nebel *et al.* 2006). Among seawater fishes generally, K^+ undergoes recycling at the gill by means of a K^+ channel.

The dynamic nature of NKA can be traced to its varied protein subunits. Activation of specific subunits in euryhaline teleosts allows acclimation to higher or lower salinities. Some of these factors have been identified during transition to different salinities by the milkfish (*Chanos chanos*), Mozambique tilapia, and certain salmonids, all euryhaline species (Hwang *et al.* 2011). Such functional "isoform switches" presumably alter NKA activity, depending on the direction of acclimation demanded (i.e. toward lesser or greater external ionic strength). For example, raising the ionic strength of water holding a euryhaline fish increases mRNA levels of gill isoform NKCC1a and protein levels of NKCC1, indicating participation in Na^+ and Cl^- secretion of branchial ionocytes during acclimation.

Because enzymes like NKA span the width of epithelial membranes, they are interfaces between the blood and the water outside, actively transporting Na^+ out of cells and K^+ into them (Lin *et al.* 2004). Any increase in up-regulation of gene expression indicates a response to internal or environmental change (Huang *et al.* 2008 and references, Wright and Wood 2009), as when euryhaline species move into habitats of varying temperature and salinity. At such times the gill might generate more ionocytes for enhancing excretion or uptake of ions.

Fishes living in waters of all types actively transport Ca^{2+} and Cl^-, although perhaps not Na^+ by freshwater teleosts (Li *et al.* 1995). Some fishes (e.g. salmonids) increase the number of ionocytes in soft waters, presumably as an aid to ionic uptake, and this apparently has the negative effect of reducing gill capacity for gas exchange as determined by a measurable decline in *critical swimming speed* (U_{crit}), the maximum burst speed to exhaustion in body lengths per unit time (Brauner and Rombough 2012 and references). The trade-off also has been noticed in other freshwater species (e.g. Matey *et al.* 2008, Sollid *et al.* 2003). For water-breathers, any compromise that diminishes oxygen uptake might be accompanied by compensatory measures like an increase in V_R, f_H, V_S, or the blood's oxygen-carrying capacity (Chapter 5). Air-breathers such as tarpon could ameliorate respiratory stress by raising the level of $\dot{M}O_{2air}$ or V_{ab}.

6.4 Acid-base regulation

Internal acid-base balance is achieved through osmo- and ionoregulation. Metabolic processes produce acids, generating H^+ and CO_2. In the absence of regulation the internal milieu would become irreversibly acidic, resulting in loss of osmotic and ionic homeostasis. As a counteractive measure, CO_2 is converted to H^+ and HCO_3^- by the simplified reaction:

$$CO_2 + H_2O \leftrightarrow H^+ + HCO_3^- \quad (6.3)$$

Fishes maintain acid-base homeostasis (i.e. they prevent undue fluctuations in alkalinity and pH of the blood and tissues) through differential excretion of H^+ and HCO_3^- across the gills (Hwang *et al.* 2011). A water-breathing fish forced into burst exercise in a low-pH environment increases its uptake of HCO_3^- from the water to help offset HCO_3^- loss, in the process experiencing a sharp reduction in the outward transfer of CO_2 and H^+ (van den Thillart *et al.* 1983). Fishes have generally low PCO_2 and HCO_3^- concentrations, which limit the buffer capacities of their internal fluids, further restricting modulation of internal pH by regulating V_R. Hyperventilation does little good. Instead, internal pH must adjust by the much slower process of ion exchange with the water. In seawater and hypersaline waters, acidic compounds generated metabolically are released via the apical Na^+/H^+ exchanger (NHE), as described by Claiborne *et al.* (1999).

Acid-base regulation in freshwaters involves different NHEs and perhaps acid excretion by H^+-ATPase (HA) with a Na^+ uptake and H^+ excretion electrical linkage, but this second process (assuming it exists) appears to be species-specific, and HA might even be involved in base excretion in ionocytes of seawater fishes (Hwang *et al.* 2011 and references). The most widely accepted model at present for Na^+ uptake and excretion – but uptake in particular – in freshwaters is also by the electrically neutral NHE.

6.5 Ammonia excretion

Nitrogenous waste in most aquatic vertebrates is excreted at the gills (Weihrauch *et al.* 2009). Nearly all teleosts are *ammonotelic,* releasing waste nitrogen from amino acid metabolism as ammonia, a potent neurotoxin whether retained internally (Ip and Chew 2010) or taken up from the environment (Spotte and Anderson 1989).

As discussed elsewhere (Chapter 5), oxygen's low solubility and low concentrations in water, in combination with water's high viscosity, make its extraction difficult. Weihrauch *et al.* (2009: 1716) pointed out that fishes are already "hyperventilated" with respect to waste gases like CO_2 and NH_3, and "in many cases no additional ventilatory energy needs to be invested by water-breathers to effectively excrete nitrogenous waste *to* the water, beyond that used to take up oxygen *from* the water."

Fishes excrete ammonia either as un-ionized NH_3 (a gas) or the protonated form (NH_4^+) (I use the general term *ammonia* when referring to its two forms in combination or in a generic context). A principal mechanism of release is outward diffusion through *paracellular* (between-cell) junctions or by *transcellular* (through-cell) means across cell membranes. In some situations there can be ion substitution of NH_4^+ (Evans 2010a, Evans *et al.* 2005, Wilkie 2002, Wright 1995). This can occur both basolaterally and apically in conjunction with the electrically neutral exchangers Na^+/K^+-ATPase and Na^+/H^+-ATPase or $Na^+/K^+/2Cl^-$ co-transport (Evans 2010a). Substantial apical $Na^+/H^+(NH_4^+)$ exchange is unlikely in freshwaters but could be important in seawater (Wilkie 2002).

Nakada *et al.* (2007) found evidence that Rhesus (Rh) proteins in pillar and pavement cells and ionocytes of the gills are ammonia transporters, mediating its excretion. One of these, Rhag glycoprotein (the "g" stands for glycosylated) is

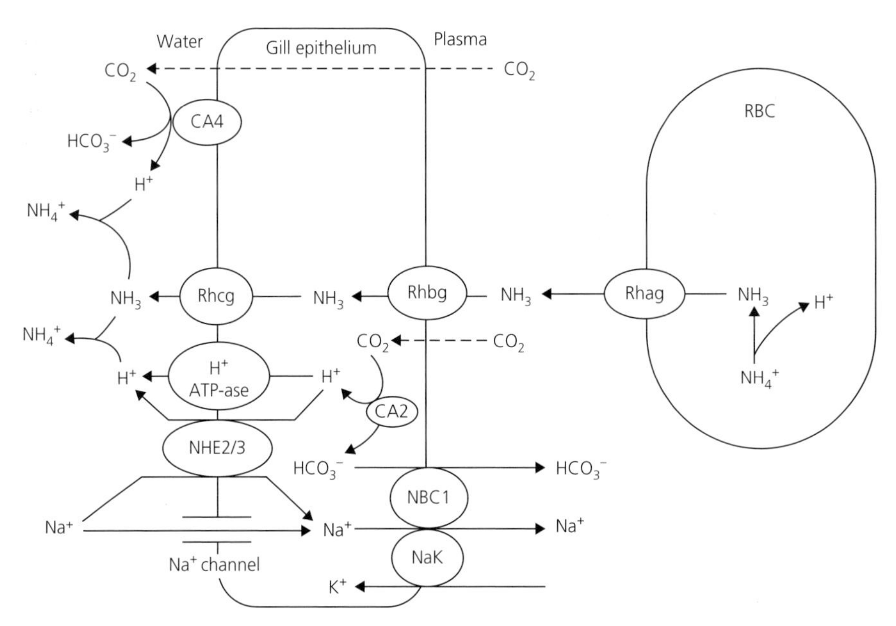

Fig. 6.3 Model illustrating how Rh glycoproteins – Rhag (in erythrocyte membranes), Rhbg (in basolateral membranes of gill epithelial cells), and Rhcg (in apical membranes of gill epithelial cells) – might facilitate ammonia excretion from blood to water in gills of freshwater teleosts. Other symbols: CA2 and CA4, carbonic anhydrase genes 2 and 4; NHE2/3, Na^+ and H^+ exchangers; RBC, red blood cell (erythrocyte). Source: Wright and Wood (2009: 2307 Fig. 2).

associated with the erythrocyte membrane (Fig. 6.3). In the illustration, NH_4^+ inside a freshwater fish erythrocyte is deprotonated, and Rhag then mediates diffusion of NH_3 across the cell membrane into the plasma. Two other glycoproteins (Rhbg and Rhcg) are thought to be involved in ammonia transport. Not discussed here is urea transport and the possibility that any of these glycoproteins might also facilitate CO_2 transport (Weihrauch *et al.* 2009).

In stenohaline freshwater fishes, and also euryhaline species that have adapted to freshwater habitats, the major pathway of ammonia excretion is by diffusion of NH_3 into the environment down its partial pressure gradient (PNH_3) originating inside red blood cells (Salama *et al.* 1999, Weihrauch *et al.* 2009). The gradient itself is defined by the unbalanced concentrations of ammonia inside and outside the fish and sustained ultimately by the pH at the gill boundary layer, the thin dynamic film where excretory gill cells meet the external environment (Salama *et al.* 1999). The acid dissociation constant (pK) of ammonia in water of ≈ 9.5 (Equation 6.4)

$$NH_4^+ + H_2O \leftrightarrow NH_3 + H_3O^+ \quad (6.4)$$

indicates that NH_4^+ is the predominant form at pH values near and slightly above neutral. In addition to pH, the shifting of Equation 6.4 is influenced to a lesser extent by temperature and ionic strength of the solution (e.g. Spotte and Adams 1983).

A current model of how a sufficiently steep PNH_3 gradient can be sustained assumes acidification of water at the water-gill boundary layer. The boundary layer must remain acidic enough to form NH_4^+ as NH_3 diffuses outward, thus serving as an instantaneous "proton trap" and perpetuating a low external PNH_3 relative to the intracellular PNH_3. This could be accomplished by (1) hydration of CO_2 to yield H^+ and HCO_3^-, or (2) active transport of H^+ via apical H^+-ATPase. The exact mechanism is unknown, although the first is potentially more important (Weihrauch *et al.* 2009, Wilkie 2002).

Any effectiveness of CO_2 hydration and acidification of the boundary layer is likely reduced in well-buffered freshwaters and even more so in brackish waters and seawater. This can be demonstrated empirically: adding small amounts of a buffer to poorly-buffered freshwater raises the pH and inhibits or stops ammonia excretion by any fishes living in it (Weihrauch *et al.* 2009). Adding ammonia to water in excess of the internal PNH_3 reverses the gradient, and ammonia enters the fish. The fish must then respond by permitting its internal PNH_3 to rise until it once again exceeds that of the water outside. Meanwhile, it must avoid neurotoxicity while re-establishing normal internal acid-base balance, now upset by the alkalosis resulting from excessive NH_3.

Ammonia excretion in seawater undoubtedly involves a combination of NH_3 and NH_4^+, most of the sum excreted passively (Weihrauch *et al.* 2009, Wilkie 2002). An increase in boundary layer pH should theoretically level the gradient and reduce ammonia excretion, and this has been shown empirically to happen (Salama *et al.* 1999). Although a PNH_3 diffusion gradient describes how the majority of ammonia is released into freshwaters, it falls short as an explanation when applied to waters of high buffer capacity and ionic strength. Acidification of the boundary layer is unlikely, even considering the involvement of CO_2 excretion and lack of apical proton pumps, which in seawater are rendered unnecessary by relentless sodium loading. The steep inward Na^+ gradient is leveled to the extent necessary for maintaining homeostasis by NHEs.

The "leaky" gills of seawater fishes allow a convenient means for paracellular transport of NH_4^+; alternatively, substitution of NH_4^+ by other monovalent transporters (e.g. Na^+ and K^+, Cl^-) are available, but probably render unlikely any link between CO_2 and ammonia excretion in seawater (Fig. 6.4). The tight junctions between ionocytes and their associated smaller accessory ionocytes loosen on transfer of a freshwater-acclimated fish to seawater.

Glycoproteins (Rhag, Rhbg, Rhcg1, Rhcg2) are expressed in gills of the Japanese puffer (*Takifugu rubripes*) and also must be considered, as demonstrated by Nakada *et al.* (2007). Pavement cells (PVCs) dominate the different cell types found in the gills of seawater fishes (Evans *et al.* 2005) and would seem the most important group involved in ammonia transport, with ionocytes in a comparatively smaller role (Weihrauch *et al.* 2009). Of the glycoproteins, Rhesus (Rh) protein Rhcg1 might have the least effect on ammonia excretion by seawater fishes.

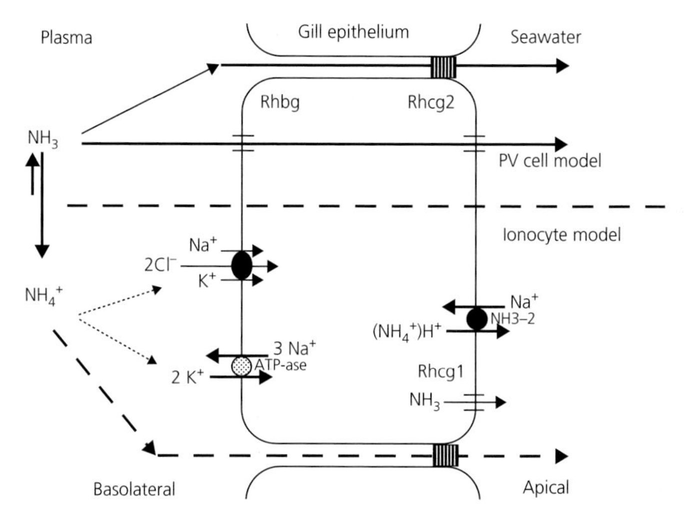

Fig. 6.4 Model of ammonia excretion by seawater teleosts, illustrating both probable transport mechanisms (active and passive) by transcellular and paracellular paths. Because seawater is well buffered, acidification of gill water is probably not involved in ammonia excretion. Epithelial pavement cells (PVCs) are the most likely sites of excretion. The glycoproteins Rhbg and Rhcg2 are likely restricted, respectively, to basolateral and apical membranes of PVCs, in which case NH_3 enters the cytosol (aqueous component of the cytoplasm) via a basolateral Rhbg and leaves by means of the apical Rhcg2. Evidence that NHE2 Na^+/H^+ exchanger 2) is expressed in the gills of many seawater fishes supports the hypothesis that apical Na^+/NH_4^+ exchange also contributes to branchial ammonia excretion. However, because NHE2 proteins are restricted mainly to ionocytes, which cover a small proportion of gill epithelium, their contribution to the excretion of total ammonia might be small. Ammonia can incidentally enter ionocytes by displacing K^+ on the branchial $Na^+/K^+/2Cl^-$ co-transporter or via Na^+/K^+-ATPase. These two mechanisms are not mutually exclusively. Thus apical Rhcg1, Na^+/NH_4^+ exchange, or both could serve as "relief valves" promoting removal of ammonia entering ionocytes via these basolateral transport systems. Source: Weihrauch *et al.* (2009: 1720 Fig. 2).

Atlantic tarpons routinely feed in lagoons near seabird colonies, where runoff from guano deposits charges the water with ammonia and higher oxidized forms of inorganic nitrogen (Vega-Cendejas and Hernández 2002). Nonetheless, we can imagine tarpons and other fishes trapped together in shallow isolated pools of deteriorating water quality. Degradation of organic matter produces ammonia, which then accumulates if the habitat is hypoxic or anoxic and nitrification (a largely aerobic process; Santoro 2016) is unable to proceed. The trapped fishes also release ammonia, and in small volumes of water without means of exchange the cumulative effect can be substantial. Ammonia can also collect from agricultural runoff and other sources of pollution during rainy periods. For the trapped fishes, if the concentration gradient outward becomes reversed blood ammonia rises, as just discussed, at which point every fish must let its internal ammonia drift upward while simultaneously attempting to maintain normal acid-base balance. *The risk of neurotoxicity from rising ammonia levels is therefore enhanced in well-buffered waters, because acidification of the water-gill boundary layer is unlikely.*

The habitat just described is probably hypoxic, perhaps dangerously close to anoxic, but tarpons can breathe air. For them, obtaining O_2 and excreting CO_2 is less serious than the potential neurotoxic effects of elevated ammonia in the environment, signs of which include hyperactivity, convulsions, coma, and death (Wilkie 2002 and references). Air-breathing, however, does not relieve nitrogen retention and a need to void ammonia, and an ABO might be of little use in this regard.

Although NH_3 is highly water-soluble, hydrating it in air by trapping a proton is more difficult than it is under the surface; unless, that is, tarpons possess a mechanism for detoxifying ammonia by converting it to less harmful urea as some other air-breathing fishes can do (Weihrauch *et al.* 2009 and references). A few species, like the giant mudskipper (*Periopthalmodon schlosseri*), an obligate air-breather, and the climbing perch (*Anabas testudineus*) can indeed excrete ammonia into air during periods of emergence, but we have no evidence of this capability by tarpons.

6.6 Euryhaline transition

The mechanisms just described apply at the molecular level to all teleosts, but considerable species variation occurs in their expression. For an aquatic organism to adapt to the changing ionic strength of its environment a set of preexisting mechanisms and the capacity to implement them must be in place (Mancera and McCormick 2000). The effects of extended osmo- and ionoregulatory interruption can be fatal by disturbing cell volume, in the process altering the density and composition of intracellular contents (Evans 2010b and references). Acute hyposmotic or hyperosmotic stress causes cells to swell or shrink, respectively, as they gain or lose water. Capacity and timing of the response are therefore essential to re-establishing homeostasis.

Euryhaline teleosts have evolved several fast-response mechanisms to minimize the delay. In some species ionocyte apical crypts close within 30 min of transfer from seawater to freshwater, shutting down ion secretion and passive loss of ions, and the process is reversible if the external ionic strength suddenly increases (Sakamoto *et al.* 2001 and references). The production and activation of ionocytes for regulation of osmotic effector proteins, including Na^+/K^+-ATPase, become necessary as ionic strength of the water rises. In many truly euryhaline teleosts tested so far, this occurs within 2–3 days following transfer from hyposmotic to hyperosmotic conditions (Mancera and McCormick 2000 and references), although in at least one, a killifish (the mummichog, *Fundulus heteroclitus*), a 70% increase in NKA activity detected within 3 h of transfer from freshwater to seawater fell to starting levels after 12 h and rose again after 3 d (Mancera and McCormick 2000). Direct transfer of mummichogs from seawater to freshwater stimulated a reduction in Cl^- excretion (both active and passive) and the blanketing of epithelial tissue by relatively impermeable pavement cells, all events occurring within 30 min (Marshall 2003).

Transfer of the euryhaline Mozambique tilapia from freshwater to brackish water of S_P 25 stimulates a rise in NKA activity within 1 h (Weng *et al.* 2002). Girling *et al.* (2003) performed an experiment to test adaptation to lower saline waters. Juvenile greenback flounders (*Rhombosolea tapirina*) acclimated to water of S_P 33 and transferred directly to dilutions of S_P 3, 7, and 16, responded with substantial drops in plasma osmolarity during the first four hours and stabilization within 24 h. Ionoregulation by specimens placed directly in freshwater failed.

Rapid acclimation to a rise in environmental salinity by mummichogs and other euryhaline species takes place in two steps (Mancera and McCormick 2000). During the first, gill permeability increases quickly to accommodate Na^+ and Cl^- secretion and a moderate increase in blood osmolality. The second step, which occurs several days later, is characterized by enhanced NKA activity, production of additional ionocytes (or greater activity of existing ones), net increases in Na^+ and Cl^- excretion, and restoration of plasma osmolality.

The initial stimulation in anadromous species (e.g. salmonids) might be delayed 7 d or so (Mancera and McCormick 2000 and references). The shift to waters of different ionic strength is often seasonal and migratory, involving life-cycle changes (e.g. smolting), and is strongly hormone-driven. Adaptation is ontogenetic and relatively slow, and my emphasis will not be on these fishes.

True euryhaline species like the mummichog, in contrast, can experience large fluctuations in ionic strength with every tidal change, in which case physiological adaptations are based on activating mechanisms already in place and poised to react, such as dynamic transport proteins and epithelial tissues. The striped mullet (*Mugil cephalus*), for example, shows an increase in gill Na^+/K^+-ATPase within 3 h of transfer from water of S_P 1 to S_P 45 (Hossler 1980). When transferred directly from seawater to freshwater the Mozambique tilapia displays a transient drop in plasma Na^+, Cl^-, and osmolality starting at 3 h, and NKA activity diminishes (Lin *et al.* 2004). Starting at 3 h post-transfer the numbers and sizes of certain ionocyte subtypes change, and most such modifications are finished by 24 h. In this species NKA expression and activity are proportional to the external environment's ionic strength (Lee *et al.* 2003, Lin *et al.* 2004 and references). Chloride-secreting ionocytes multiply when most species are transferred to hyperosmotic waters, the apical areas and accessory ionocytes intruding into the cells themselves and forming complexes (Sakamoto *et al.* 2001 and references).

A fish entering waters of ionic strength different from that to which it is adapted undergoes hyposmotic or hyperosmotic stress, and it copes with the new habitat by passing through two phases, the first acute and rapid (*critical adaptation*), the second slower to develop and ultimately sustained (*regulatory adaptation*). Both are accompanied by "remodeling" of the gills during which the cells are reconfigured to adapt (Kang *et al.* 2013 and references, Mancera and McCormick 2000). *In vitro* studies by Mancera and McCormick (2000) hinted that in mummichogs the first phase might actually precede enzymatic activity, being driven instead by the increased plasma osmolality that in turn induces quick activation of gill Na^+/K^+-ATPase. When transferred abruptly from freshwater to seawater, euryhaline teleosts undergo a rapid increase in ion permeability of the gills, increased Na^+ and Cl^- excretion, and an increase in plasma ion concentrations. On entering the regulatory adaptation phase a few days after transfer, Na^+/K^+-ATPase rises. Ionocytes already present show enhanced development or appearance of new cells. Net concentrations of Na^+ and Cl^- increase. Homeostasis of ions in the plasma recovers and drifts to original baseline levels.

The locations where ionocytes occur often correlate negatively with the water's ionic strength (Chen *et al.* 2004). Ionocytes of seawater-adapted fishes tend to be most numerous on the bases of gill filaments, the troughs of interlamellar spaces, and occasionally at the bases of the lamellae (e.g. Chen *et al.* 2004, Khodabandeh *et al.* 2009, Mir *et al.* 2011, Saadatfar and Shahsavani 2008, Sakamoto *et al.* 2001, Tano de la Hoz *et al.* 2014). Ionocytes directly on lamellae, in contrast, tend to be common in freshwater-adapted teleosts, usually disappearing when the external environment becomes hyperosmotic and indicating a role in ion uptake in dilute habitats (Sakamoto *et al.* 2001 and references). Ouattara *et al.* (2014) reported that ionocytes of freshwater-adapted black-chinned tilapias (*Sarotherodon melanotheron heudelotii*) occupied basal sites on the gill filaments when exposed to S_P values <35. At higher values (up to S_P 90) they were present on the lamellae too. Hybrid tilapias (*Oreochromis mossambicus* x *O. urolepis hornorum*) adapted previously to freshwater but transferred to saltier water of S_P 42 developed ionocytes at the bases of their gill filaments 7 d post-transfer (Sharaf *et al.* 2004). Eiras-Stofella and Fank-de-Carvalho (2002) found abundant ionocytes on gill filaments in wild-caught madamangos (*Cathorops spixii*), but fewer on the lamellae. Ionic strength of the water at the capture site was not mentioned.

Proliferation and sometimes numbers of ionocytes usually correlate positively with ionic strength (e.g. Khodabandeh *et al.* 2009, Sardella *et al.* 2004), but not always. Lee *et al.* (2003) saw no difference in densities of ionocytes in the gills of freshwater- and seawater-adapted Mozambique tilapias. Gills of tilapias in freshwater developed all three subtypes. However, one or another subtype dominated depending on ionic strength, and their morphologies were reversible within hours following transfer of the fish to waters of different ionic strengths. Seawater-adapted fish displayed only subtype III ionocytes (Lee *et al.* 2003, van der Heijden *et al.* 1997). Mozambique tilapias die during the critical phase of direct transfer from freshwater to seawater, probably of dehydration, but survive if first acclimated in brackish water for 24 h, and mortality is not an issue during reverse transfer from seawater to freshwater (Lee *et al.*, 2003 and reference). According to Stickney (1986) they can ultimately adjust to hypersaline waters of S_P 120. During acclimation, gill NKA activity increases to cope with ionic loading and dehydration.

6.7 Endocrine factors

The traditional view of endocrine control of osmo- and ionoregulation by teleosts posited that the corticosteroid hormone cortisol, secreted by the interrenal glands, drives seawater adaptation, particularly ionocyte differentiation and function and stimulating NKA activity, and that prolactin controls adaptation in the reverse direction (i.e. toward freshwater adaptation). This picture has now been expanded. Importance of the *growth hormone/insulin-like growth factor 1 (GH/IGF-1) axis* in ionoregulation has also been demonstrated, with *growth hormone* (GH) having a major influence on IGF-1 secretion and cortisol participating in ion uptake (Evans 2010b and references, McCormick 2001 and references, Sakamoto *et al.* 2001 and references). That IGF-1 increases NKA activity has been shown; it also causes ionocytes to multiply and raises tolerance to waters of higher ionic strength (Sakamoto *et al.* 2001 and references). Finally, IGF-1 dose-dependently modulates mRNA abundance in claudin isoforms (Bui *et al.* 2010). Cortisol works synergistically with both prolactin and IGF-1. In salmonids, injection of GH and cortisol increases NKA activity and salinity tolerance and in mummichogs enhances the capacity for ionoregulation in saline waters (McCormick 2001 and references, Sakamoto *et al.* 2001 and references).

Ionocytes and stem cells are the preferential location of cortisol receptors. Treatment of salmonids with GH raises their number, which correlates positively with NKA activity, an indication according to McCormick (2001: 785) "that the regulation of cortisol receptors is physiologically relevant." These receptors are stimulated by GH, increasing sensitivity of interrenal tissue to *adrenocorticotropin* (or *adrenocorticotropic hormone*, ACTH) in coho salmon.

Cortisol is also important in adaptation from high- to low-saline water, enabling ion uptake and osmoregulation by stimulating water movement across the gut in freshwater fishes (McCormick 2001 and references). Injecting specimens of the European eel (*Anguilla anguilla*), rainbow trout, Mozambique tilapia, and brown bullhead (*Ictalurus nebulosus*) increases the surface area of ionocytes and influx of both Na^+ and Cl^- (McCormick 2001 and references). Cortisol injection of gilthead seabream (*Sparus auratus*) increased their ionoregulatory capacity during exposure to low salinity (Mancera *et al.* 1994).

Prolactin undoubtedly promotes ion uptake and inhibits ion secretion in numerous and diverse species of teleosts (McCormick 2001 and references), and in the brown trout (*Salmo trutta*) counteracts activities of GH on salt secretion (Seidelin and Madsen 1997).

Mancera and McCormick (2000) suggested that at the molecular level these events are dependent on transcriptional and translational processes. Two osmotic transcription factors have been identified in Mozambique tilapias during acclimation from freshwater to seawater (Fiol and Kültz 2005, 2006): factor 1 (OSTF1) and the tilapia homolog of transcription factor 2B (TF2B). They might link cortisol with transcriptional regulation of ion transport in the teleost gill. Fiol and Kültz (2005: 927) called them "critical elements of osmosensory signal transduction in euryhaline teleosts that mediate osmotic adaptation by means of transcriptional regulation." Their flux is rapid and transient during hyperosmotic stress. Levels of mRNA rise sixfold for the first factor and fourfold for the second, peaking two hours following transfer to seawater. Protein levels increase 7.5 times for OSTF1 and 9 times for TF2B, peaking 4 h after transfer. Ka Fai Tse (2014) identified some of the same and different factors from other teleosts, for example the Japanese eel, medaka, and zebrafish (*Danio rerio*).

Cyclooxygenase type 2 (COX2) expressed in mammalian kidney cells produces prostaglandins that assist in hormone secretion and osmo- and ionoregulation to prevent dehydration and ion loading. In teleosts, COX2 expressed in branchial ionocytes perform similar functions and might aid in survival during osmo- and ionoregulatory stress caused by sudden encounters with waters of substantially different ionic strengths. The change in COX2 expression was tested by Choe *et al.* (2006) during experiments in which mummichogs acclimated long-term to freshwater or seawater were transferred abruptly to the other solution and monitored over 24 hours.

Specimens acclimated long-term were used to determine baseline levels of mRNA expression of COX2, CFTR, NKCC, and NKA. Results showed COX2 being expressed more strongly (3.1 times) in seawater-adapted than in freshwater-adapted mummichogs, as are CFTR (2.0 times) and NKCC (1.6 times) (Fig. 6.5), and that COX2 is regulated in parallel with these two critical transporters responsible for mediating Na^+ and Cl^- secretion. Direct transfer from freshwater to seawater induced a sizeable (3.4 times) and rapid increase in levels of COX2 mRNA, which returned to baseline by eight hours and remained there (Fig. 6.6a). Also occurring was a fast, large (2.8 times), and stable increase in CFTR mRNA expression that peaked at 8 h (Fig. 6.6b). The increase in NKCC1 mRNA was slow and moderate (1.9 times) (Fig. 6.6c). Levels of NKA1 mRNA levels had risen just slightly (1.4 times) by 24 h post-transfer (Fig. 6.6d).

Transfer in the opposite direction (seawater to freshwater) produced a 2.6 times increase in COX2 mRNA that fell to near-baseline by 8 h but continued to fall to 25% of pre-transfer values by 24 h (Fig. 6.6a).

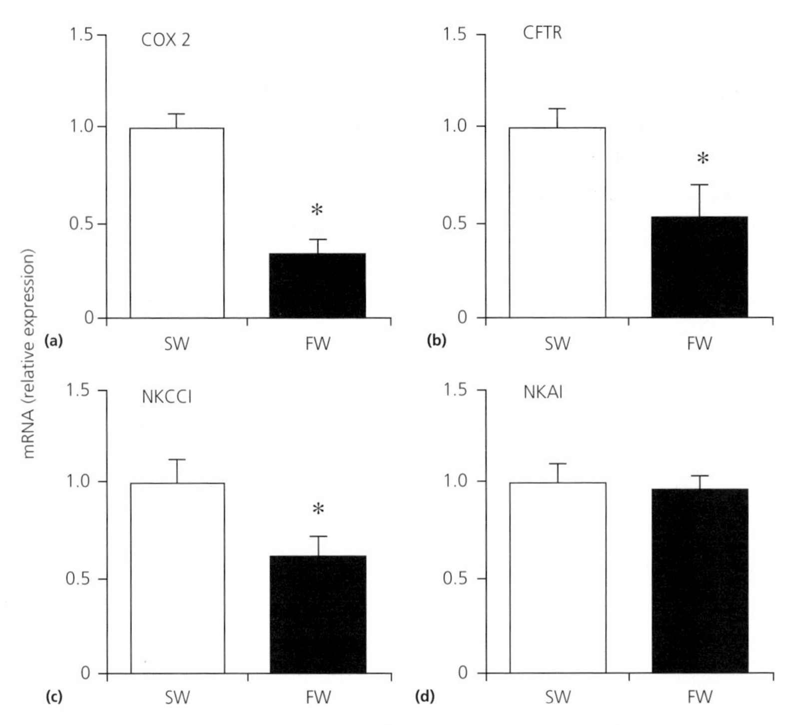

Fig. 6.5 Relative expression of (**a**) COX 2, (**b**) CFTR, (**c**) NKCC1, and (**d**) NKA1 mRNA, as measured by quantitative real-time PCR in gills of mummichogs (*Fundulus heteroclitus*) acclimated to seawater (SW, clear bars) and freshwater (FW, black bars). Values are $\overline{x}$ (± SEM); * = $p < 0.05$. Note that expression of NKA1 mRNA did not differ between the two solutions. Source: Choe *et al.* (2006: 1704 Fig. 5).

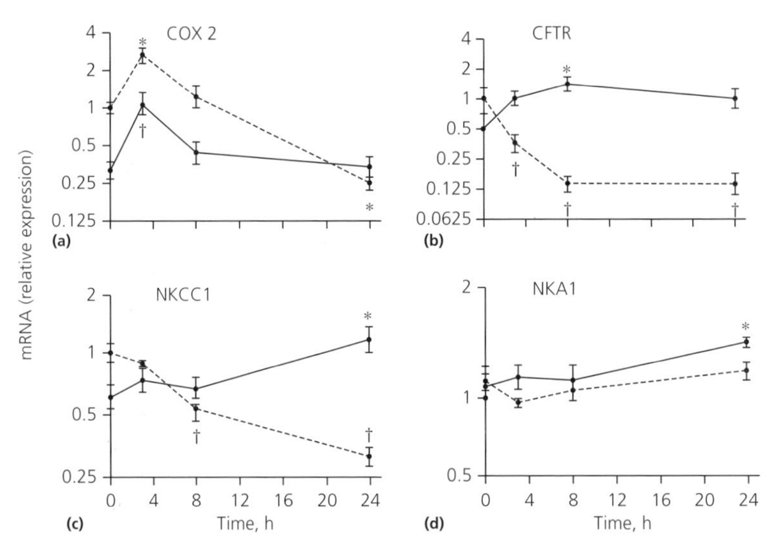

Fig. 6.6 Relative expression of (**a**) COX2, (**b**) CFTR, (**c**) NKCC1, and (**d**) NKA1 mRNA measured by quantitative real-time PCR in gills following acute transfer from freshwater (FW, dashed lines) to seawater (SW, solid lines), or the reverse. The different starting expression levels between FW and SW were standardized to the relative acclimation means of Fig. 6.5. Values are $\overline{x}$ (± SEM), $n = 5$ or 6, * = $p < 0.05$ for FW → SW transfer, † = $p < 0.05$ for SW → FW transfer. Source: Choe *et al.* (2006: 1704 Fig. 6).

Unlike the reverse transfer, this one resulted in a rapid, large, and stable decline in CFTR mRNA levels that reached 14.2% of pre-transfer levels by 8 h (Fig. 6.6b). Also occurring was a rapid and sizeable decrease in NKCC1 mRNA that reached <33% of pre-transfer levels by 24 h (Fig. 6.6c). There was no significant change in NKA1 mRNA (Fig. 6.6d). Thus COX2 expression enhances survival of branchial ionocytes in teleosts, the same function it serves in protection of analogous cells in mammalian kidneys. Its expression in fishes is regulated by the ionic strength of the external environment.

6.8 Eggs and larvae

The ovarian fluid and cytoplasm of fish eggs prior to spawning are nearly isosmotic with the female's plasma, which in turn is regulated according to the ionic strength of the surrounding water. The eggs of marine teleosts are therefore hyposmotic upon discharge, requiring osmo- and ionoregulation to begin immediately. It seems intuitive that a tiny tarpon egg with its high surface-to-volume ratio (calculated as a sphere) should suffer nearly instantaneous desiccation. In the words of Alderdice (1988: 200): "That is, given equal rates of water permeation across the plasma membrane, smaller eggs with their smaller volume would tend to come into a steady state with the external medium more rapidly than larger eggs." However, membrane permeability actually depends on rates of diffusion through both the internal cytoplasm and external medium (Hansson Mild and Løvtrup 1985). Nonetheless, how does an egg, its developing embryo, and eventually the newly hatched larva, maintain life-sustaining osmotic and ionic gradients while lacking the fully developed integuments, gills, esophagus, gut, and kidneys of adults?

Marine fish oocytes at spawning contain a supply of water, stored mainly in the yolk, to overcome loss by diffusion to the hyperosmotic external environment (Fyhn *et al.* 1999), a situation that becomes increasingly precarious with rising temperature (Alderdice 1988: 197–199). From the start, fish eggs entering the ocean confront the same problems of desiccation and salt loading endured by adult fishes, and most also regulate at ≥300 mOsm (Alderdice 1988). The only structure capable of osmo- and ionoregulation at fertilization is the plasma membrane, a two-layered lipid matrix interspersed by patches of protein (Alderdice 1988: 182).[3] Water transport occurs across the hydrophobic layer, ions through ion "channels" across the proteinaceous

[3] As Alderdice (1988: 203–204) pointed out, "The term 'vitelline membrane' also is used in the nonteleost literature for the outermost egg membrane." To avoid confusion Alderdice referred to it instead as the plasma membrane, a convention I follow here.

"islands" by passive or active transport (Alderdice (1988: 184). These processes are modulated by the plasma membrane's permeability to water and capacity to transfer ions.

Osmo- and ionoregulation begin at *epobily* (internal cell movement that commences at gastrulation) in the developing embryo. Before this, from expulsion into the sea to fertilization an egg functions in a passive regulatory state under control of the plasma membrane, a state Alderdice (1988: 216) called "resistive maintenance of the integrity of the egg proper." Permeability of the teleost egg's plasma membrane diminishes rapidly after fertilization, attaining its minimum at ≈24 h (Alderdice 1988: 238). This process, known as *water hardening*, helps maintain osmotic balance (Guggino 1980a, 1980b). The blastula with its blastoderm represents a slight tightening of water and ion flow, and both processes become notably more restricted with gastrulation and the appearance of ionocytes. Further control is gained as ionocytes develop on the yolk-sac membrane (Sucré *et al.* 2011).

During early development the integument is the primary site of both gas exchange and osmo- and ionoregulation. Prior to adult structures appearing to modulate these functions, water loss is prevented in part by the low permeability of the embryonic epithelia. The adult equivalent of drinking seawater is accomplished in the embryonic mummichog and its congener the Bermuda killifish (*Fundulus bermudae*) through external pores contiguous with the developing pharynx (Guggino 1980a, 1980b); drinking rate in embryonic teleosts is influenced by the action of cortisol (Lin *et al.* 2000).

Nothing is known about the osmo- and ionoregulation of the tarpon egg, embryo, and yolk-sac larva, but information from "conventional" forms offers insight. Molecular mechanisms driving these functions during ontogeny are no doubt the same, and only timing in the appearance and maturity of critical organs and structures is likely to differ.

The European seabass, a euryhaline teleost, might be a suitable proxy (Lorin-Nebel *et al.* 2006 and references). Many features of its life-history parallel the tarpon's. Adults occupy coastal seas and brackish regions, but tolerate waters from fresh to hypersaline (Lorin-Nebel *et al.* 2006 and references). Their eggs hatch in the sea, the larvae drifting and swimming to inshore areas then moving into estuaries and lagoons. The need and direction of osmo- and ionoregulatory processes shift with ontogenetic stage as tissues develop and organs form.

Lorin-Nebel *et al.* (2006) followed NKCC through early development of the European seabass, although the antibody used did not distinguish between secretory and absorptive isoforms (the presumed basolateral NKCC1 and apical NKCC2). They found NKCC to be immunolocalized in the embryonic and prelarval stages and localized in the apical area of digestive tract cells and basolaterally in the integumentary ionocytes. From the standpoint of the timing of

NKCC and gill development, a few opercular cells were evident at 1 d post-hatch (dph), some immunoreactive cells being visible along the branchial slits with stained opercular and branchial cells evenly distributed. At 6 dph, epithelium of the gill chamber revealed a few immunopositive cells at the junction with the operculum, and ionocytes were present on the budding branchial filaments.

In late embryos, NKCC had already appeared in the intestinal and integumentary ionocytes, those of the integument probably performing osmo- and ionoregulatory functions in pre-larvae prior to development of the gills, intestine, and kidney (Lorin-Nebel *et al.* 2006). Gill lamellae have not yet formed, and the principal function of the gills is probably ion exchange, not gas exchange, as also suggested for other species investigated (e.g. Li *et al.* 1995; Rombough 1999, 2007). Most teleost pre-larvae can osmoregulate, and this ability increases with later development (Varsamos *et al.* 2005 and references). Before the gills appear in early teleosts, most osmoregulation is conducted across the integument, which contains large numbers of ionocytes high in Na^+/K^+-ATPase activity (Varsamos *et al.* 2005 and references).

Gill filaments and lamellae in the European sea bass could be distinguished at 30 dph, but only the filaments had ionocytes, as shown by basolateral NKCC staining. The gill chamber and epithelium of the opercula contained immunoreactive cells that took stain similarly, as did sections of the gut and the ducts and dorsal section of the urinary bladder. Sucré *et al.* (2011) later discovered ionocytes in the gill-slit integument, which they presumed to be precursors of the eventual gill ionocytes.

Li *et al.* (1995) noticed that ionocytes develop on the gill filaments of Mozambique tilapia larvae before the lamellae appear. They proposed that ionoregulation and not gas transfer was the original evolutionary function of gills, citing the ontogenetic development of today's larval teleosts as evidence. Brauner and Rombough (2012) and Rombough (2007) concurred, further developing the hypothesis and noting that ionoregulation might also relate to acid-base balance or ammonia excretion. Others (e.g. Fu *et al.* 2014) continued to provide empirical confirmation. Anatomical evidence is reinforced by high Na^+-ATPase and K^+-ATPase activities, which are greater in ionocytes of larval fishes than adults and more developmentally advanced than the diffusing capacity of gas-transfer tissues (Li *et al.* 1995). Shih *et al.* (2008, 2013) showed that nitrogenous waste as ammonia is excreted by skin epithelial cells (ionocytes) of larval zebrafish through an acid-trapping mechanism involving H^+-ATPase and Rhcg1 with the yolk sac being the site of highest excretion (Fig. 6.7). A similar process occurs in fishes during later ontogenetic stages once the gills have formed (Chapter 1).

Whether tarpons and other primitive air-breathers fit within these hypothetical boundaries has not been established, but the possibility is strong considering their gills still function to regulate gases and ions and to control acid-base balance. Perhaps more relevant in an evolutionary sense is that all air-breathing fishes – even obligate air-breathers – spend some part of their early lives as water-breathers (Brauner and Rombough 2012).

Evidence for the early ionoregulation hypothesis can be stated briefly. Prior to appearance of the gills – that is, starting with the embryo and continuing through the yolk-sac stage into early larvae – gas exchange, osmoregulation, and ionoregulation are functions conducted through the skin. Ionoregulation, for example, first occurs via integumental ionocytes (Horng *et al.* 2009, Hwang *et al.* 1994), arising only later in gill tissues (Li *et al.* 1995, Schreiber and Specker 1999). Gas exchange in the absence of gills is still efficient during early developmental stages because embryos and larvae have large surface areas (*SA*s) relative to their mass and metabolic rate, this last parameter measured as the rate of aquatic oxygen uptake ($\dot{M}O_{2w}$). As Brauner and Rombough (2012) pointed out, body *SA* is roughly proportional to body length squared (L^2), $\dot{M}O_{2w}$ to body volume (L^3). With growth, values of the $SA/\dot{M}O_{2w}$ ratio decline in approximate proportion with body weight (W) to

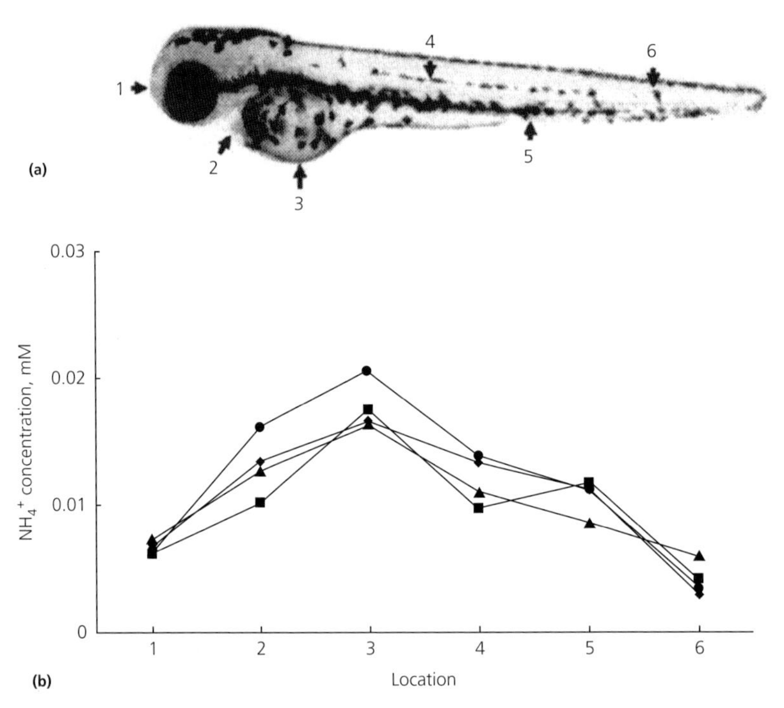

Fig. 6.7 Change in total ammonia (as NH_4^+) excreted along a concentration gradient (ΔNH_4^+) at the skin of a zebrafish larva: (**a**) The six locations where ammonia was measured with an ion-selective electrode: 1 snout, 2 pericardial cavity, 3 yolk sac, 4 trunk, 5 cloaca, 6 tail. (**b**) Mean ΔNH_4^+ concentration at the six locations (n = 4 larvae). Source: Shih *et al.* (2008: C1626 Fig. 1).

the two-thirds power ($W^{0.67}$). In addition, diffusion distance increases as the skin thickens, lessening gas transfer effectiveness. At this time the developing gills, increasingly convoluted and branched and thus not restricted to the surface area limits of smooth skin, can continue gaining *SA* at exponential values exceeding $W^{0.67}$.

CHAPTER 7
Ecology

7.1 Introduction

You would think that a fish as physically spectacular as the Atlantic tarpon would have been an object of intense ecological study, but like most aspects of its biology this too has been largely ignored. That neither species is the focus of a sustained commercial fishery anywhere in the world probably accounts for the general disinterest. We humans are most likely to fund research on species we prefer to eat, and tarpons are largely excluded from that category (Chapter 8.5).

Still, I was surprised at the overall scarcity of ecological information when beginning the brief essay comprising these chapters. That initial response faded to disappointment upon discovering how little information is actually there. The Pacific tarpon typically appears as a name in regional species lists, often without further comment. Where either species spawns is still to be determined. Most of what we know of tarpon ecology applies to post-metamorphic larvae and juveniles, no doubt because shallow inshore sites are easier and cheaper to study than offshore blue waters. These reports tell us three things: tarpons (1) are carnivorous; (2) have broadly varied diets, prey-switching as they grow; and (3) are seemingly casual about habitat choices. Tagging efforts are revealing a little about adult Atlantic tarpons in the western North Atlantic and Caribbean (Section 7.6), although comparable work in the eastern North Atlantic and South Atlantic has yet to be done. Seasonal movements of Pluto are more fully described and mapped than those of adult Pacific tarpons.

7.2 Habitats

Atlantic and Pacific tarpons can live indefinitely in freshwaters (e.g. Alikunhi and Rao 1951, Gunter 1942, Howells and Garrett 1992, Raj 1916: 252, Thomas 1887). Atlantic tarpons routinely occupy inland locations many miles from the sea, such as Lake Nicaragua and Panamá's Río Chagres. Breder (1925a: 140)

Tarpons: Biology, Ecology, Fisheries, First Edition. Stephen Spotte.

reported both small and large Atlantic tarpons underneath the Gatun spillway in Panamá, 10 km up the Río Chagres from where it meets the Caribbean Sea. They ranged from 254–1880 mm TL (to 38.5 kg) and "were seen constantly leaping near the foot of the dam." The habitat was presumably freshwater (Breder 1925a: 139). Chacón Chaverri (1993, Appendix D) reported adult tarpons >70 km upstream in the Río Colorado of Costa Rica, noting that their sizes did not differ significantly from coastal specimens. Babcock (1936: 19–20) listed several freshwater locations in the Western Hemisphere where tarpons could be found, and Breder (1944: 219) reported having seen them in the Río Tampaon, México, >160 km inland. Breder also mentioned several small specimens living in a freshwater aquarium at the Bingham Oceanographic Laboratory of Yale University (New Haven, Connecticut). At the time of his writing they had been there 6 y without evidence of physiological stress.

Many adult Atlantic tarpons might be permanent residents of Lake Nicaragua, 190 km up the Río San Juan from the Caribbean. After examining an otolith of one specimen, Brown and Severin (2008: 271) wrote that it had, "appeared to enter freshwater after the first year of life and remained there from that point on." The fish measured 1570 mm FL and was clearly of adult length. Another large specimen from Lake Nicaragua and one from the Río San Juan "showed similar patterns of marine life early and what appeared to be exclusively freshwater life later...." The otolith of still another, this one caught in the Gulf of Mexico, indicated the first year had passed in freshwater, subsequent years in saline waters.

Fishermen told Gunter (1938) of having gigged tarpons and sharks in the Black River near Jonesville, Louisiana >435 km inland from the Gulf of Mexico, and Haldeman (1892: 112) stated, "During September, 1879, I saw large numbers of Silver Fish eight or ten miles up the Apalachicola River [of Florida], and am told that that was not an unusual occurrence." He wrote of tarpons ascending Florida's Homosassa River and several Texas rivers too. Atlantic tarpons have occasionally been released into freshwater reservoirs for recreational fishing in the US (Howells and Garrett 1992), and more recently Thailand.[1]

Juvenile and adult tarpons can also survive indefinitely in saline waters of variable ionic strength, often able to make abrupt changes into freshwater and back again without permanently harmful effects. Breder (1944: 234) summed it by stating, "Small tarpon can easily withstand a transfer from straight fresh water to sea water of more than normal density or vice versa." These direct transfers ordinarily occur without mortality (Randall and Moffett 1958). Nonetheless, transfer from seawater to freshwater might be more stressful than the reverse procedure, at least temporarily. Moffett and Randall (1957: 6) moved two seawater-acclimated Atlantic tarpons (80 mm FL) directly to aged tap water:

[1] http://www.palmtreelagoon.co.uk/fish_species_–_tarpon.html; http://www.fishsiam.com/species/atlantic-tarpon-87.html. Downloaded 05 July 2015.

"In about one minute the fish displayed obvious distress; one lay on its side on the bottom of the ... container and the other was upside down. After several minutes, however, they resumed normal orientation and activity. Twenty-four hours later they were returned to sea water at which time they did not exhibit any signs of discomfort."

Roberts (1978) reported catching Pacific tarpons in the Palmer River, a tributary of Papua New Guinea's Fly River, 905 km inland. Coates (1987: 530) recorded them 530 km upstream from the mouth of New Guinea's Sepik River, while noting that "areas above this have not been sampled to any significant extent." In a survey of ichthyofauna of the Fitzroy River and its tributaries in Western Australia, Morgan *et al.* (2004) caught Pacific tarpons far inland. The distance was not stated, but from their map some of the locations appear to be 300–400 km from King Sound, the nearest estuarine habitat.

Kulkarni (1992) recaptured adult Pacific tarpons released as juveniles in two Indian freshwater lakes (Valvan and Shirota) in July 1939. The presence of healthy Pacific tarpons after 53 years is surely proof they can survive indefinitely in freshwaters.

The earliest mention I found of Atlantic tarpon habitats is Dampier (1906: 118), writing about the Bay of Campeche on México's east coast. He stated: "These Fish are found plentifully all along that shore from Cape Catoch to Trist, especially in Clear Water, near sandy Bays, but no where in muddy or rocky Ground." Dampier's observations, however, were limited to locations the indigenous people found convenient to establish their fisheries, and he was evidently writing about adult tarpons. Furthermore, discussing habitat "preference" borders on unrealistic in the tarpon's case.

Killam *et al.* (1992: ix) stated that Atlantic tarpons occupy lagoons, ditches, canals, tidal tributaries, and open coastal waters; Zale and Merrifield (1989: 6) added salt marsh and mangrove ponds, tidal creeks, beaches, rivers, and mosquito-control impoundments. Lakes, creeks and streams, estuaries, floodplains, mangrove swamps, sloughs, mudflats, sandflats, saltpans, reservoirs, sawgrass marshes, coral and oyster reefs, bayous, ponds, potholes, and karst sinkholes (called *cenotes* in México and Central America) could be added to these lists and they would still be incomplete. In other words, Atlantic tarpons can survive in practically any nearshore or inshore aquatic habitat within their natural range where temperature and food are not limiting.

The same applies to Pacific tarpons (e.g. Bishop *et al.* 2001; Ley 2008: 5). Losse (1968: 81) reviewed the earlier literature and concluded that Pacific tarpons mainly inhabit "estuaries, lagoons, bays and mangrove areas, often together with *Elops*. The species was rarely found out to sea and apparently shows a preference for euryhaline conditions." Adults are common inhabitants of lagoons that form on Australian floodplains (Perna *et al.* 2011). Munro (1967: 41) limited the habitat to "Harbours and rivers ...." When not confined to a restricted body of water, tarpons everywhere are typically coastal, but also occur offshore (Hammerschlag *et al.* 2012) and far inland (e.g. Breder 1925a, 1944: 218; Coates 1987; Ferreira de Menezes and Pinto Paiva 1966; Kowarsky and Ross 1981; Roberts 1978).

Larval (Stage 1) Atlantic tarpon leptocephali are exceptional, occurring only in offshore pelagic waters (Killam *et al.* 1992: 3–5). According to Rickards (1968: 223), habitats into which larval tarpons move when migrating inshore are "generally shallow, brackish pools of dark-colored water connected to the sea only during periods of extreme high water." Atlantic tarpon larvae are not particular about environmental conditions including salinity, water clarity or composition, pH, or temperature, having been collected in coastal habitats ranging from turbid to clear and varying widely in temperature, pH, and salinity (Erdman 1960, Evermann and Marsh 1902: 80, Harrington 1958, Harrington and Harrington 1961, Moffett and Randall 1957, Simpson 1954, Zale and Merrifield 1989, Zerbi *et al.* 2005). Often these waters are darkly pigmented (e.g. Breder 1933, Rickards 1968, Wade 1962), I presume with tannic and fulvic acids and other humic compounds, and ionic strength shifts depending on tides and rainfall (Storey and Perry 1933). Pacific tarpons over a large size range also have been caught in freshwaters both landlocked and flowing, muddy and clear. Bishop *et al.* (2001: 23) caught juvenile Pacific tarpons in the Alligator Rivers area, Northern Territory, Australia, mainly over bottoms of mud and clay, and sand to a lesser extent, and in areas with submerged vegetation.

That fishes occupy locations preferentially is indisputable; otherwise, their distributions would be random and unpredictable. In assessing abundance and habitat preferences of juvenile Atlantic tarpons, Clark and Landry (2002) sampled >100 sites along the Texas coast from Galveston Bay to Corpus Christi Bay. Ten of these were monitored monthly to evaluate recruitment, retention, and growth. The catch totaled 92 specimens of 17–206 mm SL including eight recaptures. Juveniles were most numerous in August and present in minimal numbers from September through November. Nearly all were caught in shallow marshes having little tidal exchange. The ranges of three variables: S_P 21–35, dissolved oxygen 1–9 mg/L, and temperature 16–31°C.

Killam *et al.* (1992: ix) wrote that juvenile Atlantic tarpons "prefer backwater mangrove and marsh areas with soft muddy bottom sediments and little or no submerged vegetation." This statement and others like it are probably unfalsifiable; too many reports of "preference" are conflicting. Randall and Moffett (1957: 145) found juveniles in pools in the Florida Keys where "the bottom was so rocky, or contained so much brush, that a seine could not be used ...." Mercado Silgado (2002: 97) noted that collection sites in Colombia where he captured early metamorphic tarpons "were often characterized by high levels of sulfates, low pH, and near anaerobic water." Mercado Silgado (1971: 12) caught leptocephali "*en una area pequeña, de aguas turbias, salobres y Ilanas de fondo fangoso.*" [in a small area of cloudy, brackish, and calm water over a muddy bottom.]

Ferreira de Menezes and Pinto Paiva (1966: 89) claimed that juveniles in northeastern Brazil preferred shallow lagoons with "lots of aquatic grasses." Erdman (1960: 146) caught 11 Atlantic tarpon larvae of 24–26 mm SL in surface waters of the Río Añasco, Puerto Rico, "in muddy freshwater near a small stand of cat-tails." Simpson (1954: 72) caught two Atlantic tarpons of 37 mm SL (4.8 mm TL) and 7.7 mm SL (9.9 mm TL) in a slough in the Aransas National Wildlife Refuge: "The collecting station had a sandy-mud bottom with no vegetation in the water, although there were clumps of reeds along the shore." Water depth was ≈ 60 cm. Zerbi *et al.* (2005: 488) captured juveniles at Boquerón, Puerto Rico over bottoms that were "extremely soft, black, anoxic mud with strong [*sic*] hydrogen sulfide smell and rotting vegetation." Other nearby locations had hard mud bottoms and clearer water.

Water flow does not appear to affect survival, juveniles of both species having been found in both still and moving waters. Obviously, neither does bottom type, nor the type or density of emergent and submerged vegetation. Habitats where Pacific tarpons occur are just as varied, the species having been reported from waters both clear and muddy, fresh and salty, with pH values ranging from 5.2–9.1 and temperatures of 23–34°C (Merrick and Schmida 1984: 53 and references). In conclusion, tarpons seem largely indifferent to their living conditions.

As hinted earlier, sometimes the habitats an individual fish has occupied over time can be inferred indirectly. Many fishes deposit new layers daily around each otolith, their sum constituting a measure of age. Daily deposition has been validated in both species of tarpons (Chen and Tzeng 2006; Crabtree *et al.* 1992; Cyr 1991: 10, 13; Tsukamoto and Okiyama 1993). Age determination is done by sectioning an otolith through its core and counting the layers under a microscope. The inorganic elements, having been derived from the external environment, leave chemical traces trapped within the matrix, providing information about the composition of the water in which the fish has been living.

Strontium (Sr^{2+}) and calcium (Ca^{2+}) are divalent cations with similar ionic radii, meaning that strontium can be incorporated into an aragonite otolith as a substitute for calcium. Because the concentration of strontium in seawater is ≈ 100 times greater than in most freshwaters ($\approx 8.7 \times 10^{-5}$ *M* vs. $\approx 9 \times 10^{-7}$ *M*) the Sr^{2+}/Ca^{2+} ratio in a tarpon otolith (examined using an electron probe microanalyzer) reveals not only the fish's age but the nature of its habitat (Campana 1999), water temperature having exerted no discernible effect (Chen *et al.* 2008).

Tarpons move effortlessly through locations ranging from fresh to hypersaline, and their otoliths are therefore records of both time and place (Chen *et al.* 2008, Shen *et al.* 2009). More recently, Woodcock and Walther (2014) showed that chemical signatures from scale edges of Atlantic tarpons caught in Florida and Puerto Rico are consistent with movement across salinity gradients. Specifically, $\delta^{13}C$ values related positively to Sr^{2+}/Ca^{2+} ratios and revealed patterns of movement between freshwaters and saline waters. These findings mean that fishes do not need to be killed to gain such information, which is necessary to obtain otoliths. A scale suffices instead.

The bimodal distribution of Sr^{2+}/Ca^{2+} otolith patterns in some Lake Nicaragua tarpons reported by Brown and Severin (2008) infers considerable time spent in freshwaters and coastal seas, but comparatively little in brackish areas. Other specimens examined spent substantial time in brackish waters, possibly moving rapidly between freshwaters and ocean waters. Still another otolith, this from a juvenile, hinted at a first year lived in a hypersaline habitat. Pacific tarpons of 1–3 y in western Taiwan also showed a range of habitat use based on otoliths (Shen *et al.* 2009). Some lived mostly in freshwaters, others in brackish areas, and still others shifted back and forth. Juveniles of 4–5 y returned to the sea.

Based on extensive sampling, Cyrus and Blaber (1992) recorded 128 species of teleosts in Embley Estuary of Australia's Northern Territory and evaluated their distributions based on turbidity and ionic strength. Variation in both factors depended on time of year and correlated negatively. During the wet season, heavy rainfall and runoff flush stream and river waters into the estuary raising its turbidity and lowering its ionic strength. Pacific tarpons occupy the length of the estuary during the wet season and start of the dry season, after which they presumably move into Albatross Bay. The estuarine water turns clearer and saltier as the dry season progresses, but concluding that tarpons by their absence find such conditions distasteful would be conjecture based entirely on association. The authors did not suggest this, including the Pacific tarpon in the category of "seasonal marine species."

These and similar reports from around the world are evidence of the high tolerance and nonspecific requirements of tarpons, but their purely descriptive nature renders such observations of either species less than instructive. Viewed from a wide-angle perspective, they merely illustrate that easily identified and measured environmental factors have no discernible effect on the fitness of tarpons.

Tarpons are hardly alone among fishes capable of adapting to hypoxic conditions, a trait common in teleosts indigenous to tropical freshwaters (Chapter 5.7). Not all possess ABOs. Junk (1984: 225) reported that of 120 species from a small *várzea* lake near Manaus, a Brazilian city in central Amazonas at the confluence of the Rio Negro and Rio Solimões, 40 occurred routinely "under pronounced hypoxic conditions." Of these, 10 could take up atmospheric oxygen and 10 more "were able to increase the [size of the] lower lip to utilize the oxygen rich surface layer of water or had other adaptations already known." Of the remaining 20 species, Junk knew of no specific adaptations, "but we may presume that some exist." Chapman *et al.* (2002) found a similar high tolerance of hypoxia by several water-breathers in Lake Nabugabo, Uganda, a satellite of Lake Victoria, and in nearby Lake Kayanja, enabling them to delay ASB.

Poor water quality other than depleted oxygen is not always an issue either, hydrogen sulfide having been reported from different habitats where metamorphosing and juvenile tarpons occur (e.g. Beebe 1927, 1928: 70; Beebe and Tee-Van 1928; Breder 1933; Chacón Chaverri 1994; Chacón Chaverri and McLarney 1992; Cyr 1991: 34; Franks 1970: 35; Mercado Silgado 2002; Randall and Moffett 1958; Rickards 1968; Zerbi *et al.* 2005), indicating anoxia from anaerobic activity (Dunnette *et al.* 1985). Such conditions usually persist in stagnant waters or result from weak tidal exchange or fluvial flow. Sulfide in anoxic sediments originates from many sources, including putrefaction and sulfate reduction resulting from the combined activities of proteolytic and sulfate-reducing bacteria (Dunnette *et al.* 1985).

Geiger *et al.* (2000) exposed juvenile Atlantic tarpons to total sulfide concentrations of 0–232.1 μ*M* (0–150 μ*M* H_2S) in water of S_P 22 in stepped increases over the pH range 6.8–7.8. The water was aerated and presumably maintained at normoxia. Values of V_R showed negative correlation with total sulfide concentration, opercular activity nearly stopping at the highest levels (Fig. 7.1). Some coughing was seen, although no fish died. At sulfide levels > 230 μ*M*, f_{ab} reached 60 breaths/h, higher than for juveniles in water made nearly anoxic and containing no sulfide. The correlation between V_R and pH over the range 4.25–8.75 was positive but weak. Ventilation rates were 20–30 breaths/min at pH levels of 4–5. In more basic water (pH 7–8) V_R rose but was much more variable (22–54 breaths/min).

High saline conditions are not uncommon in the tropics, especially during times of drought when the volume of evaporated water exceeds the volume entering from freshwater streams and rivers, or where the outflow of a river is restricted and estuarine water and seawater do not mix freely (Packard 1974). The Casamance River in Sénégal terminates in a lengthy estuary. Because of the

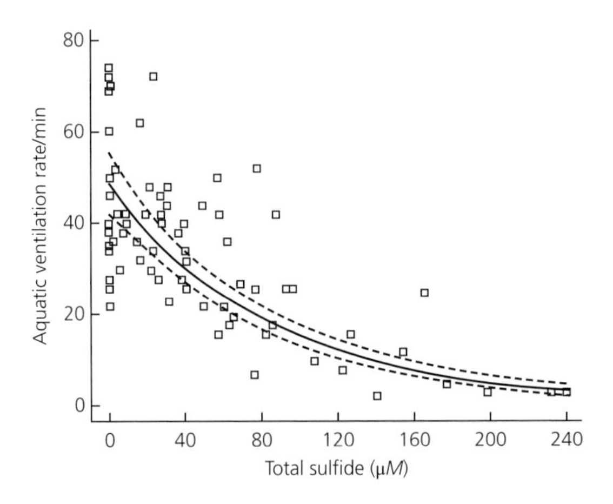

Fig. 7.1 Association between aquatic ventilation rate (V_R) as opercular movements/min and total sulfide concentration (μ*M*) at 25°C (presumably pooled data for pH values 6.8–7.8). The curve is described by the exponential regression equation $V_R = e^{3.877-0.011x}$, in which x = μ*M* H_2S (n = 73, r^2 = 0.677, CI = 95%, $p < 0.001$). Source: Geiger *et al.* (2000: 186 Fig. 2).

intensely hot climate, low rainfall, and high evaporation rate, ionic strength in some years actually increases inland from the mouth, and the water turns hypersaline with distance from the sea, instead of becoming more dilute as might be expected. Between March 1984 and April 1985, S_P values of 90 (2.6 times saltier than seawater) were recorded at Mankono, >200 km inland, 80 at Sedhiou, ≈25 km farther downriver, and ≈54 at Ziguinchor, approximately 50 km from the river's mouth (Albaret 1987). The Casamance River contains a rich fish fauna: 75 species of 39 families, among which is the Atlantic tarpon (Albaret 1987: 309). However, the number of tarpons caught, their ontogenetic stages, and practical salinity determinations at capture sites along the estuary evidently were not recorded.

Pacific tarpons inhabit waters of variable ionic strength after metamorphosis, easily shifting back and forth between brackish and freshwaters. In western Taiwan the general pattern was for the smallest to prefer freshwaters, then moving downstream into increasingly saltier waters with age, until at 4+ y, none remained in freshwater and most had entered the sea. However, saltpans and tidal pools along the Norman River estuary in Queensland are 2.7 times saltier than seawater, and young Pacific tarpons (22 mm FL, 25 mm TL) have been collected at these places in December (Russell and Garrett 1983). Shen *et al.* (2009) caught 19 specimens offshore adjacent to the mouth of Tadu Creek with otolith ages of 2–5 y. All were still juveniles (i.e. none had mature gonads).

7.3 Predators of tarpons

Adult Atlantic tarpons have few natural predators. American alligators (*Alligator mississippiensis*) have been mentioned (Wade 1962), and Babcock (1936: 64–65) reported porpoises (presumably Atlantic bottlenose dolphins, *Tursiops truncatus*) feeding on 15 lb (≈6.7 kg) tarpons near a boat basin in Tampa Bay, Florida, but sharks take the greatest toll (Fig. 7.2).

Clark and von Schmidt (1965) retrieved tarpon parts from stomachs of bull sharks (*Carcharhinus leucas*) and great hammerheads (*Sphyrna mokarran*). According to Nelson *et al.* (1992: 84–94), bull sharks and Atlantic tarpons are sympatric in estuaries throughout the Gulf of Mexico. If one species is abundant the other usually is too, and in areas where bull sharks are rare (e.g. Baffin Bay, Texas), so are tarpons. These findings in some respects contradict those of Hammerschlag *et al.* (2012), who found little overlap in habitat use between bull sharks and Atlantic tarpons in southern Florida, including the southeastern Gulf of Mexico, despite similarities in diet. Bull shark numbers were high all year but peaked in winter; tarpons were most numerous in late spring, based on catches in the recreational fishery. Tarpons sometimes gathered at upriver locations where bull sharks were absent. When traversing deep open waters where both

Fig. 7.2 Atlantic tarpon killed by sharks off Coquina Beach, a popular tourist destination at Bradenton Beach, southwestern Florida. The attack occurred about 10:30 am on 2 July 2015, and the remains soon washed ashore. Source: Photograph courtesy of Ryan and Jill Castline and *The Islander*, Anna Maria Island, Florida. Reproduced with permission.

sharks and food were abundant, tarpons moved rapidly in straight trajectories, behavior inconsistent with foraging. On reaching shallow waters containing complex physical structures their movements became slow and meandering, indicative of foraging.

Juvenile tarpons in bays and estuaries are often large enough to survive predation by other fishes. According to Rickards (1968) they do this by rapidly outgrowing juveniles of sympatric carnivores like the ladyfish and spotted weakfish (*Cynoscion nebulosus*), although not necessarily members of their own species; Moffett and Randall (1957) captured a Florida Keys tarpon of 189 mm FL with a smaller juvenile measuring 70 mm FL in its stomach. Rickards (1968: 226), who based his findings on studies of tarpons in Georgia salt marshes, wrote that "consumption by other fish would seem to account for only a small loss from the [juvenile] tarpon population." He added that competition for food by these other species could have affected tarpon survival. A night heron (the bare-throated tiger heron, *Tigrisoma mexicanum*)[2] has been seen preying on juvenile Atlantic tarpons in Costa Rica, along with other herons (*garzas*) of unstated species (Chacón Chaverri 1994).

[2] Common name *martín peña* in New World Spanish.

Farmer (1997) speculated that the rise of aerial predators (pterosaurs and birds) perhaps stimulated the evolution of swim bladders and loss of lungs in air-breathing fishes. She noted how members of certain species seem wary when surfacing, and that many are nocturnal. To Domenici *et al.* (2007), air-breathing increases exposure to aerial predators, resulting in some fishes surfacing more often in sheltered locations. The risk appeared to be offsetting because hypoxia deadens the escape response, and not to surface also has detrimental consequences. Such wariness has an empirical basis and is apparent in obligate water-breathing fishes too.

That estuaries and lagoons serve as refuges from predation is therefore false if accepted as a universal statement because any safety they offer is relative. Piscivorous birds were thought to be the principal predators of juvenile Atlantic tarpons at Sapelo Island, Georgia (Rickards 1968) and in estuaries and marshes along the upper Texas coast (Clark and Landry 2002). Beebe (1927) suggested the same was true in an isolated lagoon he seined in coastal Haiti, and Zerbi *et al.* (2005) speculated that bird predation influenced habitat shifts by juvenile tarpons in Puerto Rican lagoons. Babcock (1951: 49) speculated that young Atlantic tarpons isolated in vernal ponds along the Florida coasts would be quickly devoured by fish-eating birds as the waters evaporated. In a study conducted at West Kleinemond Estuary in South Africa, up to 80% of Cape stumpnoses (*Rhabdosargus holubi*) were eaten by wading and diving birds (Blaber 1973).

Tarpons of both species tend to avoid floodplains, preferring main channels and lakes where the water is deeper (Bishop *et al.* 2001, Breder 1925a: 189, Coates 1987, Davis *et al.* 2014). Breder (1925a: 189) reported that when the level of the Río Chagras below the Gatun Dam spillway was lowered, suddenly the tarpon "at once rush to mid-stream" where the water flows at maximum speed, evidently to keep from becoming trapped in the shallows. These fish were of adult size and doubtfully afraid of piscivorous birds. Harrington and Harrington (1982) wrote that Atlantic tarpons living at locations subjected to seasonal drying emigrate to more stable habitats, ostensibly to keep from becoming confined in the ever-shrinking ponds and marshes. Davis *et al.* (2014) reported that tarpons in the Ross River estuary chose deeper tidal pools (>85 cm deep) and kept away from shallow pools without speculating on a reason, which again could simply be to prevent becoming trapped.

Even the threat of aerial predators imposes a physiological cost on vulnerable fishes. Captive specimens of the largemouth bass (*Micropterus salmoides*) (200–425 mm TL), exposed for 30 s to simulated predation using models of a great blue heron (*Ardea herodias*) and an osprey (*Pandion haliateus*), demonstrated bradycardia, its magnitude correlated negatively with size of the fish (Cooke *et al.* 2003). Recovery times to homeostasis in fish of all sizes were ≈40 min for $\dot{Q}$ and f_H and ≈30 min for V_S (Cooke *et al.* 2003). Smith and Kramer (1986) used a stuffed and mounted great blue heron as a simulated predator to test responses of captive Florida gars (*Lepisosteus platyrhincus*) subjected to hypoxia (6.24 kPa). Intervals between breaths of air rose 118% in the heron's presence, while intervals between water breaths (presumably measured as pumps of the opercula per unit time) fell 13% (i.e. air-breathing frequency, f_{ab}, declined; aquatic breathing rate, V_R, increased).

7.4 Environmental factors affecting survival

Fishes are sensitive to numerous environmental factors that affect physiological processes (e.g. changes in pH, ionic strength of the water, temperature, the presence of toxicants) or influence feeding and prey capture. Turbulence, for example, affects the feeding of fish larvae negatively by making prey capture more difficult (Huebert and Peck 2014). Depending on their state of development, fishes can often select preferred habitats and avoid others when suitable choices are available (Serrano *et al.* 2010). Gray snappers (*Lutjanus griseus*) occupying mangrove habitats in southeastern Florida's Biscayne Bay routinely experience S_P values in the range 0.4–40.3 and temperatures of 12.3–35.6°C along a narrow depth range of 17–145 cm (Serrano *et al.* 2010). Given choices in the laboratory, they preferred salinities within the range 9–23.

This capacity to choose suggests that during tidal changes euryhaline species might linger in water having an ionic strength to which they are already adapted, perhaps making forays across the boundary into more stressful regions but exiting before major anatomical changes become necessary (Chapter 6.6). Whether fishes actually make behavioral decisions involving such trade-offs is unknown, although doing so would obviate any need of initiating long-term physiological and anatomical adaptations.

Tarpons, although generally inured to bad water quality, are especially susceptible to cold. Atlantic tarpons of all ages seem more tolerant of higher than lower water temperatures. Rickards (1968) reported that juveniles at Sapelo Island, Georgia tolerated water of 36.0°C. The lowest temperature at which they were captured was 16°C. Juveniles confronted by water temperatures consistently below optimum developed more slowly, if they survive at all. Cyr (1991: 37), for example, reported that when a cold front moved over Jack Island in southeastern Florida and air temperature dropped to –4°C, apparently no first-year tarpons survived.

Snelson and Bradley (1978) reviewed the history of 15 Florida fish kills caused by cold from 1856 through 1962 and described a later event in 1977. These incidents occurred on average about once per decade. Tarpons have been among the victims in all of them. Storey (1937: 10–11 Table I) listed tarpons among fishes in southwestern Florida waters "always hurt" by sudden declines in water temperature, although specific temperatures were not provided. Atlantic tarpons have featured prominently in reports of North American winter fish kills (e.g. Breder 1944: 232–233, Cyr 1991: 37, Griswold 1922: 9–10, Haldeman 1892: 114, Howells 1985, Howells and Garrett 1992, Overstreet 1974, Randall and Moffett 1958, Smith 1896, Storey 1937, Storey and Gudger 1936, Wade 1962: 592, Willcox 1887), usually dying when water temperature declines to ≤15°C (Galloway 1941, Gilmore 1977, Howells 1985, Moffett and Randall 1957, Rickards 1968, Springer and Woodburn 1960). In laboratory experiments Randall and Moffett (1958) found juveniles able to survive temperatures changes down to 18°C and to 12.8°C if acclimated gradually over 2 weeks. However, delayed mortality after several hours was often observed at the lowest temperatures (Moffett and Randall, 1957: 7 Table 2). Florida anglers have long recognized that tarpons become torpid and stop feeding in cold water. Griswold (1913: 100–101, 1921: 27–28) wrote, "They do not show on cold days; but if the water is sixty-eight degrees [≈20°C], or warmer, you can see them, and can fish for them with some hope of success."

Susceptibility is heightened when the temperature decline is abrupt. Smith (1896: 173) recorded the deaths of many tarpons of 13.6–72.6 kg in Biscayne Bay, southeastern Florida during February 1895 when air temperature diminished suddenly to –3.3°C. Many still alive were "floating belly up in a stupefied or benumbed condition, and it would appear that the immediate cause of death [of the others] was drowning or asphyxiation." Durbin C. Tabb (personal communication in Wade 1962) saw dead tarpons in Everglades National Park, Florida when the water temperature fell from 24.0°C to 11.0°C in a few hours. Rickards (1968: 225) reported mortality of juvenile tarpons held in outdoor tanks for laboratory experiments when the water temperature dropped overnight from 21°C to 12°C.

Howells (1985) noticed changes in behavior of Atlantic tarpons subjected to low temperature. They began swimming restlessly instead of hovering quietly above the substratum. Several hours before dying the dorsal coloration of some changed from light blue-gray to deep steel-blue. Whether equilibrium was lost varied by individual, some losing it for extended periods prior to death, others succumbing without loss of equilibrium. Feeding continued to at least 16.2°C.

Tabb and Manning (1961) reported that a December 1957 cold front killed many tarpons and other species in Coot Bay and Whitewater Bay, Florida, when temperatures in shallow brackish waters reached 12–14°C. Storey and Gudger (1936: 641) repeated an eyewitness account of a fish kill at Sanibel in southwestern Florida over 29–30 December 1894 when air temperature at nearby Ft. Myers declined to –4.4°C. Within a day, "The banks of the Caloosahatchee [River] became 'white with tarpon,' and the Sanibel beaches were lined with all sizes and kinds of fishes from alongshore minnows to huge tarpon." A later cold spell at Sanibel 12 December 1934, when the Ft. Myers air temperature read 1.7°C, killed all juvenile tarpons in a brackish pond that had not been open to the Gulf of Mexico for 2 y. On a strip of shore ≈30 m long, 68 dead specimens of ≈360–610 mm washed up. The other fishes survived, indication of the tarpon's cold-sensitivity.

Overstreet (1974) described a massive fish kill in bayous of Jackson County, Mississippi in January 1973 following 8 d in which air temperature dropped steadily to –4.4°C. He reasoned that under severe temperature stress estuarine fishes perhaps die from disruption of normal osmo- and ionoregulation when enzymatic control of these processes is compromised to the point of becoming nonfunctional. The tarpons he necropsied seemed thin, and although Overstreet did not suggest a specific cause of this condition, dehydration comes to mind and gives credence to his suggestion. Atlantic tarpons might actually seek warmer waters during cold periods. They are known to aggregate around heated power plant effluents in winter (Snelson 1983: 190).

Tarpons captured outside the subtropics are sometimes thin, but this could be from lack of food or other reasons, not necessarily cold. Fowler (1928: 608), for example, recorded "Six poor, thin examples, between Barnegat Inlet and Atlantic Highlands [New Jersey] in August, 1927." From my experience, sea temperatures in August along that part of the New Jersey shore should be 21–23°C, or 2–4°C below what might be considered ideal for tarpons but warm enough to sustain them, and certainly not cold enough to be debilitating. Hickey *et al.* (1976: 187) reported the fate of an adult male trapped in a pound net off eastern Long Island 14 July 1974. An accompanying photograph shows it to be in good condition, and there was no doubt it was up to fighting strength: "The fish leaped and thrashed violently and finally exhausted itself after becoming entangled in the funnel of the net."

Atlantic tarpons tolerate high temperatures, but f_{ab} rises at 40.7°C, and delayed mortality is common after several hours (Moffett and Randall 1957: 7 Table 1). In this series of experiments, tarpons recovered more often when subjected to cold and then returned to the original acclimation temperature than when the procedure was reversed, indicating that although low temperatures are both debilitating and finally lethal, exposure to unusually high temperatures might be ultimately more damaging.

One report I found suggests a mass fish kill in southwestern Florida evidently following an outbreak of red tide before the causative agent (the dinoflagellate *Karenia brevis*) of these periodic episodes was discovered. According to Taylor (1919: 7) the event was accompanied by an "odorless but exceedingly irritating gas" forcing many people to seek medical assistance for sore throats and "colds." Waters in the affected areas had become "dark-colored," and fishes died in the thousands, including Atlantic tarpons.

Tarpons, like other fishes, sometimes die during droughts when inshore waters dry up. Suriname in northern South America lies between 6° and 2° north latitude and has two wet and two dry seasons. The short rainy season is December into early February. This is followed by a short dry period encompassing the remainder of February and all of March, then comes a long rainy season extending from April through July and a long dry season lasting from August through November (Mol *et al.* 2000). The Bigi Pan area is a tropical estuary of ≈ 530 km^2 featuring intertidal mudflats, mangroves, and brackish lagoons. The lagoons are shallow (<80 cm deep), seawater entering by means of small creeks on spring tides. Water flows in the reverse direction during rainy periods. The lagoons, including Bigi Pan Lagoon, the largest (≈7 km^2), are 2–5 km behind a coastal barrier and occupy ≈10% of Bigi Pan. In the long dry season the smaller ones can dry out completely. Depending on surface evaporation and rainfall the S_P ranges from 5–40. During El Niño-induced droughts, such as that occurring in the long dry season of 1997/1998, even Bigi Pan Lagoon dried up, killing all fishes living there. As observed by Mol *et al.* (2000), tarpons were last to succumb. These investigators also reported that in May 1998, a month after the first major rains, the lagoons held only juveniles of all commercial fishes including tarpons, which were < 50 mm TL. In an ecological context droughts are not only deadly, they impede migrations of many species (e.g. Titus and Mosegaard 1992).

Hypoxia is a potential detrimental factor, but as discussed in Chapter 5, tarpons routinely survive hypoxic conditions that drive away or kill obligate water-breathing species (Kulkarni 1983, Townsend 1992, Townsend *et al.* 1992), even when oxygen falls to 1.14–2.28 kPa (Townsend *et al.* 1992). Fish kills often occur during periods of hypoxia or anoxia abetted by photosynthesis and respiration of aquatic plants, and at such times upwelling caused by flooding or wind brings anoxic water to the surface, depleting oxygen throughout the water column (Bishop 1980).[3] Thus mixing, which alleviates stratification and logically should oxygenate the water column, induces the opposite effect. Such events are seldom fatal to tarpons (Bishop 1980). Coates (1987) noted that large mats of the floating fern *Salvinia molesta* had formed in many oxbows of the Sepik River of northern Papua New Guinea, turning their waters hypoxic. Among the few fishes able to tolerate conditions underneath the mats and in waters >15 m deep was the Pacific tarpon.

Bishop *et al.* (2001: 22) reported that juvenile Pacific tarpons in the Alligator Rivers area, Northern Territory, Australia experienced seasonal shifts in surface temperatures of 23–34°C and dissolved oxygen levels of 1.9–9.7 mg/L (surface waters) and 0.2–7.4 mg/L (bottom waters). Also, across seasons, water clarity ranged from clear to turbid (visibility 4–270 cm) and pH values from 5.3–9.1 (surface waters) and 5.2–7.1 (bottom waters), further evidence of the tarpon's general hardiness. Tarpons were not among the species killed during a hypoxic event in Leichardt Billabong, Northern Territory in the early-wet season of 1978–1979 (Bishop 1980).

[3] A similar event known as a *friagem* ("big chill") occurs in the Brazilian Amazon when an influx of cold air causes mixing of the water column, turning it hypoxic, more acidic, and rich in hydrogen sulfide (Sloman *et al.* 2009).

7.5 Gregariousness

Atlantic tarpons are social, sometimes forming monospecific groups (Breder 1959), notably when migrating. Whether schooling is obligate or facultative is undecided. Anyanwu and Kusemiju (2008: 121) described "fry, fingerlings, and subadults" in Nigerian waters as moving in shoals. That tarpons are often seen in groups is not evidence of schooling behavior, and my personal observations indicate a stronger trend toward aggregation of a few individuals, often just two or three, rather than many swimming in the synchronous formation characteristic of such fishes as herrings and mackerels. Large numbers of Atlantic tarpons gather at places like Boca Grande Pass in southwestern Florida just prior to spawning offshore, and their presence has stimulated a sizeable recreational fishery. These might be better termed seasonal aggregations than schools. Atlantic tarpons also tend to gather around artificial structures (Layman and Silliman 2002). For example, they often aggregate underneath docks and large piers (personal observation) and when disturbed move off together in unison. Such sightings are particularly common near docks where fishermen clean their catch or tourists throw food to them (Chapter 8.2). In such situations it seems the main attraction is the structure itself or what it offers, not the presence of other tarpons. Hildebrand (1963: 116) wrote:

> *The Tarpon does not school, at least not habitually like the menhaden, for example. When it does congregate in considerable numbers, as at the foot of the spillways of the dams in the [Panamá] Canal Zone, it does so presumably because food is abundant there. In other places it sometimes congregates in pursuit of schools of fish.*

These statements conform with my conclusion that any schooling behavior is facultative.

Munro (1967: 41) evidently considered the Pacific tarpon to be a schooling species, Perna *et al.* (2011) mentioned similar behavior of Pacific tarpons feeding in open waters, and Pollard (1980: 54) described juveniles schooling in estuaries. However, Coates (1987: 532) seemed equivocal after extensive sampling, noting that his single large catch from a gill-net set in the Sepik River yielded only small specimens (150–250 mm), which "were probably a school of migrating fish briefly entering the floodplain." Otherwise, Pacific tarpons there were usually caught in "low numbers, normally in ones or twos." He continued: "If larger fish (>250 mm) do school within the river then these groups probably tend to be small." Floodplains are not a predominant habitat of Pacific tarpons in the Sepik River region. In a report on fishes of southern Africa, Mann (2000: 66) described the Pacific tarpon as mainly solitary and nowhere abundant.

Diel patterns of movement by both species have been described only superficially. Ley (2008: 13) reported that of 208 Pacific tarpons netted over an extended sampling period, 84% were caught at night. Capture rates were highest just after sunset, tapering off overnight, which would seem to make them *crepuscular* (i.e. most active at dawn and dusk, both periods called *twilight*).

7.6 Seasonal movements

What we commonly call "seasonal migrations" of tarpons are more accurately described as seasonal departures and appearances. A reasonably complete description of a species' migration implies not merely geographic knowledge of arrival and departure points, but the routes traveled between locations. For tarpons this information is painfully incomplete.

Randall and Moffett (1968: 139) wrote that Atlantic tarpons in the Gulf of Mexico (presumably adults), "undertake a northerly migration during the warm months and move southward when the waters begin to cool." Movement is in the reverse direction as the water warms, and they "arrive in considerable numbers at Aransas Pass, Texas, about March, and appear along the Louisiana coast early in May, where they remain until October." Tarpons were noticeably more abundant near Port Aransas in summer 1972, correlating with a simultaneous spike in surface sea temperature (Moore 1975). The rise in temperature conformed with a pattern first apparent in 1967 and counteracted a trend of declining tarpon numbers starting in the early 1960s.

Atlantic tarpons (both juveniles and adults) are widespread in southern Florida (Wade 1962), occupying canals, airboat trails, lakes, both coasts, and often moving far into the Everglades and Big Cypress Swamp (Kushlan and Lodge 1974). Tabb and Manning (1961: 606) thus considered the species a permanent resident of northern Florida Bay, its abundance peaking from August through January. Juveniles of 40–150 mm (SL?) were abundant from June through October in Coot Bay Pond and similar bodies of water nearby in the Florida Everglades. Juveniles gather in large numbers during August and September to feed on cyprinodont fishes and the grass shrimp *Palaemonetes intermedius*. "This feeding 'jubilee' corresponds with the end of the rainy season and an attendant oxygen depletion which forces the small fish and prawns from the flooded swamps and into the open margins of bays."

Randall and Moffett (1958: 139) speculated that Atlantic tarpons indigenous to the Gulf of Mexico might spend winters relatively inactive in deep water, returning to shallower inshore locations with warming spring temperatures: "This would convey the impression of northerly movement, for those in the more northern localities would appear progressively later in the year." They emphasized that such a hypothesis does not disprove long migrations, only that evidence to the contrary was lacking. Genetic and tagging studies are now filling the blank spaces in our knowledge of when and where tarpons go.

That Atlantic tarpons on either coast of Florida move offshore *en mass* during the cold months is untrue. Were it so their winter gathering places at sea would probably be known, and they would not be present in such numbers inshore during winter fish kills. Some Atlantic tarpons obviously inhabit rivers, bayous, lagoons, and estuaries in winter, which anglers have known for years. Griswold

(1921: 26), for example, stated, "Winter fishing for tarpon is river fishing ...." On the next page Griswold (1921: 27) wrote: "The tarpon that come to the rivers, bayous, and inlets of our coast in April and May in great numbers leave in the autumn, supposedly for the warmer waters of the Gulf Stream; but some fish remain in the deep rivers of the east coast of Florida all winter." And, of course, on the west coast too.

The few assessments of genetic variation in Atlantic tarpons point to limited population structuring (Ward *et al.* 2008). Haplotype diversity is controlled by several processes, including mutation, recombination, and demography. Blandon *et al.* (2002) found reduced haplotypes in tarpons from the eastern Atlantic and Pacific coast of Panamá, nor was there much genetic divergence among sampling locations in the western Atlantic and Caribbean. Of these, divergence was higher among specimens collected in the Gulf of Mexico, but a consistent geographic pattern was not evident. García de León *et al.* (2002) discovered what they considered separate populations of tarpons in Texas and the western Gulf of Mexico. Ward *et al.* (2005) applied microsatellite analysis to fish from 15 locations from the Gulf of Mexico and Chetumal, Quintana Roo, México in the western Caribbean, finding those from the Gulf genetically distinct. Samples from Brazil and west Africa diverged genetically, as did samples from

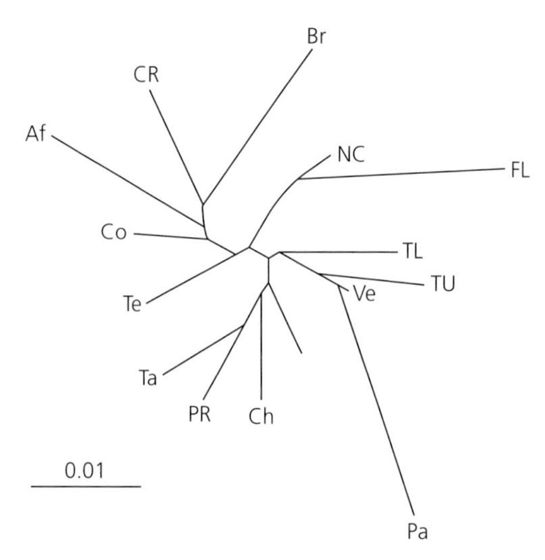

Fig. 7.3 Unrooted neighbor-joining (N-J) tree showing structure for 15 collections of Atlantic tarpons using pairwise Cavalli-Sforza and Edwards chord distance matrix. Bootstrap support for all nodes $<50\%$. Collection site designations: United States – North Carolina (NC), Florida west coast (FL), Louisiana (LA), Texas at two locations designated "upper" and "lower" (TU, TL); México – Tampico (Ta), Tecolutla (Te) Veracruz (Ve), Chetumal (Ch); Costa Rica (CR); Panamá, Pacific coast (Pa); Puerto Rico (PR); Colombia, Caribbean coast (Co); Brazil, state unspecified (Br); Nigeria, western Africa (Af). Louisiana (LA) apparently was inadvertently omitted in the source publication and thus is not shown here. Total $n = 328$. Source: Ward *et al.* (2008: 139 Fig. 10.2).

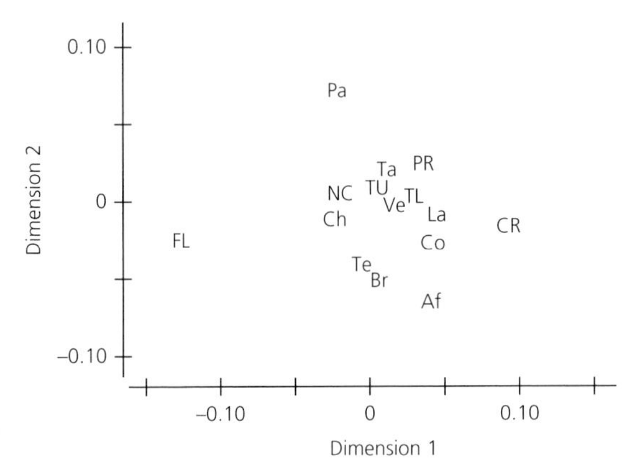

Fig. 7.4 Plot of first two dimensions of a multi-dimensional scaling analysis of genetic affinity of Atlantic tarpons from 15 locations. Collection site designations: United States – North Carolina (NC), Florida west coast (FL), Louisiana (LA), Texas at two locations designated "upper" and "lower" (TU, TL); México – Tampico (Ta), Tecolutla (Te) Veracruz (Ve), Chetumal (Ch); Costa Rica (CR); Panamá, Pacific coast (Pa); Puerto Rico (PR); Colombia, Caribbean coast (Co); Brazil, state unspecified (Br); Nigeria, western Africa (Af). Louisiana (LA) apparently was inadvertently omitted in the source publication and thus is not shown here. Total n = 328. Source: Ward *et al.* (2008: 139 Fig. 10.3).

Panamá, Costa Rica, and Florida (Fig. 7.3). The remainder – particularly from the Caribbean and Gulf of Mexico – clustered tightly, in contrast with earlier findings of García de León *et al.* (2002) and Ward *et al.* (2005). Multi-dimensional scaling (Fig. 7.4) yielded a tight clustering of most sampling locations in dimension 1 with only Florida and Costa Rica standing apart. Dimension 2 separated samples from the Pacific coast of Panamá in one direction and the Nigerian sample in the opposite direction, and showed the African sample to be most closely related to the Brazilian fish. The rest grouped tightly along both dimensions.

As these two figures demonstrate, Atlantic tarpon populations can so far be distinguished genetically only at the peripheries of their distributions. These identifiable "subpopulations" (I use the term hesitantly considering the small sample sizes and limited sampling locations) are represented by samples from the Costa Rican Caribbean, Florida, Panamá's Pacific coast, and the Gulf of Guinea (Nigeria). The Panamanian "subpopulation" is especially interesting because it was established sometime after opening of the Panamá Canal in 1914 (Hildebrand 1939); that is, relatively recently. The span of its genetic diversity (Table 7.1), as evaluated by the mean number of alleles per locus, encompassed 3.33–5.67, all above the median. Other "subpopulations" on the periphery of distribution are North Carolina (H_S = 0.47) and Nigeria (H_S = 0.37). Values of heterozygosity ranged from H_O = 0.36 (Puerto Rico) to H_O = 0.56 (Brazil). Within-population genetic diversity showed a range of H_S = 0.31 for Costa Rica to H_S = 0.50 for tarpons from the Pacific coast of Panamá.

Table 7.1 Assignment test of individual Atlantic tarpons to their 15 source populations. Collection site designations: United States – North Carolina (NC), Florida west coast (FL), Louisiana (LA), Texas at two locations designated "upper" and "lower" (TU, TL); México – Tampico (Ta), Tecolutla (Te) Veracruz (Ve), Chetumal (Ch); Costa Rica (CR); Panamá, Pacific coast (Pa); Puerto Rico (PR); Colombia, Caribbean coast (Co); Brazil, state unspecified (Br); Nigeria, western Africa (Af). Total n = 328. *Source:* Ward *et al.* (2008: 140 Table 10.4).

Location	n	$\bar{X}_{alleles}$	H_o	H_s
Veracruz MX (Ve)	31	4.50	0.44 (0.11)	0.45 (0.27)
Tecolutla MX (Te)	15	3.83	0.43 (0.09)	0.45 (0.28)
Tampico MX (Ta)	20	4.67	0.46 (0.13)	0.37 (0.23)
Texas "lower" (TL)	22	5.50	0.38 (0.10)	0.40 (0.25)
Texas "upper" (TU)	39	5.67	0.48 (0.10)	0.49 (0.29)
Louisiana (LA)	24	3.67	0.41 (0.11)	0.37 (0.23)
Florida (FL)	20	3.50	0.48 (0.11)	0.49 (0.29)
North Carolina (NC)	25	4.17	0.51 (0.10	0.47 (0.28)
Puerto Rico (PR)	32	4.00	0.36 (0.12)	0.35 (0.22)
Chetumal MX (Ch)	19	3.33	0.55 (0.11)	0.47 (0.28)
Costa Rica (CR)	17	3.33	0.40 (0.13)	0.31 (0.21)
Colombia (Co)	16	3.50	0.40 (0.08)	0.44 (0.27)
Brazil (Br)	9	3.33	0.56 (0.13)	0.42 (0.27)
Africa (Af)	16	4.17	0.43 (0.14)	0.37 (0.23)
Panamá, Pacific coast (Pa)	23	4.17	0.49 (0.07)	0.50 (0.30)
Total	328			

An important aspect of such assessments is how closely the results predict assignment of individuals tested to their source populations (Table 7.2). These data showed a range in assignment capability of 6.5% (Veracruz) to 62.5% (Nigeria) and demonstrated assignment in all cases except Veracruz and the "lower" Texas coast. As Ward *et al.* (2008: 137) pointed out, "Collection sites that are not believed to have substantial permanent tarpon populations (e.g., Texas, Louisiana, and North Carolina) have relatively low correct assignment values, supporting the hypothesis that tarpon in these samples may be migrants from other regions."

Returns from conventional tagging programs established for tarpons have been poor. For example, Wiggers (2010: 42) reported five recaptures of 289 tarpons tagged by anglers in South Carolina between 1982 and 2009, although four were recovered from South Carolina waters, providing slim evidence of philopatry. The longest time at large was a fish tagged at Port Royal Sound 23 August 1993 and recovered 24 March 1995 after 578 days. It measured 1905 mm TL (presumably after recapture).

Luo *et al.* (2008b: 276) pointed out that such programs are restricted in what they reveal. The data retrieved "allow only limited inferences on movement behavior and migrations, restricted to the start and end points of the time at liberty." I add that the same restriction holds true of the present undertaking

Table 7.2 Assignment test of individual Atlantic tarpons to their 15 source populations. Collection site designations: United States – North Carolina (NC), Florida west coast (FL), Louisiana (LA), Texas at two locations designated "upper" and "lower" (TU, TL); México – Tampico (Ta), Tecolutla (Te) Veracruz (Ve), Chetumal (Ch); Costa Rica (CR); Panamá, Pacific coast (Pa); Puerto Rico (PR); Colombia, Caribbean coast (Co); Brazil, state unspecified (Br); Nigeria, western Africa (Af). Total n = 328. Source: Ward *et al.* (2008: 140 Table 10.4).

Location	Ve	Te	Ta	Tl	Tu	LA	FL	NC	PR	Ch	CR	Co	Br	Ni	Pa
Ve	2			1	1	1	1	1			1			2	
Te	4	8	1	2	1	2	2	2	3	3	2	1		1	1
Ta	4		11	1	3	1		3	1	2	1				1
TL	1			2		1		2							
TU	1			1	9			1		1					1
LA	1	3		2	1	9			4	1	1				2
FL			2	1	2		11			1		2			1
NC					4		2	9	1	1				1	4
PR	4	2	2	3	9	3	1	2	17			2	1		
Ch	2		1		2	1	1		1	9	1	2	2		
CR	3		2	2	2	2			2		10	2		1	1
Co	3	2		4		2		3	1		1	5	1	1	
Br				1	1			1				1	4		
Af	2		1	1	2	1	1			1		1	1	10	
Pa	4			1	2		1	1	2						12
%C	6.5	53.3	55.0	9.1	23.1	37.5	55.0	36.0	53.1	47.4	58.8	31.3	44.4	62.5	52.2

to involve catch-release Atlantic tarpon anglers in DNA sampling.[4] Both methods provide starting and endpoints while revealing little about the path of travel and what occurs along the way. They tell us, in other words, where a tarpon went, but nothing about its specific route and what it did along the way.

These gaps are being closed. Luo *et al.* (2008b) published a detailed report describing results of tagging Atlantic tarpons using satellite-based, pop-up archival transmitting tags (PATs). When applied successfully a single tag yields thousands of data points charting an individual fish's movements in real time, and offering remarkable insight into how its behavior correlates with abiotic factors like location, depth, temperature, salinity, turbidity, and level of illumination. Conventional tags and DNA findings will certainly fill other empty spaces in our knowledge of tarpon biology, but for gaining insight into the timelines of what tarpons really do, PATs are unrivaled. Among the findings was the first evidence of four and perhaps five migration patterns of tarpons in the western North Atlantic and Caribbean based on successful application of 25 PATs.

Pattern 1: Southward migration in autumn and winter along the US southeastern coast. A 36.4 kg tarpon tagged at Savannah, Georgia 21 September 2001 was tracked ≈460 km south to near Sebastian Inlet on Florida's east coast where the tag popped off 4 November 2001 as scheduled. Data collected over the 43.5 d of attachment comprised ≈64 000 minute-by-minute recordings of temperature, depth, and levels of illumination. The fish usually stayed in water of 21.7–27.2°C ($\bar{x}$ = 26°C) and depths of 0–25 m ($\bar{x}$ = 10 m). Its rate of travel was ≈11 km/d. Its southerly passage became more rapid as cold fronts passed overhead. The PAT of a fish tagged at Hilton Head, South Carolina 6 September 2001 popped off 3 January 2002, also near Sebastian Inlet.

Pattern 2: Northward migration in spring and summer along the US southeastern coast. A PAT applied 18 May 2002 at Islamorada in the Florida Keys was recovered 4 August 2002 at Merritt Island, Florida 50 km north of Cape Canaveral. This tarpon had traveled north ≈440 km in 79 d and provided 112 761 data points for time, temperature, depth, and subsea illumination. Another tarpon, also tagged in the Florida Keys at Key Largo 1 June 2006, traveled the same distance north, but in 26 d at a speed of $\bar{x}$ = 16.9 km/d.

Pattern 3: Winter migration from the northern Gulf of Mexico to the Florida Keys. A fish tagged 4 September 2003 at Venice, Louisiana was tracked to Key West. Buoy data from the National Oceanographic and Atmospheric Administration (NOAA) indicated that water temperature fell to <24°C around 10 December that year in the vicinity of Venice. Had the fish started its migration on that date, it would have traveled 1160 km in 60 d ($\bar{x}$ = 20 km/d).

[4] http://miami-dade.ifas.ufl.edu/documents/TarponFactSheet.pdf; http://www.projecttarpon.com/DNAresearchflorida.html; http://myfwc.com/media/1389022/DNA-project-overview.pdf. Downloaded 6 July 2015.

Pattern 4: Northward summer migration from the western Gulf of Mexico (e.g. Veracruz, México) to the northern Gulf area of Texas and Louisiana. In May 2003 and June 2004 Luo and colleagues tagged 15 adult tarpons ≈3–5 km offshore of the Antigua River north of Veracruz. Five tags failed to transmit data; another fell off near the attachment site after 61 d. The remaining nine detached 230–1730 km north of Veracruz, indicating movements of 8–33 km/d, or $\bar{x}$ = 17.2 km/d (see Luo *et al.* 2008b: 282–283 Table 18.2).

Pattern 5 (tentative): Northward movement through the southern Lesser Antilles and indicating some intermixing with tarpons in the northern Caribbean Sea. Two of several PATs applied 6 August 2006 near Trinidad (Trinidad and Tobago) released prematurely after a month, both north of the original attachment site. One was recovered 165 km away near Grenada, the other 290 km from Trinidad near St. Vincent (St. Vincent and the Grenadines).

Diel patterns in such data are grab-samples of natural history at its most fascinating. These tarpons swam through water temperatures of 16–34°C and at depths to 88 m (Luo *et al.* 2008b: 282–283 Table 18.2), a tag staying attached in one instance for 121 d. The authors presented their results for two fish, and I summarize the highlights here.

Tag T-03 was the PAT mentioned above as applied 21 September 2001 at Savannah, Georgia and recovered 4 November after 44 d. This fish dived to a maximum depth of 25 m ($\bar{x}$ = 10 m) and displayed intermittent vertical movements in the water column. It mostly stayed shallow during the day and ventured deeper at night, but was active at all hours during full moons. It strayed through temperatures of 21.5–27.1°C ($\bar{x}$ = 24.8°C). Activity increased when visibility was ≈15 m.

A 55 kg tarpon tagged on 10 May 2004 at Veracruz, México (PAT T-24) released its tag 8 September 2004 on schedule after 121 d. The tag was found the same day on Port Aransas Beach, Texas having generated >177 000 data points for depth, temperature, and level of illumination. This fish dived to a maximum of 48 m ($\bar{x}$ = 6.3 m). During 5 d in mid-June 2004 it descended to deeper water at dawn, where it stayed until dusk when it rose near the surface for the night with only a few deeper descents. This behavior was the opposite of T-03's. Tarpon T-24 experienced a wide temperature range (20.7–33.7°C, $\bar{x}$ = 27.7°C, maximum daily range 5°C). The fish continued northward with warming waters, reaching the coastal lagoons in late July. Its pattern, in combination with patterns of others, showed a strong preference for 26°C with seasonal preferences of 24–26°C (spring and autumn) and 28–30°C (summer).

Pacific tarpons in the Alligator River area, Northern Territory, Australia are distributed most widely from the late-wet into the early-dry season, when they disperse across the floodplain, sandy corridors, and backflow billabongs (Bishop *et al.* 2001: 18).

Kowarsky and Ross (1981) studied fish movements past the Fitzroy River barrage at Rockhampton, Queensland by placing a fish trap just above the weir and sampling it 73 times between August 1978 and October 1979. The barrage is 59.6 km from the river mouth, and the gates were lifted or lowered according to stream flow, which in turn was controlled by rainfall. Six Pacific tarpons were included in the catch, all of them in March or April. They ranged from 139–370 mm TL (i.e., juveniles to potential breeding size), although reproductive status was not mentioned.

7.7 Feeding and foods

Suction feeding is the most widely used method of prey capture by teleosts (e.g. Grubich 2001, Holzman *et al.* 2012, Lauder 1982, Oufiero *et al.* 2012 and references, Tran *et al.* 2010), often combined with *ram-feeding* that involves forward motion of the body, jaws, or both (e.g. Norton and Brainerd 1993). Most species combine suction and ram-feeding, but rely more on one method than the other. Ram suction feeding is thus considered a continuum (Norton and Brainerd 1993, Oufiero *et al.* 2012, Tran *et al.* 2010), which led to devising of a *ram-suction index*, RSI (Norton and Brainerd 1993) to evaluate a species' position along the continuum:

$$RSI = \frac{\left(D_{predator} - D_{prey}\right)}{\left(D_{predator} + D_{prey}\right)} \tag{7.1}$$

in which $D_{predator}$ is the distance the predator swims during the strike and D_{prey} the distance the prey moves toward the predator during the strike as hydrodynamic forces suck it in. Norton and Brainerd (1993) discovered that RSI values for small-mouthed fishes fall closer to the suction end of the continuum; large-mouthed predators revealed the opposite pattern.

The feeding sequence of a typical teleost can be divided into two phases, each with a series of stereotyped maneuvers occurring rapidly in tandem (e.g. Grubich 2001). These are suction (intake of prey and filling of the buccal cavity with water) and compression (swallowing of prey and emptying of the buccal cavity of water). The suction sequence involves (1) opening of the jaws, (2) closing of the opercula to seal the buccal cavity, (3) expansion of the buccal cavity abetted by elevating the head, (4) depression of the hyoid and lateral expansion of the suspensorium. The sum of these movements forces a negative pressure gradient locally causing water and prey to be sucked into the mouth. The compression phase involves (1) progressive closing of the jaws posteriorly, (2) protraction of the hyoid and suspensorium adduction, and (3) opening of the opercula allowing water to exit from the buccal cavity. The combined events seem to me so highly conserved as to border on an instance of exaggerated water-breathing (Chapter 5.2).

To generate suction a fish rapidly expands its buccal cavity, which initiates intake of the water in front of its mouth (Lauder 1982). Some predatory fishes shorten the distance to their prey by chasing it down and then capturing it using suction.

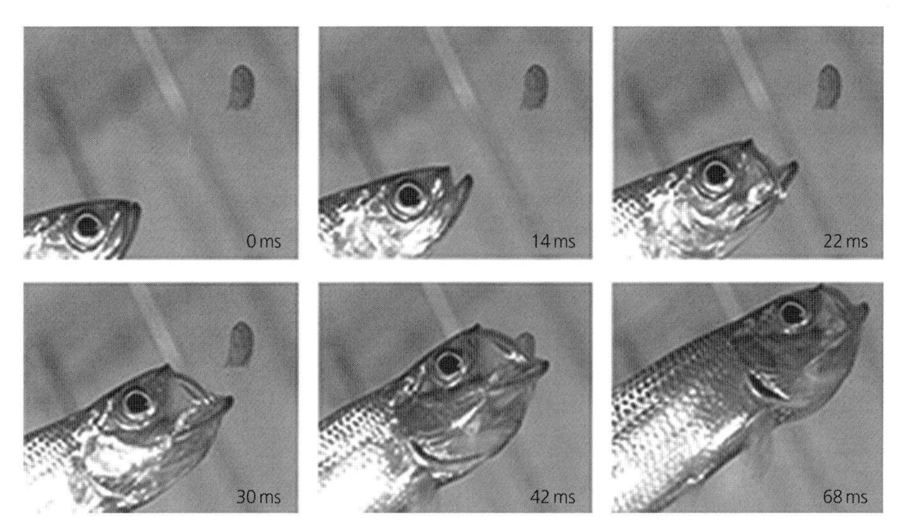

Fig. 7.5 Kinematic sequence of prey capture behavior by juvenile Pacific tarpons. Scale bar = 1 cm. Source: Tran *et al.* (2010: 76 Fig. 1).

Others lie in wait or slowly stalk targeted prey, relying on a quick lunge when the target organism approaches within suction distance (Tran *et al.* 2010 and references). By either method timing of the gape is critical. A fish swimming rapidly or lying in ambush with its mouth fully open would be unable to generate suction upon contact with the prey. Complicating matters, target organisms often use evasive maneuvers when aware of pursuit.

Tran *et al.* (2010) assessed ram feeding and prey capture kinematics using captive juvenile Pacific tarpons (135–142 mm SL, n = 3). For their purposes *time to peak gape* (TTPG) was measured as the time necessary for gape to increase from 20 to 95%. Ram speed was calculated as closing distance to the prey from start of the strike to immediately before the prey disappeared into a tarpon's mouth. Because the capture sequence in most fishes employs both suction and ram motion the study was intended to evaluate which is relatively more important to Pacific tarpons, as measured by the ram-suction index. In 42 capture sequences (Fig. 7.5) strikes were initiated as far away as 4.7 cm from the prey ($\bar{x}$ = 1.9 ± 0.15 SE cm). Prey items (pieces of frozen crustaceans) were captured in an average of 25.35 (±1.47 SE) ms.

The kinematic sequence began with peak rotation of the lower jaw, followed quickly by peak cranial elevation and peak displacement of the hyoid (Fig. 7.6).

Tran *et al.* (2010: 79) wrote that hyoid displacement sustained peak values beyond those of other kinematic variables, a characteristic of less derived teleosts (Hernández *et al.* 2002, Lauder 1982). The tarpons started opening their mouths early in the gape cycle but nonetheless sucked in prey at 95% peak gape. Ram speed ranged from 0.19–1.38 m/s ($\bar{x}$ = 0.65 m/s). Speed obviously varied, although in general "the farther away tarpon initiated their strike,

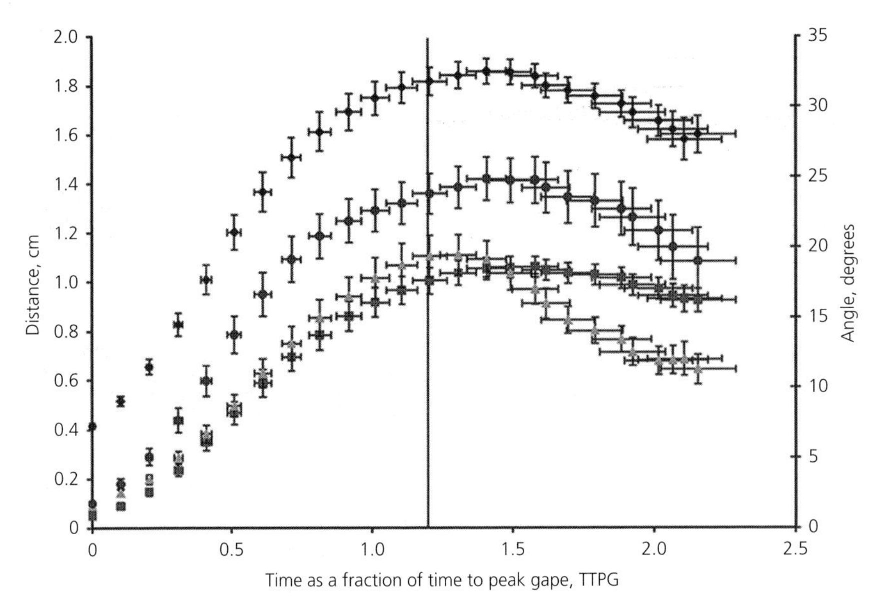

Fig. 7.6 Kinematic profile of cranial movements observed during prey capture by juvenile Pacific tarpons. Data points represent means of three fish. Time is scaled to time to peak gape (TTPG) to account for variation in time of prey capture. Scaled time, $T_{(0)}$, is the point at which 20% of peak gape is attained at the start of every trial. Vertical error bars are ± SE of scaled time. The single vertical line represents mean time to prey capture from all trials. Symbols: gape, filled circles (black); lower jaw rotation, filled circles (gray); cranial elevation, filled triangles; hyoid depression, clear squares. Source: Tran *et al.* (2010: 76 Fig. 2).

the faster they swam towards the prey ...." Ram speed correlated positively with initial predator-prey distance ($r^2 = 0.72$, $F_{1,45} = 89.46$, $p < 0.001$) and the distance at which the strike started ($r^2 = 0.79$, $F_{1,45} = 108.27$, $p < 0.001$). The pattern of this last association is shown in Fig. 7.7. Peak gape and most other variables were unaffected by changes in ram speed, which in turn had no effect on suction distance ($r^2 = 0.24$, $F_{2,45} = 0.04$, $p = 0.85$).

According to the definition used by Tran *et al.* (2010), ram-feeding involves the predator overtaking and swallowing prey using rapid acceleration of its whole body; the predator, in other words, moves while the prey remains stationary. In ram-suction feeding, however, both predator and prey move toward each other, the latter at the crucial moment. In fishes using *ram-biting*, a common method of sharks, the predator relies on a high-velocity lunge (alternatively, rapid acceleration) to close the distance to the prey, and the sequence ends with a hard bite.

These experiments identified juvenile Pacific tarpons as ram-suction predators, relying on both forward movement (≈6 *l*/s) and buccal expansion. In comparison with other species, juvenile Pacific tarpons tested faster in the sequence than the largemouth bass, slightly slower than juvenile barracudas (family Sphyraenidae), and markedly slower than pickerels and pikes (family Esocidae). Tran *et al.* (2010: 81) wrote: "When tarpon initiated their strike from further [*sic*]

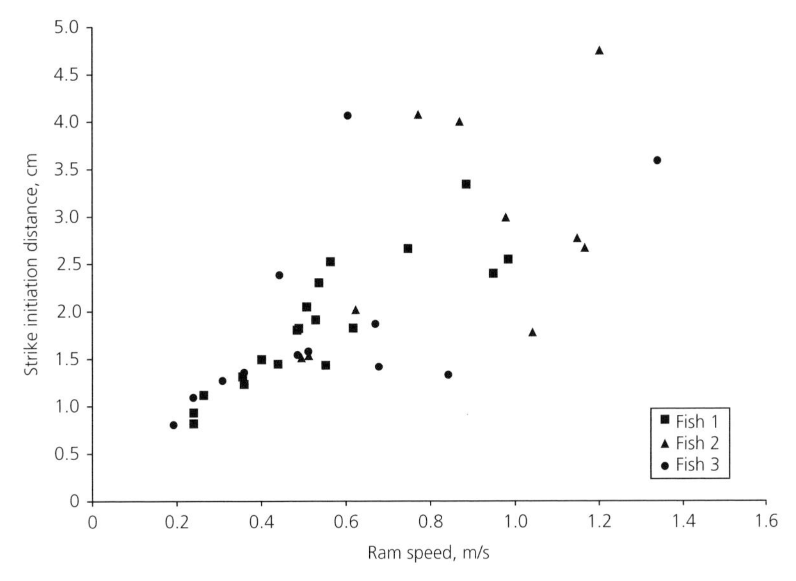

Fig. 7.7 Relationship between strike initiation distance and ram speed (m/s). Strike initiation distance is the distance from a fish's mouth to the prey at start of the strike (i.e. when 20% of peak gape is reached). Each data point represents a feeding trial, and individuals are shown in different symbols. Note how they display variation in range of ram speed and how more variable strike initiation distances are observed at higher ram speeds ($r^2 = 0.79$, $n = 3$). Source: Tran *et al.* (2010: 77 Fig. 4).

away, they achieved higher ram speeds but longer overall strike times revealing that [they] were unable to fully compensate for the increase in distance by swimming faster." This could affect both prey selection and capture success, especially in clear waters. Turbid conditions might offer feeding advantages to young tarpons, but this is conjecture.

Oufiero *et al.* (2012) found that the extent of jaw protrusion of a species during a strike correlates positively with attack speed, but not suction capacity, and that species using high-speed attacks have low capacity to generate suction. In contrast, low-speed predators demonstrate a full range of suction capacities, making the ram-suction continuum more complex than previous work had indicated. The benefit of upper jaw protrusion during the strike is that it has the effect of increasing attack speed because the mouth moves more rapidly toward the prey, in the process increasing the hydrodynamic forces exerted on it (Oufiero *et al.* 2012 and references). Tarpons do not use jaw protrusion (Tran *et al.* 2010) when attacking, and their attacks occur quickly, features that place them toward the ram-feeding end of the continuum.

Juvenile Atlantic tarpons use ram-suction feeding too. Their feeding kinematics are scarcely distinguishable from the Pacific tarpon's. However, the timing of some steps in the suction-compression phases differ, as do methods of approaching prey. Pacific tarpons approach targets head-on; Atlantic tarpons tend to initiate

Fig. 7.8 Schematic representation of a juvenile Atlantic tarpon, illustrating the postero-dorsal cranial inflection. Source: Guigand and Turingan (2002: 46 Fig. 1).

attacks from beneath their prey. Guigand and Turingan (2002) noted that the Atlantic tarpon, like the Pacific species, demonstrates an exaggerated postero-dorsal cranial movement (i.e. it rapidly tilts its head back) during buccal expansion and intake of prey (Fig. 7.8), but this maneuver seems more exaggerated than the Pacific tarpon's. Attacks are explosive, followed by a lower compression phase when the head then rotates antero-ventrally, not down past the midline but returning to horizontal.

Kinematic displacement profiles of most teleosts are symmetrical, or "bell-shaped," indicating that the suction and compression phases are of similar duration. The Atlantic tarpon's is different: the first phase occurs more rapidly than the second (Fig. 7.9). This is partly because the tarpon keeps swimming with its mouth still open. Continued flooding of the buccal cavity would slow closing of the jaws while also facilitating transport of prey to the esophagus (Guigand and Turingan (2002).

These descriptions of feeding kinematics would have thrilled Louis L.Babcock, life-long angler, tarpon *aficionado*, and amateur marine biologist. He was particularly taken with learning exactly how a tarpon feeds. On 25 July 1928 a 5 ft tarpon (≈1514 mm) was caught by commercial fishermen in Sandy Hook Bay, New Jersey. The specimen, which had been injured, was nonetheless taken to the New York Aquarium. There, at last, he hoped to observe how a tarpon attacked its prey (Babcock 1936: 62). Charles H. Townsend, the aquarium director, had alerted Babcock of the specimen's arrival, although both men were ultimately denied the opportunity of watching it eat. To Babcock's dismay a note with bad news arrived in the next day's mail:

Dear Mr. Babcock
July 26, 1928

We regret to announce the death of Tarpon atlanticus [*sic*] who passed away this morning refusing the consolation of nourishment. *Requiescat in pace*.

Yours regretfully,
C. H. Townsend

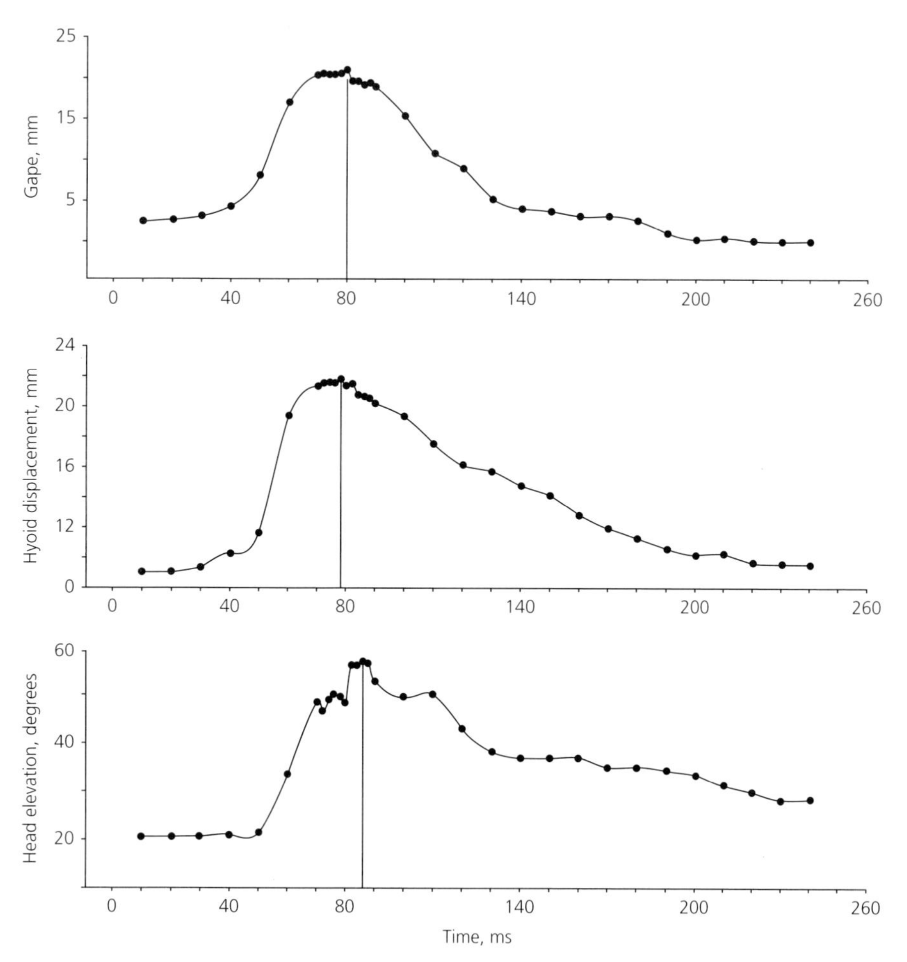

Fig. 7.9 Representative kinematic profiles of gape, hyoid, and head displacement during prey capture by Atlantic tarpons. None of the three profiles is symmetrical along the abscissa, indicating that the velocity of buccal expansion and opening of the jaw are much faster than the velocity of buccal compression and closing of the jaws. Source: Guigand and Turingan (2002: 49 Fig. 3).

Not yet mentioned is suction feeding by fish larvae. As pointed out by Hernández *et al.* (2002: 226), the standard definitions "assume that a rapid abduction of bony cranial elements results in reduced pressure within the buccal chamber, which brings in a bolus of water that will continue moving through the mouth and out the opercular openings." This works for large fishes, but tiny ones have difficulty overcoming water's viscosity. "In a sticky medium fluid will not continue to flow once skeletal elements have stopped moving." The situation is aggravated by incomplete development of muscles controlling abduction and small opercular openings, which in combination tend to reverse fluid flow as the hyoid returns to its resting position. Relative hyoid depression correlates with

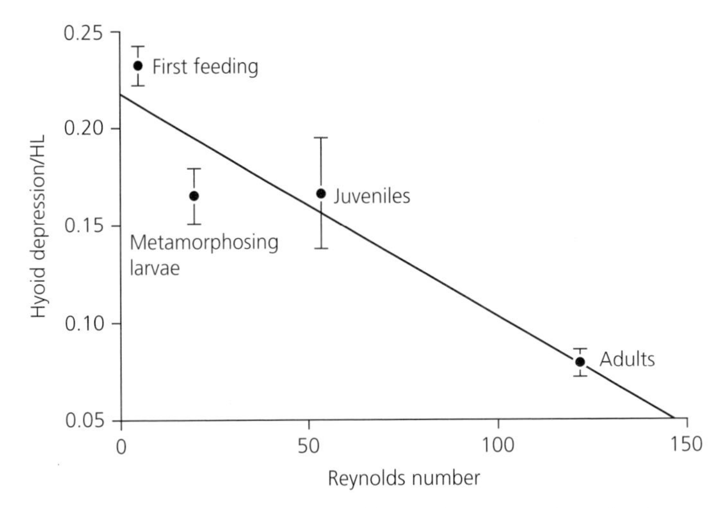

Fig. 7.10 Relative hyoid depression by degree of head lift (HL) correlates significantly with Reynolds number in suction feeding by zebrafish and decreases with ontogeny. When adjusted for body length (TL), relative hyoid depression is greatest in first-feeding larvae, declining in older stages. Reynolds number and relative hyoid depression are related inversely: first-feeding larvae (3–4 mm TL); metamorphosing larvae (6–7 mm TL); juveniles (10–12 mm TL); adults (25–27 mm TL). Number of trials representing the mean values, form of the variance, and sample size unstated. Regression equation: $y = 0.001x + 0.217$, $r^2 = 0.88$. Source: Hernández *et al.* (2002: 226 Fig. 3).

Reynolds number and decreases with ontogeny, as demonstrated in larval zebrafish (Fig. 7.10).

A partial list of organisms eaten by Atlantic and Pacific tarpons is available in Appendix E). Atlantic tarpons are often classified as piscivorous (e.g. Rueda and Defeo 2003), but this is not strictly true because they also eat many other organisms. What they are is carnivorous. The species was classified as a pelagic omnivore by da Silva de Almeida (2008: 58 Table 3), who gave no reason for doing so. Ferreira de Menezes and Pinto Paiva (1966: 89) stated that juvenile Atlantic tarpons maturing in coastal lagoons feed on insects, but after attaining 150 mm FL switch to small crustaceans and fishes. Beebe (1927) and Beebe and Tee-Van (1928) reported that a single127 mm tarpon seined from an isolated lagoon in Haiti contained the remains of 68 water boatmen (*Trichocorixa reticulata*) in its stomach.

Babcock (1951: 50) believed that young tarpons swept into pools isolated from the sea became stunted from lack of food. He placed specimens of ≈310–490 mm in "a stagnant, dark-colored, land-locked pool" and noticed how they "failed to grow appreciably." He speculated that in the open sea they might have reached 40–50 in. (≈1020–1270 mm) over the same period. Possible stunting in isolated bodies of water was proposed earlier by Breder (1944: 243). To my knowledge, evidence of stunting for either species of tarpon resulting from isolation has not been reported.

According to Chacón Chaverri (1994), juvenile Atlantic tarpons of 15–90 mm SL feed principally on small crustaceans (e.g. copepods) and insects, graduating to fishes at >270 mm SL, although Vega-Cendejas and Hernández (2002) found that fishes were not important dietary items to juveniles >270 mm SL at Isla Contoy, México, where amphipods (*Corophium* spp.) made up 65% of total stomach contents by weight. Their sample comprised just six specimens that ranged from 330–420 mm SL.

Juvenile Atlantic tarpons appear to be opportunistic predators of small invertebrates and fishes. Rickards (1968) reported that almost 96% of food items recovered from Tarpons <125 mm SL in Georgia salt marshes consisted of ostracods, grass shrimps (*Palaemonetes* spp.), and fishes, principally the mosquitofish, *Gambusia affinis*. Ostracods disappeared from the diet between August and November after a series of late-August storms. In Rickards' study fishes were the main prey item of juveniles of all sizes (Rickards 1968: 232 Table 7). Fishes composed ≈80% of the total stomach contents examined in tarpons of all lengths, but nearly all were mosquitofish. Mummichogs and striped mullets were consumed only by the largest juveniles. No copepods were recovered, ostracods apparently having been substituted by the smallest tarpons (<125 mm SL). Rickards also mentioned that although grass shrimps were always available they did not become part of the stomach contents until early September, evidently because the tarpons were still too small (<76.0 mm SL) to eat them.

Harrington and Harrington (1960) found copepods to be the dominant food in 442 stomachs of Indian River Lagoon, Florida tarpons of 16.0–75.0 mm SL, and the same was true of 231 specimens up to 150 mm SL from the same area (Harrington and Harrington 1961).

Ferreira de Menezes (1968b: 146 Table 1, Appendix F) listed stomach contents of 178 juvenile Atlantic tarpons (70–470 mm FL) captured in brackish waters, 118 within the municipal districts of Fortaleza and Acaraú, State of Ceará, northeastern Brazil, and 60 more from mangroves in the municipal district of Acaraú. She also examined 78 stomachs of adults (510–1900 mm FL) from coastal waters of Acaraú. She divided her findings into categories of decreasing importance labeled primary, secondary, and occasional foods. For juveniles these were: insects – hemipterids (mainly Notonectidae) and dipteran larvae; fishes – principally Poecilidae and various larvae; crustaceans – cladocerans, copepods (mainly Cyclopoidae), and decapods; algae – diatoms and chlorophytes (particularly *Spyrogira* spp.) and unicellular forms (cells of *Chlorella* spp. were abundant); and finally higher plants – exclusively *Halodule* (= *Diplanthera*) cf. *wrightii*. Juveniles were opportunistic, depending on food availability in the habitat. On salt flats the foods, in decreasing order of importance, were: crustaceans, insects, fishes, algae; in the mangroves: fishes, insects, crustaceans.

Primary foods for adults were fishes (e.g. families Carangidae, Clupeidae, and Scombridae). Secondary foods comprised crustaceans (e.g. family Portunidae). Occasional foods were mollusks and plants.

Moffett and Randall (1957) presumed that small juvenile Atlantic tarpons are filter feeders, straining water through their unusually long and numerous gill rakers, although theirs is the only report I found mentioning this possible feeding method, and the authors offered no supporting evidence other than noting that a Florida Keys fish of 78 mm FL had gill rakers up to 3 mm long, and the longest gill raker of a specimen of 570 mm FL measured 18 mm. The stomach contents of the 78 mm specimen yielded an estimated 3968 copepods (*Cyclops* spp.) of 0.4–0.8 mm TL.

Mercado Silgado and Ciardelli (1972: 156) wrote: "*En experimentos efectuados sobre leptocéfalos del Estado II, Fase VII, se pudo comprobar que a estos animales les es imposible tomar alimentos, ya que, como el primer autor pudo demostrar, al presionar la vejiga natatoria y tratar de dirigir una burbuja de aire hacia el ano ésta no pudo salir por no existír la abertura anal. Esta experiencia, como se anotó anteriormente, demuestra que en esta Fase los leptocéfalos no pueden tomar alimentos sólidos ya que se verían imposibilitados de excretar los productos catabólicos, los que tenderían a acumularse en el intestino y deberían poder observarse con el microscopio.*" [In experiments on leptocephali stage II, phase VII, it was found that these animals are unable to take food because, as the first author was able to show, an air bubble introduced into the gut could not be forced out by pressing the swim bladder because there is no anal opening. This experience, as noted above, shows that in this phase leptocephali cannot eat solid food because they would be unable to excrete the catabolic products, which tend to accumulate in the intestine and should be visible with a microscope.]

They continued: "*Para determinar la Fase en que los leptocéfalos empiezan a tomar alimentos, se introdujeron números exactos de nauplii de* Artemia salina*, larvas de mosquitos u otros insectos en los acuarios y se cubrieron con angeo u otro material que impidiera la introducción de elementos extraños dentro del acuario; al día siguiente se contaron los nauplii de* Artemia salina *o las larvas de mosquitos remanentes. De esta manera se determinó que los leptocéfalos toman sus primeros alimentos en la Fase VIII del Estado II. En los estados más avanzados se usaron larvas de crustáceos,cladóceros y larvas de mosquitos* Culex *y otros invertebrados acuáticos. Posteriormente se utilizó, para su alimentación, peces de la familia Poecilidae, vernacularmente denominados "pipones"*." [To determine the phase when leptocephali start taking food, exact numbers of *Artemia salina* nauplii, mosquito larvae, or other insects were introduced into the aquariums, which were covered with *angeo* [canvas or coarse linen] or other material that would prevent the introduction of foreign elements; the next day the nauplii of *Artemia salina* and remaining mosquito larvae were counted. Thus it is determined that leptocephali take their first food in phase VIII of Stage II. Crustacean larvae, cladocerans, *Culex* mosquito larvae, and other aquatic invertebrates were used [to test] the more advanced stages. Used subsequently as food were fish of the family Poecilidae, locally called "pipones".]

I could not find specific information about age at first feeding by Pacific tarpons. Tzeng *et al.* (1998) mentioned rearing leptocephali in both freshwater and seawater and feeding them brine shrimp nauplii, but did not state specifically the size, age, or developmental stage when they started feeding. Noble (1973) reared post-metamorphic and early juvenile Pacific tarpons in the laboratory, reporting they ate only live prey. Even freshly killed prey eaten readily when alive were rejected, and the flesh of prawns and other food items was ignored even by starved tarpons. Their stage of development at first feeding evidently was not recorded.

Pacific tarpons are most likely carnivorous too, although Pusey and Arthington (2003: 9), citing their own unpublished data, wrote that "the consumption of fruit by Australian freshwater fish is common .... In the Wet Tropics, many large fish, such as *Hephaestus fuliginosus* [sooty grunter], *Hephaestus tulliensis* [khaki grunter], *K.* [*Kuhlia*] *rupestris* [jungle perch] and *Megalops cyprinoides* (tarpon), consume the fruits of many riparian species during the wet season ...." Others have reported captive Pacific tarpons readily consuming commercial fish pellets (Seymour *et al.* 2004), which they could scarcely recognize as prey. The same can be said of hatchery-reared salmonids, which are also carnivorous.

Food selection by carnivorous fishes is not necessarily dependent on prey-like prior attributes (e.g. wriggling or swimming movements, coloration, reflectivity, size, or shape). What remains to be determined is whether tarpons of either species actually do eat fruit or even ingest it accidentally and, if so, what percentage of the diet it composes in nature and whether frugivory can sustain normal growth. Coates (1987) found root hairs of the floating fern *Salvinia* sp. (probably *S. molesta*) among the ingested items of Pacific tarpons from the Sepik River system, concluding they were swallowed incidentally while feeding, evidence that Pacific tarpons forage directly *in* submerged vegetation, although doubtfully *on* it.

Perna *et al.* (2011) reported increased numbers of Pacific tarpon in Australian floodplain lagoons after the mechanical removal of alien aquatic vegetation that had served as refuge for prey fishes, principally the eastern mosquitofish (*Gambusia holbrooki*). One such plant, the water hyacinth (*Eichhornia crassipes*), had formed extensive surface mats, their decaying infrastructure of stems and roots providing a suitable substratum for encroachment of para grass (*Urochloa mutica*). Removing both species opened sizeable sections, hastening the immigration of large piscine predators including tarpons.

According to Coates (1987), prey items of Pacific tarpons in the Sepik River did not vary noticeably throughout the year. Pacific tarpons feed mainly on invertebrates of small to medium size, and on small fishes. Although the sizes of food items tend to increase as tarpons grow, they still eat a variety of organisms (Coates 1987, Roberts 1978), and Coates reported seeing tarpons jumping out of the water to catch flying insects. Allen and Coates (1990: 53) examined the stomachs of 142 Pacific tarpons caught in the Sepik River system and recovered a variety of food items, including small fishes. However, in their opinion the species "is not an important piscivore."

Pacific tarpons are opportunistic predators like the Atlantic species. Prey selection depends on their state of maturity (to a large extent, body size), which determines the size range of organisms that can be captured and swallowed. Prey availability also changes seasonally. Pandian (1968) assessed these factors in Pacific tarpons captured in backwaters of the Cooum River, Chennai (formerly Madras), India. Prey species recorded in Pandian's report are included here in Appendix E, essential results in Table 7.3. Prey-switching was evident. The smallest tarpons feed mainly on crustaceans, graduating to insects and then fishes. Most prey-switching was attributed to growth, but ease of capture caused partly by changes in prey behavior, in combination with seasonal availability, were also thought to be contributing factors. Cataño and Garzón-Ferreira (1994) examined stomach contents of juvenile Atlantic tarpons (range 120–530 mm FL, n = 582) in the Ciénaga Grande de Santa Marta region of the Colombian Caribbean. Specimens collected from February–December 1990 showed evidence of prey-switching. The tarpons were largely piscivorous, feeding predominately on small mullets (*Mugil* cf. *incilis* from March–July, but mainly mollies (*Poecilia* cf. *gilli*) from September and February.

Based on stomach analyses, Chacón Chaverri (1994, Appendix G) concluded that Atlantic tarpons of Stages 2–4 feed selectively, demonstrating a preference for the biggest available particles. However, this statement was never tested as a hypothesis (i.e. that particle size alone was the basis for selection), nor was potential prey at the sampling sites assessed to determine whether young tarpons choose prey of certain types. His conclusions are no doubt correct if interpreted as opportunistic feeding. Tarpons, like other generalist predators, eat what the habitat makes available. Chacón Chaverri's findings show them shifting to larger prey with increasing body length, which is to be expected, and which has been observed at other locations (e.g. Harrington 1966). His sample of 261 specimens collected within the Gandoca-Manzanillo National Wildlife Refuge, Costa Rica revealed that from commencement of feeding at ≈15 mm SL to ≈90 mm SL, volumetric stomach contents were dominated by copepods (32%). Notonectids (backswimmers) made up 12%, and unidentified ("waste") material counted for 25%. Miscellaneous plant material and insects other than notonectids

Table 7.3 Prey-switching in Pacific tarpons according to body length (SL) during October–December 1963 in backwaters of the Cooum River, Chennai, India. Percentages are by volume. Source: Pandian (1968, 1973 Table 3).

Mean SL, mm	Range	*n*	Crustaceans, %	Insects, %	Fishes, %	Miscellaneous, %
83	43–99	10	52.2	11.6	18.1	17.1
125	100–149	32	42.9	36.25	11.3	9.5
170	150–249	75	33.3	21.6	40.5	4.6
277	259–303	4	34.9	11.5	53.75	–

(e.g. odonates, tricopterids) constituted the remainder. However, at 90–270 mm SL fishes were the predominant prey (30%) followed by notonectids (20%).

Young Atlantic tarpons evidently forage from the surface to the benthos (Austin and Austin 1971). These authors examined stomachs of 10 juvenile Atlantic tarpons from a mangrove habitat in western Puerto Rico. Three were empty. The fish ranged from 162–200 mm (whether FL, SL, or TL was not reported). The remainder contained (by volume) 83% insects and 17% crustaceans (anomurans and penaeid shrimps). Specifically which anomurans was not mentioned. Mojarra fry (*Eucinostomus* sp. and *Diapterus* sp.) were numerous, yet did not appear in the stomach contents.

CHAPTER 8
Fisheries

8.1 Introduction

Just 12% of North Americans are thought to have never participated in recreational angling (Pelletier *et al.* 2007 and references), and the US alone has an estimated 10 million saltwater angling enthusiasts (Shiffman *et al.* 2014 and references). Worldwide, recreational anglers include ≈ 12% of Earth's population, and they catch > 47 billion fishes each year, releasing approximately two-thirds, many of which die anyway (Cooke and Cowx 2004; Shiffman *et al.* 2014 and references). With so many people involved, almost no fishery is unaffected.

Part of Chapter 8 builds on physiological principles outlined in Chapters 5 and 6, providing a practical platform on which to set those theoretical and empirical findings. They pertain to how a fish caught on hook and line or in a net responds to the resultant stress factors and injuries, which in turn form the basis of proper handling procedures in catch-release fisheries. Tarpons are easily reared in captivity, although aquaculture has been slow to develop except at an artisanal level. Commercial fisheries have apparently had a negative impact at some locations, perhaps caused by the removal of excessive numbers of adults (e.g. northeastern Brazil). Populations of both species worldwide might be declining, but too few data are available to verify or falsify this notion.

8.2 Recreational fisheries

The US Atlantic tarpon recreational fishery, which extends from Virginia to Texas, contributes > $6 billion annually to the national economy according to the nonprofit Bonefish and Tarpon Trust,[1] with Florida benefiting most (Fedler

[1] http://www.bonefishtarpontrust.org/tarpon-research/tarpon-research.html. Downloaded 7 July 2015.

Tarpons: Biology, Ecology, Fisheries, First Edition. Stephen Spotte.

Fig. 8.1 Team tournament fishing for Atlantic tarpons in southwestern Florida. The boat has been painted with the names and logos of the team's commercial sponsors, and a television camera crew films the action. Source: ©NcBateman1 | Dreamstime.com 5545151.

2009, 2011; Killam *et al.* 1992: 3–1). Tarpon tournaments are marketed as exciting sporting events similar to big-time baseball and football, in which contesting teams of anglers pursue the fish while others participate as spectators or watch on television. Beer manufacturers are frequent sponsors (Fig. 8.1), augmenting the promise of big fish, pleasant weather, and sexy women (Fig. 8.2). These events are designed for *trophy fishing*, defined by Shiffman *et al.* (2014: 318) as targeting "the largest members of a species with the goal of obtaining an award with perceived prestige." The purpose is to catch a record specimen that afterward is certified by a central authority.

The first Atlantic tarpon caught by rod and reel was landed on Florida's east coast. Of this event, Heilner (1953: 219) wrote:

> *Undoubtedly the first tarpon ever taken on rod and reel was taken by Samuel H. Jones of Philadelphia in March, 1884, in Indian River Inlet, Florida. The fish weighed 172½ pounds [78.2 kg] and was taken trolling on a Buell spoon. The honor is generally supposed to have belonged to W. H. Wood who caught a fish in Surveyor's Creek, Florida, on March 25, 1885, but Mr. Jones' fish antedates this by one year.*

Some authors subsequently failed to credit Jones. Fritz (1963: 89), for example, wrote: "In 1885, a New York sportsman, W. H. Woods [*sic*], fishing with a rod and reel, hooked and landed a huge silver tarpon. His feat was given wide publicity – and a new sport was born." Kaplan (1937: 96–98) traced the history of the controversy.

Fig. 8.2 Tarpon tournament fishing as a sporting event, complete with beer advertisements painted directly on the boats (Fig. 8.1) and pin-up girls. ©Lantern Press. Reprinted with permission.

Although the Atlantic tarpon has become one of the world's premier recreational fishes since the mid-1880s, its sobriquet "silver king" rings false. "Silver queen" seems more pointed. Mature individuals are sexually dimorphic, and the truly big specimens – including those representing most angling records – are probably females. From a large sample of Florida tarpons, Crabtree *et al.* (1995) reported that females ranged from 1193–2040 ± 141.5 mm FL ($\bar{x}$ = 1677 mm FL, n = 322) and were significantly longer than males: 901–1884 ± 130.2 mm FL ($\bar{x}$ = 1447 mm FL, n = 125). The biggest specimens examined had come from the recreational fishery.

Despite the Atlantic tarpon's large size, high endurance, and burst strength, technological advances in fishing equipment place it at a disadvantage. Materials science has made possible lightweight, durable rods of glass or carbon fibers. These are manufactured to exacting standards of power and action. Today's fishing lines are mass-produced to equally narrow tolerances of weight, breaking and knot strength, sinking velocity, and resistance to abrasion and ultraviolet light.

The subsurface visibility of fishing lines correlates positively with diameter, and some thinner products have refractive indices near that of water, making them functionally invisible to an approaching fish. Hooks are available in bewildering choices of size, shape, composition, hardness, and corrosion resistance; the variability in artificial lures seems infinite. Not lagging behind has been the enhanced efficiency of reels, which Griswold (1922: 20) decried long ago: "With the invention of the reel drag the science of tarpon fishing received a *coup de grâce*. It is no longer fishing but 'coffee grinding' and the fish have no chance whatever."

Anglers who pay attention to experimental findings also have advantages over their quarry. The work of Guigand and Turingan (2002) is a good example. They demonstrated in laboratory experiments that Atlantic tarpons attack prey from beneath, meaning that surface lures should be the most effective. They also found that tarpons approach prey slowly and are easily startled, at which point the feeding sequence is interrupted or terminated. A quiet, smooth approach is best. Tarpons seem less attracted to fast-moving prey. Finally, the fact that the compression phase of prey-capture is slower than the suction, or "strike" phase (Fig. 7.9) suggests that tarpons can be lost if the hook is set too early.

The Atlantic tarpon is a key component in the recreational fisheries of other countries bordering the Caribbean (Cruz-Ayala 2002, Dailey *et al.* 2008, Fedler and Hayes 2008, Garcia-Moliner *et al.* 2002). Its value to western African countries is poorly documented (Anyanwu and Kusemiju 2008), although the all-tackle world record fish weighing 286 lb 9 oz (≈ 130 kg) was caught 19 March 2003 at Rubane, Guinea-Bissau. Until then the record fish had been shared by another western African fish of 283 lb (≈ 128.4 kg) from Sherbro Island, Sierra Leone and one weighing the same from Lake Maracaibo, Venezuela. These two specimens were landed, respectively, in 1991 and 1956 (Dailey *et al.* 2008).

The Pacific tarpon is sometimes considered a recreational fish in Australia (Ley 2008: 6) and elsewhere, and occasionally advertised as such on recreational fishing and travel websites and publications, but the much smaller size makes it a less attractive target. Pollard (1980: 54) described it as "A very hard-fighting sport fish ...." The all-tackle record is ≈ 2.99 kg (6 lb 9 oz) from Tide Island, Queensland, Australia, 14 May 2000. Tarpons of both species have occasionally been released into impoundments for anglers (e.g. Howells and Garrett 1992; Thomas 1887), a practice that seems to hold little interest today, except perhaps in southeast Asia, where business groups in Thailand have imported Atlantic tarpons to stock recreational fishing lakes and reservoirs (see Preface).

Tarpons are also objects of passive entertainment, serving as spectacles in the equivalent of finny petting zoos. At Robbie's Marina at Islamorada in the Florida Keys, dozens of large, semi-tame tarpons gather underneath the docks. For a modest fee a tourist may buy a few freshly-thawed bait fishes (thread herring, *Opisthonema oglinum*, on a day I was there) and feed them (Fig. 8.3).

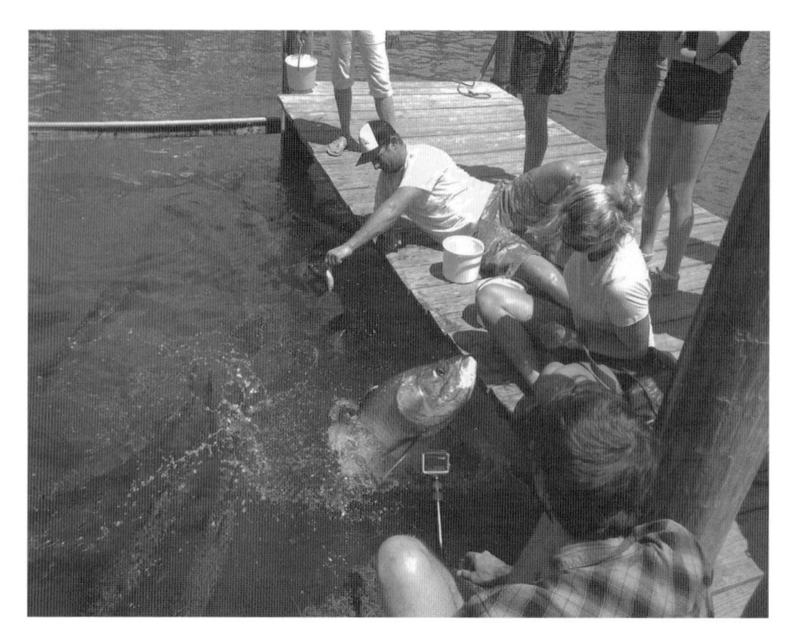

Fig. 8.3 Feeding the tarpons at Robbie's Marina, Islamorada, Florida. Source: Stephen Spotte.

8.3 Handling

Handling is activity that involves touching a fish *in any way,* including while it struggles at the end of a line, a kind of handling I consider *distant touch.* Just because human hands have not become involved directly, their influence on the fish is apparent from the instant the hook is set. After much observation and experimental work we know with certainty that angler behavior and choice of equipment affect the survivorship of fishes caught and released (e.g. Cooke and Suski 2005, Pelletier *et al.* 2007 and references). I define *handling time* in the recreational fishery as extending from the instant a fish is hooked and secured until the moment of its capture and retention or capture and release. If this seems unusually long, consider that the physiological effects of stress arise after variable time-lags, then continue to increase along a continuum. Stress is applied at the instant of hooking, and a sequence of stress responses continues to play out during the struggle, some stages in full force before the fish has been brought close enough to be touched by hand.

Any partitioning of this sequence goes unrecognized by the fish's nonspecific stress response. For example, plasma cortisol values were measured in snappers (*Pagrus auratus*) stressed by angling, longlining, and trawling with or without removal from the water and subsequent confinement. The investigators (Pankhurst and Sharples 1992: 345) wrote: "There was no difference in plasma cortisol between fish that were serially sampled or bled only once, indicating that increases in cortisol concentrations were due to capture stress and not the handling protocol." Put simply, when you hook a fish, everything that happens afterward is handling.

Short- and long-term homeostasis are both compromised by handling, often severely. That a recently caught fish swims away when released is not evidence of its ultimate survival. *Delayed*, or *latent*, *mortality* caused by capture stress often occurs hours – even days – later. That most fishes are adapted to tolerate at least some burst activity is indicated by dominance of the anaerobic myotomal muscle mass, which composes 80–95% of the swimming musculature (Driedzic and Hochachka 1978) and ≈30% of a fish's total weight (Skomal 2007).

However, the capacity to *sustain* burst activity is sorely tested once a fish has been hooked, which makes how long it is "played" important when keeping track of handling time. A prolonged fight at the end of a line is obviously detrimental and compromises survival prospects. This is nonetheless the objective of many anglers. Pelletier *et al.* (2007: 765) wrote: "It is impossible to ignore the fact that extended fighting is glamorized by the angling media and sought after by many anglers." They continued: "The angling media (e.g., fishing television shows) could catalyze this shift [in 'mindset'] by spreading the message about the importance of reducing playing times." Good advice, although unlikely to be accepted anytime soon by tarpon tournament anglers and their commercial sponsors.

To minimize both short-term and long-term mortality of tarpons brought close to the boat ("brought to leader" in angling terminology), four rules should be followed rigorously:

Rule 1: Play a tarpon for the shortest time possible to prevent exhausting it.

Rule 2: Never touch the fish when it reaches the boat, *even to measure it*. Disturbing the mucus enveloping a fish's surfaces or abrading or puncturing the skin increases exposure to potential disease-causing microorganisms from human hands, the boat and fishing gear, and the water. Rule 2 includes never using a net of any kind – including a knotless net. All of them strip away mucus, abrade the skin (sometimes with loss of scales and exposure of damaged skin), fray the fins, and require the fish to be lifted out of the water.

Rule 3: Never lift a tarpon out of the water (Fig. 8.4). Doing so causes the gills to collapse in addition to risking skin abrasion, loss of mucus, and frayed fins.

Rule 4: Avoid sloshing *any* fish back and forth underwater, "walking" it (Fig. 8.5), or otherwise attempting to "resuscitate" it prior to release.

Pelletier *et al.* (2007) queried every Canadian territorial and provincial natural resource agency and its equivalent in every state in the US about handling methods recommended in its recreational fisheries, including advice on how a newly unhooked fish should be "resuscitated." The results were to drag the fish back and forth in the water (63%), hold it facing upstream (14%), move it slowly forward (11%), pull it around in an S-shaped pattern (3%), and shift it side to side (3%).

All these manipulations would seem to have little or no ameliorative effect and might even be detrimental. So is "resuscitating" a tarpon for several minutes

Fig. 8.4 This illustrates inappropriate catch-release technique. Avoid handling a tarpon and lifting it out of the water. Source: ©lschneider | Dreamstime.com 29192490.

by attaching a Boga grip tool to a its mouth and pointing its head into the current, pulling it around in circles or figure-eights, or towing it slowly behind a boat. Enhancing oxygen uptake – which appears to be the principal objective – is doubtfully achieved, and the additional handling time is can hardly be beneficial. Guindon (2011: 19) noted: "In the case of [Atlantic tarpons], the capture event, boat-side handling, distance towed and weigh-in procedures would all contribute to confounding any towing specific effects, but results ... support the idea that towing is a form of excessive handling that may exacerbate observed stress responses and merits further investigation."

Although the initial flood of epinephrine (see below) heightens lamellar perfusion (Wendelaar Bonga 1997), additional gill lamellae have not had time to recruit. As mentioned previously in a mechanical context (Chapter 5.2), *the amount of oxygen taken up from each volume of water passing over the gills does not change with increases in water flow during exercise and hypoxia, and this holds true regardless if the hypoxia is environmental or functional (i.e. internal) in origin.*

Fig. 8.5 Another questionable catch-release procedure. No empirical evidence indicates that "walking" a tarpon prior to release or attempting to otherwise "resuscitate" a tarpon or any other fish is beneficial. Source: ©Flyfishingnation | Dreamstime.com 25557883.

Increasing volume flow artificially is therefore unlikely to be effective either. Oxygen transfer is passive, and diffusion between water and plasma is slow (Chapter 5.2). Because of oxygen's low solubility and diffusivity,[2] augmentation of gas transfer is by gill ventilation, except in species relying on ram ventilation. Pumping water is strenuous, consuming 10–20% of the metabolic activity of a resting fish (Chapter 5.2, 5.5, 5.6), percentages that rise during exercise. I see little value in dragging a fish around in an effort to simulate ram ventilation when the physiological benefit of doing so is untested and the procedure's effect on survivorship unknown. To tease out just the effect of "resuscitation" on oxygen uptake and survivorship requires an experimental design capable of evaluating each stage in the handling sequence separately and then assessing the data for interactive effects.

A fish just brought to leader is likely fatigued and distressed (Section 8.4). Little stored energy remains in its muscle tissues to be mobilized for powering opercular movements. Its plasma is undergoing severe acidosis. I mentioned previously that PCO_2 and the concentration of HCO_3^- are low in fish plasma (Chapter 6.5). Buffer capacity is thus limited, further restraining the modulation

[2]The diffusion coefficient in water is 2×10^{-9} m^2/s.

of internal pH by regulation of V_R. Hyperventilation is of little help. Internal pH must instead adjust by the much slower process of ion exchange with the water. Restoration of acid-base balance requires dehydration of plasma HCO_3^-, after which CO_2 can diffuse outward across the lamellar epithelia. None of these processes is enhanced by dragging a fish open-mouthed through the water for several minutes. Finally, any "resuscitation" procedure extends handling time, and the presumed benefit, especially to species not ordinarily dependent on ram ventilation, is doubtful enough to demand careful testing before becoming accepted practice in recreational fisheries.

These comments can be the basis for evaluating handling procedures currently used in the Atlantic tarpon recreational fishery. Removing tarpons from the water is now illegal in Florida, a useful improvement and one that makes Rule 3 above superfluous in state waters, although not necessarily elsewhere. A graph for estimating the weight of a tarpon based on its fork length and girth was published by Ault and Luo (2013: 115 Fig. 5) and also can be seen on the Internet.[3] Its objective is to provide an acceptably accurate and precise estimate of weight without having to directly weigh the fish or lift it above the surface. No such procedure shortens how long a fish is played (Rule 1 above), an issue that still needs addressing in tarpon recreational fisheries.

I agree with Shiffman *et al.* (2014) that length, and not weight, ought to be the criterion used in trophy fishing. In the case of Atlantic tarpons, weight can instead be estimated from length and girth (Section 8.3, Chapter 2.1), although even a rapid girth measurement requires added manipulation of the fish and prolongs handling time. Neither weight nor length as a single measurement meets the definition of size as defined here: a fish's weight at the time its length is measured (Chapter 2.1). However, as emphasized elsewhere, length is largely unaffected by body condition if angling records are to be based on just one indicator (Chapter 2.4).

Both length and weight provide important information to fishery managers and ecologists, although not necessarily to anglers practicing catch-release. I feel strongly that direct weighing of *all* fishes caught by angling and destined for release, and all proxies of weight including girth measurement, ought to be eliminated from recreational fisheries for two reasons. First, determination of weight, whether direct or indirect, extends handling time, if only briefly, often requires removing a fish from the water (if weighing is direct), and usually involves touching it. Second, weighing a fish is a less reliable procedure than measuring it. Scales require calibration to guarantee accuracy and precision, and proxies of weight introduce statistical errors of estimation. The only possible error a tape measure introduces is restricted to the user's competence, and almost anyone can learn to measure a fish.

[3] http://www.bonefishtarpontrust.org/btt-publications/tarpon-weight-calculator.html. Downloaded 7 July 2015.

Because growth in fishes is largely indeterminate (i.e. they continue growing throughout life), length relative to weight is less affected by body condition (Chapter 2.3). A fish's age, sex, nutritional and reproductive status (e.g. starved or well-fed, ovaries immature vs. packed with roe), and even whether it has eaten recently and its gut is empty or stuffed with food, affect how much it weighs but not how long it is. Were recreational fishing records based on length alone, handling times might be shortened and a hooked fish set free in a less compromised physiological state.

8.4 Stress effects

A stressful event for a fish can be defined as unpredictable disruption in its routine (Pankhurst 2011). This certainly includes being caught on hook and line or in a net. Severe disruption in routine precipitates behavioral changes, typically violent struggling involving burst exercise (Chapter 5.4), and this in turn triggers a cascade of physiological responses apparent from changes in pertinent blood values and often identifiable by deviant behaviors. Physiological stress responses are believed to be compensatory and thus adaptive (Wendelaar Bonga 1997 and references).

Some of these responses happen so quickly after a stress factor has been applied that gathering baseline (i.e. control) data has been a continuing problem (e.g. Perry *et al.* 1985, Skomal 2007). The issue becomes how to take a blood sample quickly and without handling – and thus stressing – the subject, and it remains a principal observer effect in physiology. Capturing and bleeding a fish rapidly does not guarantee baseline values of all appropriate variables, nor does the use of anesthetics beforehand. Certain stress responses, such as upsets in acid-base balance, occur almost instantaneously (see below), and some teleosts reveal endocrine changes even under deep anesthesia (Spotte *et al.* 1991), which induce secondary stress responses (see below).

Breder (1944: 243) wondered whether extended playing of Atlantic tarpons by anglers lowers survival, and since his time some have attempted to test it. Breder noted shark predation as a factor that increases the mortality of hooked and recently released tarpons, as did early anglers (e.g. Griswold 1921: 171, Kaplan 1937: 132), and this is still a common observation during modern tarpon tournaments.[4] Of course, where tarpons gather in large numbers, sharks do too, and some predation is inevitable.

The greatest impact on adult Atlantic tarpons in North America, separately and in combination, is predation by sharks and angling-induced death in the recreational fishery. In catch-release programs, human predation is indirect: tarpons

[4]http://myfwc.com/research/saltwater/tarpon/catch-release/tampa-bay-study; http://myfwc.com/research/saltwater/tarpon/catch-release/boca-grande-pass-study. Downloaded 13 July 2015.

occasionally die later from delayed mortality induced by physiological factors (*primary mortality*), or their loss to sharks becomes more probable because they are restrained during capture or exhausted from struggling and killed after release (*secondary mortality*). This is often the case with other species too. For example, predation on bonefishes caught by angling is greatest in the first 20 minutes post-release (Danylchuk *et al.* 2007). Catch-release programs can be considered successful only if the fishes released survive to reproduce, or at least fall within acceptable confidence intervals of longevity compared with cohorts that have never been captured. A catch-release protocol presumes, in other words, that angling does not diminish fitness (or affects it minimally), and this should be the ultimate standard by which such a program is judged.

Edwards (1998) assessed short-term survival of Atlantic tarpons released by anglers near Boca Grande Pass, southwestern Florida, using acoustic tracking of transmitter tags applied prior to release. The fish ranged from an estimated 8–68 kg ($\bar{x} \approx 38$ kg) and struggled on the line from 10–75 min ($\bar{x} = 22$ min). Fish were monitored until their transmitters detached and floated to the surface (range 1.1–12.1 h, $\bar{x} = 4.8$ h). Of 27 tagged tarpons, 26 survived. Edwards (1998: 1) attributed this high survivorship to "local angling techniques and handling practices." These are optimistic numbers, but not necessarily cause for celebration. Adams *et al.* (2013: 13) pointed out that at present >95% of the tarpon catch occurs in Florida waters, and that most specimens are released. Using permits issued as proxies for harvest in recent years, perhaps <1% of the recreational catch is kept (and killed) by anglers. Thus, "where *M. atlanticus* supports a catch-and-release fishery, post-release mortality likely has a greater impact on abundance than does [*sic*] harvest." Based on a study by Guindon (2011), Adams and colleagues concluded that post-release deaths from sharks was ≈13% (95% CIs 6–21%). However, the upper 95% of the CIs "may reach 28% when analysed by individual estuarine systems." Guindon's analysis pointed to an annual recreational post-release mortality of 8000–16000 tarpons in Florida alone (Fig. 8.6). I could not find this data in her report.

Short-term survival can only provide a minimum estimate of actual mortality (Edwards 1998), and this is why long-term survivorship eventually requires monitoring too, especially of foul-hooked specimens. Of the fish Edwards monitored, 19 were caught using live bait on single hooks and eight on jigs with single hooks. Hooking location was inside the mouth in the jaw (22 fish), outside the jaw (three), and outside the ocular orbit (one). Hook location for one fish was undetermined. Two fish had been removed from the water for photographs, and one of these died.

More recently, estimated short-term mortality of Atlantic tarpons also fitted with acoustic tags and caught and released in the same area was ≈20% with death defined as cessation of tag movement, and ≈10% when death was confirmed visually.[5] These percentages are neither better nor worse than catch-release data

[5] http://myfwc.com/research/saltwater/tarpon/catch-release/boca-grande-pass-study. Downloaded 13 July 2015.

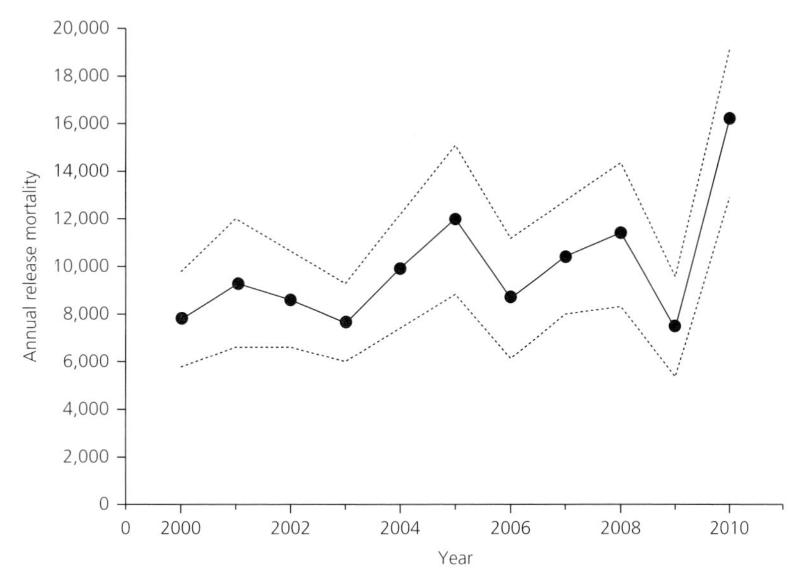

Fig. 8.6 Annual post-release mortality of Atlantic tarpons in the US recreational fisheries from 2000–2010. Dotted lines represent 95% CIs. Mean mortality was calculated using short-term catch-release deaths of 13%, representative of Florida's Gulf Coast recreational fishery and included post-release shark predation. Also see Guindon (2011). Source: Adams *et al.* (2013: 14 Fig. 8).

examined in a meta-analysis (n = 274) by Bartholomew and Bohnsack (2005) involving 48 freshwater and seawater species (but not tarpons). They found an overall median release mortality of 11% $\bar{x}$ = 18%, range 0–95%). Hooking location was the most important variable affecting mortality, and foul-hooked specimens (i.e. those hooked anywhere except the mouth) were most likely to die later. Other harmful variables included removing hooks from deeply hooked fishes instead of simply cutting the line, although according to Skomal (2007 and references) hooks retained in the stomachs of blue sharks (*Prionace glauca*) causes chronic systemic disease (presumably inflammatory). Even so, cutting the line and leaving the hook in place in deep-hooked specimens of common snook (*Centropomus undecimalis*) resulted in lower short-term mortality (Taylor *et al.* 2001).

Other factors that increase the incidence of catch mortality include use of J-hooks instead of circle hooks (Cooke and Suski 2004), use of natural instead artificial baits, exposure to air, and extended handling times. Hooks are manufactured to exacting specifications (Fig. 8.7). Hook type is undoubtedly an important variable, as shown in studies of other species both large and small. Hooks that are J-shaped tend to be ingested more deeply than circle hooks (Fig. 8.8), and natural baits are more likely to be swallowed than lures (Cooke *et al.* 2001, Pelletier *et al.* 2007 and references, Siewert and Cave 1990). In both cases hooks lodged deeply can cause internal organ damage and increase the handling time needed to remove them (Cooke *et al.* 2001, Siewert and Cave 1990). Hook removal is

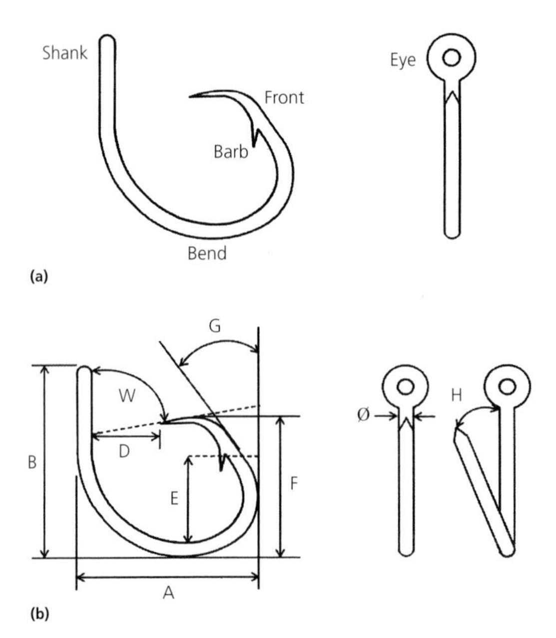

Fig. 8.7 Parts of a circle hook. (**a**) Basic components. (**b**) General specifications: A, width; B, length; D, gap; E, throat; F, front length; W, point angle; G, front angle; H, offset angle; Ø, wire diameter. Lettering conforms to hook manufacturer conventions. Specifications for a "true" circle hook: angle of point to shank must be a minimum of 90°; angle of front length must bend a minimum of 20° toward shank; front length of hook should be 70–80% of hook's total length. Source: Serafy *et al.* (2012: 375 Fig. 2).

usually impossible without increased handling, which also involves touching the fish (often gripping it tightly) and exposing it to air. Cutting the line eliminates both situations but leaves the hook inside the fish. Skomal *et al.* (2002) estimated post-release mortality at 4% of juvenile bluefin tunas (*Thunnus thynnus*) caught using circle-hooks, but 28% in those caught on J-hooks.

The introduction of conservation practices in angling has advanced considerably since Kaplan (1937: 102) wrote:

> *The tarpon's undisputed reputation for gracefully fighting to the finish gives it a distinct appeal to those who seek this mansize sport. The magnificent thrills furnished by the oscillating tarpon justifies the belief that this species was created especially to entertain salt water anglers.*

Such a homocentric perspective, often illustrated and reinforced in old angling publications by sepia-tinted photographs of dead tarpons strung up by their mandibles (Fig. 8.9), today provide a seagoing equivalent of Oscar Wilde's take on fox hunting as the unspeakable in full pursuit of the uneatable. Or Saul Bellow's obvious jab at Hemmingway's macho persona in the opening of his novel *Dangling Man* (Bellow 1944: 1–2): "The hard-boiled are compensated for their silence; they fly planes or fight bulls or catch tarpon, whereas I rarely leave my room."

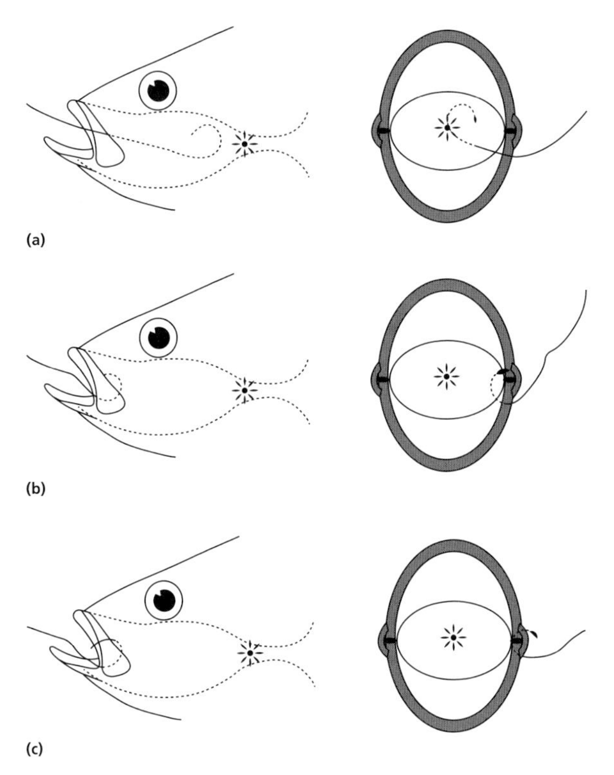

Fig. 8.8 Schematic illustrations of how a circle hook works when pressure is applied to the line (lateral and frontal perspectives). (**a**) The fish ingests the bait and starts to swim away (or the angler applies gentle pressure). (**b**) The hook is pulled to the side of the mouth, where its point catches the flesh at the jaw and pivots outward with increasing pressure. (**c**) At a critical line tension, either by the fish's swimming or the angler giving a tug, the hook slides over the jaw and rotates, becoming embedded. Unlike J-hooks, circle-hooks do not require a hard yank on the line to become "set" in a fish's jaw. Only gentle pressure is usually required, and tugging too hard is likely to cause foul-hooking. Source: Cooke and Suski (2004: 301 Fig. 2).

Prior to catch-release of tarpons becoming mandatory in Florida, recreational anglers selected for large size, ordinarily killing all or most of the fish they caught and saving the largest to be weighed and photographed. Modern anglers still select for size, the difference being that the fish are not killed, at least not purposely so. A few early tarpon anglers showed sympathy for their prey. Griswold (1922: 16) disapproved of tarpon fishing using light tackle, writing that "the charm of heavy fishing is being 'up against' a big fish and landing him as quickly and as humanely as possible." Griswold nearly always released his catch. He seems to have been an exception.

The objective of most was to subdue the fish at all cost, which included fighting it to total exhaustion. The fish was then gaffed and slid aboard. Haldeman (1892: 120) described his preferred method, devised after having lost a large specimen through careless handling by his boatman: "Since [then], I play my

Fig. 8.9 Dead tarpons strung up. Source: Griswold (1921: frontis).

fish, as usual, until completely exhausted, when I bring him to the side of the boat, and make the boatman run his hand and arm up his gills, out through the Tarpon's capacious mouth, and lift him gently into the boat." He emphasized that "hereafter I will exhaust my fish thoroughly ...." Haldeman's (1892: 124) notion of properly latching onto a tarpon was inducing it to swallow the hook: "It is generally conceded by all that a Tarpon must be well hooked in the gullet before the chances are at all favorable for his capture." These practices – fatiguing the fish, running an arm under the operculum and out the mouth, and deep-hooking – are antithetical to modern catch-release protocols. Unfortunately the first is still practiced.

Tarpons can still die on the line "while gracefully fighting to the finish," but more often are released barely able to move and suffering the effects of severely disturbed homeostasis. These fish display abnormal behavior. Some swim slowly away; others drift passively with the current, lose equilibrium, float on their sides or upside-down, or surface frequently before sinking from view (e.g. Edwards 1998). Their short-term fates are hard to monitor, any long-term effects extended over several days or weeks even more difficult, but monitoring is necessary if fishery managers expect to obtain useful data. Being able to predict mortality rates based on some measurable population factor and obtained independently of the recreational fishery would also be helpful. One is survivorship, which requires adequate data on sizes of the target species at known age (Section 8.8).

Hooking a fish elicits burst exercise, consuming energy and inducing anaerobic respiration. The initial series of physiological responses, termed *primary (neuroendocrine) responses*, is hormonal, and it stimulates a suite of *secondary responses* requiring adjustments in metabolism, respiration, acid-base balance, immune function, osmo- and ionoregulation, and other internal events. Secondary responses are identified by increases in $\dot{Q}$ and $\dot{M}O_{2w}$; unbalanced water and ion uptake, transfer, and retention with subsequent shifts in plasma osmolality; a rise in plasma metabolites (e.g. glucose, lactate); and changes in enzyme concentrations. Spikes in hematocrit and sometimes hemoglobin can occur (Suski *et al.* 2007). Secondary responses also include some or all of the nonspecific behavioral abnormalities mentioned earlier (e.g. loss of equilibrium, floating upside-down). Along the stress-death continuum, a fish suffering these physiological and behavioral traumas can be considered not merely stressed, but *distressed* (Wendelaar Bonga 1997 and references).

Many quantitative aspects of stress are species-specific (Skomal 2007 and references), in addition to depending on the particular stress applied and its intensity and duration. *However, the physiological and behavioral responses named above are nonspecific; that is, they remain the same regardless of the nature of the stress itself.* Thus playing a fish on hook and line induces the same physiological and behavioral responses as close confinement, chasing, netting, exposure to toxicants, or nearly any other deviation from its typical experiences.

Recovery time from a nonspecific stress response varies by phylogenetic group, species, and even individual. Plasma acidoses associated with exhaustive exercise are strictly metabolic in blue sharks (Skomal and Chase 2002), but their origins are both respiratory and metabolic in tarpons, the bluefin tuna (*Thunnus thynnus*), skipjack tuna (*Katsuwonus pelamis*), yellowfin tuna (*Thunnus albacares*), and some other teleosts (Perry *et al.* 1985, Skomal and Chase 2002). In the species just named, plus the white marlin (*Tetrapturus albidus*), stress level is commensurate with time spent struggling on the line (Skomal and Chase 2002). The skipjack tuna clears blood lactate quickly, recovering to normal concentrations after 50 minutes, a remarkably short time for a fish (Perry *et al.* 1985). In contrast, Holeton and Heisler (1978) found the process to take up to 36 hours in the larger spotted dogfish (*Scyliorhinus stellaris*).

Less easily identified are *tertiary (whole-animal) responses*, which occur over extended time, appear at the population level, and are manifested in some general or specific dysfunction like disturbances in metabolic scope, growth, condition, disease resistance, behavior, and survivorship. By being chronic instead of acute, any adaptive value in the response attenuates (Wendelaar Bonga 1997 and references).

The focus here is on acute stress and the primary and secondary responses induced by handling and restraint.

The teleost primary (neuroendocrine) stress response occurs in two steps separated by a short interval (Barton 2002). The first is a rise in plasma catecholamines released from the kidney's chromaffin tissue. Epinephrine is the most prevalent form of these in teleosts. The second step is a more persistent spike in plasma cortisol released to the plasma after synthesis in the interrenal tissue. Fishes facing acute stress typically experience elevated cortisol values of 30–300 ng/mL and peaking within 0.5–1 h after the disturbance (Barton 2002 and references).

Because a fish's weight comprises ≈ 30% white (anaerobic) muscle but just 3–6% blood, biochemical changes in the blood strongly mirror those occurring in the muscle tissue. In practical terms, deviations from normal of certain blood values not only correlate with a stress factor, they often delimit its magnitude (Skomal 2007 and references). Secondary responses vary by situation and species, but a spike in glucose (hyperglycemia) released from the liver in response to increased metabolic demands is notable (Wendelaar Bonga 1997 and references). The source is catecholamine-mediated glycogenolysis and followed by cortisol-mediated gluconeogenesis. Also occurring is a lactate spike from the formation of muscle lactic acid following anaerobic metabolism (e.g. Pankhurst 2011). Recovery to baseline concentrations of both factors takes hours or days, depending on species and intensity of the stress. In seawater-acclimated fishes this includes dehydration from compromised osmoregulation (Wendelaar Bonga 1997 and references). Bonefishes forced to exercise for 4 min experienced disrupted ionoregulation, shown by substantial rises in plasma Na^+, Cl^-, and Ca^{2+}, and a two-fold increase in plasma glucose, which returned to control values by 4 h post-exercise (Suski *et al.* 2007). Lactate had risen 13 times above control values by 1 h post-exercise and returned to normal after 4 h. Hematocrit, which had spiked four-fold by 1 h post-exercise, still remained significantly higher than controls 4 h later, although hemoglobin, which was twice control values at 1 h post-exercise, was the same as the controls after 4 h.

Exposing a fish to air for even a short time – especially an exhausted fish – compromises its chances of survival when released. Exercising bonefishes for 4 min followed by exposure to air for 1 min caused blood lactate to rise about 10 times higher than control values, or about four times the values of fish exercised for 4 min but not air-exposed. Fish exercised for 4 min and air-exposed for 3 min had lactate concentrations ≈ 15 times control values even after 2 h of recovery. Some of these specimens died. When Ferguson and Tufts (1992) exposed fatigued rainbow trout to air for 30 s the mortality increased 38%; exposure for 60 s raised it to 72%. Ferguson and Tufts (1992: 1157) concluded: "These results indicate that the brief period of air exposure which occurs in many 'catch and release' fisheries is a significant additional stress which may ultimately influence whether a released fish survives."

Stress resistance varies by species, and quantitative results obtained from one are not directly applicable to another.

A fish's existing physiological status sometimes exerts a feedback effect. The rise in catecholamines, for example, is more severe if accompanied by functional hypoxia. The results might then include an increase in the hemoglobin's affinity for oxygen or release of hepatic glycogen as glucose to the plasma (Wendelaar Bonga 1997 and references).

In the colorful description of Wells and Baldwin (2006: 347), fishes are "swimming machines composed of two slabs of anaerobically poised muscle attached laterally to an incompressible backbone." They continued: "This large muscle mass is poorly perfused, has few mitochondria, and low concentrations of myoglobin." Moreover, "The fast twitching, anaerobically operating muscle fibres predominate over slow oxidative fibres and their proportions reflect the athletic ability of the species ...." So described, tarpons qualify as moderately athletic (Chapter 5.5), and were this not so, then angling for them would be less exciting and exploitive.

Atlantic tarpons are admired as recreational fish in part for their leaping ability. According to Randall and Moffett (1958) they can jump 12 ft (≈3.6 m) vertically and 25 ft (≈7.6 m) horizontally. This evidently inspired some early anglers to shoot at them as they became airborne. Southworth (1888: unpaginated), after describing the dense, bony structure of tarpon scales, wrote: "A close examination of these scales readily accounts for the failure of fishermen heretofore in all their endeavors to shoot the tarpon when jumping in the air." He went on to state: "A rifle-ball will strike the scale and immediately be deflected, failing to penetrate the body." Thus did angling and hunting briefly merge into the sport of skeet-shooting hooked tarpons.

We can assess this athleticism from other features of anatomy and physiology. Tarpons have a high proportion of aerobic (i.e. red) muscle tissue. Tarpon blood has considerable oxygen carrying capacity and demonstrates a strong Bohr effect (Chapter 5.5), this last favoring unloading of oxygen from blood to tissues for sustained aerobic swimming (Wells *et al.* 1997). Furthermore, bimodal and obligate air-breathing fishes in general have higher oxygen carrying capacities than obligate water-breathers, perhaps as a response to low P_aO_2 (Johansen *et al.* 1978).

Water holds only ≈3% as much oxygen as contained in atmospheric air at sea level. We might think that because tarpons are bimodal breathers (Chapter 5.1) they can recover quickly after heavy exertion simply by inhaling air, but this assumption would be a wrong. As Heisler (1993: 20) wrote: "The switch of facultative air breathing fish from water breathing to air breathing is directly comparable to the exposure of exclusive water breathers to hyperoxia: in both cases the concentration of oxygen in the breathing medium is largely enhanced." Were tarpons mammals, hyperventilation above the surface would restore acid-base balance in minutes.

During anaerobic energy production, such as occurs when a tarpon struggles at the end of a line, lactic acid empties into the blood. A consequence of the fish's explosive activity is functional hypoxia. The dissociation of lactic acid yields H^+ and lactate, the latter carrying a negative charge. Some excess hydrogen ions are excreted from intracellular spaces into the water outside (Heisler 1993 and references). Others are buffered in the blood by mobilized HCO_3^- and non-bicarbonate buffers. However, lactate persists, is *never* transferred to the environment, and must be cleared by aerobic metabolism. One product of burst exercise, in other words, is lactate with H^+ and CO_2 being released into the blood and lowering its pH. Because the tarpon exhibits a sharp Bohr effect when its internal milieu becomes acidic (Wells *et al.* 2005: 89), it is probably incapable of saturating its blood with oxygen when plasma pH falls.

Despite rising to the surface occasionally to take in air, tarpons released after capture must rely on oxygen in the air retained in their swim bladders, but also on what can be obtained via the gills. One result of functional hypoxia is production and retention of excess CO_2, forcing repayment of the oxygen debt incurred by anaerobic respiration during the struggle and then clearing the blood of excess lactic acid. Any reduction in blood CO_2 for acid-base balance (i.e. blood pH compensation) is unlikely, simply because the amounts produced exceed what can be eliminated. In the larger spotted dogfish, an elasmobranch, H^+ produced during brief periods of anaerobic exercise is equivalent to the amounts generated during 3.5 h of resting aerobic metabolism (Holeton and Heisler 1983).

To repeat, tarpons experience a substantial Bohr effect during functional (internal, or respiratory) acidosis, which enhances oxygen turnover by releasing O_2 to the muscles as blood pH declines (Wells *et al.* 2005). This effect can also be induced directly during times of aquatic hypoxia and indirectly by aquatic hyperoxia. Adult tarpons living in seawater are unlikely to encounter either condition. The rates of both air- and water-breathing increase during environmental hypoxia. Air-breathing enables a tarpon to sustain high aerobic metabolism by transferring oxygen from its swim bladder to other tissues. The mechanism (Bohr effect) capitalizes on the low affinity of tarpon blood for oxygen, providing an advantage over many obligate water-breathers.

With either the Bohr or Root effect in play, functional acidosis reduces the oxygen available for aerobic metabolism. Hypoxia – either environmental or functional – is a strong stimulus for the induction of air-breathing, no less so during exercise (Seymour *et al.* 2004). However, air-breathing is less advantageous for a tarpon, even in normoxic water with free access to the atmosphere, as it struggles at the end of a line experiencing anaerobic acidosis.

What happens when any fish is played on hook and line? During prolonged aerobic swimming at high speed (not burst exercise), $\dot{M}O_{2w}$ typically increases 12–15 times above resting values, >90% of it shunted to aerobic muscles straining in the fish's effort to escape (Randall 1982 and references). Meanwhile $\dot{Q}$ increases 3–4 times its resting level, mainly through increased V_S. The consequences amount to a disproportionate increase in the $\dot{M}O_{2w}/V_R$ ratio from a resting value of ≈10–30 and a nine-fold reduction in water residence time in the gill cavity from ≈250 ms at rest to ≈30 ms while struggling.

As Randall (1982: 284) pointed out, although increased volume flow temporarily thins the boundary layer between gill lamellae and newly inspired water, theoretically enhancing oxygen uptake, "The net result is that the amount of oxygen removed from each volume of water as it passes over the gills does not change with increases in water flow during exercise and hypoxia...." (Or doubtfully during "resuscitation" either.) The rise in volume flow results largely from small spikes in V_R and substantial increases in V_S. With prolonged aerobic exercise come manifestations of functional hypoxia, resulting from muscular activity and functional acidosis. Breathing eventually becomes arrhythmic, forcing compensation by switching to ram ventilation, an advantageous situation for a fish swimming freely but impossible for one tethered to a hook. Then the only outcome is fatigue.

Lactic acid entering the blood of a fish can be viewed as purging it of gases (Alexander 1993). As plasma pH falls – both a cause and an effect of the rising lactic acid concentration – oxygen bound to oxyhemoglobin is released through Bohr and Root effects. In a milieu of reduced internal pH, lactic acid also releases carbon dioxide (another gas) from its bound form with plasma bicarbonate. And in the manner of any electrolyte, lactic acid renders all blood gases less soluble by the "salting out" effect, much as oxygen is less soluble in seawater than in freshwaters.

We know tarpons water-breathe during environmental normoxia, but they start surfacing to breathe more often as environmental conditions turn hypoxic at ≈6 kPa (Wells *et al.* 2007). A tarpon hooked in normoxic water often jumps several times, but mainly stays submerged. We might assume air is imbibed during those seconds of being airborne, but thinking this would be speculation. Even if true, leaving the water evidently fails to ameliorate an internal cascade of events already in progress that culminates in functional hypoxia, acid-base imbalance, and exhaustion. Air-breathing by fishes with physostomous swim bladders is a four-stage process, fine-tuned by evolution to seamlessly integrate the mechanics of breathing with the physiology of respiration (Chapter 5.3).

Strenuous aerobic swimming during capture on hook and line that is interrupted by flashes of burst exercise causes spikes in blood lactic acid and glucose, disrupts hematological stasis by elevating hemoglobin (Hb), and reduces mean cell hemoglobin concentration (MCHC). Wells *et al.* (2003) monitored these and other factors after playing juvenile Pacific tarpons on hook and line. Pacific tarpons normally have high resting levels of hematocrit (37.6 ± 3.7%) and hemoglobin (120.6 ± 7.3 g/L), but both nonetheless rise significantly to 51.9 (± 3.7%) and 142.8 (± 13.5 g/L) immediately after struggling for as little as 15 min. Swelling of the erythrocytes induces a drop in MCHC from a 24-h resting value of 317 (± 19.8) g/L to 275.1 (± 14.5) g/L. Lactate derived anaerobically from white muscles during burst exercise reached 40.67 (± 9.44) μmol/g; after 24 h of rest the value fell to 5.62 (± 3.18) μmol/g. Lactate, which is transferred quickly from muscle tissue to the blood, measured 6.66 (± 0.99) mmol/L compared with the 24-hours resting value of 0.67 (± 0.17) μmol/L, a spike of 1×10^4. *In a separate group of fish hooked and retrieved rapidly, values of these factors immediately after capture were close to resting values 24 h later.*

Those intense 15 min endured by the test group depleted tissue oxygen and left insufficient quantities for normal aerobic metabolism. Once the fight ends and the exhausted tarpon has been brought to the boat and released, its tissues must repay this oxygen debt. Although air-breathing helps restore normal hematological and blood chemistry parameters and keeps the metabolic scope stretched during strenuous exercise (Wells *et al.* 2003, 2007), *clearance of lactate from incurring an oxygen debt is accomplished by breathing water* (Wells *et al.* 2007). Water's greater density and diminished oxygen-holding capacity compared with air make the restoration of tissue oxygen a slow process. Among its many functions, aerobic metabolism is the pathway for reprocessing the excess lactate and glucose and returning them to resting values.

Elevated lactate, by disrupting acid-base status and interfering with oxygen uptake, is itself an indicator of metabolic scope (Wells *et al.* 2007 and references). The tarpon's high Bohr response is theoretical evidence that post-angling acidosis could be severe, possibly life-threatening. Would holding a post-caught tarpon, or any fish, briefly captive hasten lactate clearance? Probably not, and especially if the procedure requires removal from the water and exposure to air. Furthermore, restraint by confinement is an additional stress.

In practice, some anglers believe that post-release mortality can be reduced by temporarily keeping newly-caught fishes in live-wells prior to release, putatively buying them time to recover normal physiological and behavioral stasis. Whether more of them actually survive is unknown, and the expected return of blood values to homeostatic concentrations during a few hours of close confinement is questionable. Shultz *et al.* (2011) addressed this issue using bonefishes fatigued by simulated angling and found the procedure especially detrimental if water in the live-wells was either hypoxic or made hyperoxic by entrainment of gaseous oxygen. Lactate clearance was most efficient in normoxic seawater. They wrote (Shultz *et al.* 2011: 19): "These results suggest that anglers and tournament organizers should recover angled bonefish in normoxic seawater." This is not be surprising. The harmful effects of aquatic hypoxia and hyperoxia have been discussed (Chapter 5.6). In the case of hyperoxia, a predictable and well-documented outcome is acidosis. In reference to captive fishes, Spotte (1992: 221) wrote:

> *The most remarkable lesson to be learned from how fishes respond to environmental hyperoxia is that oxygenation of the tissues overrides any requirement to rectify simultaneous disturbances in acid-base balance. In other words, a fish placed in a hyperoxic environment "forgets" to breathe, because its tissue oxygen requirements are satisfied even though the subsequent reduction in gill ventilation rate causes a rise in* $PaCO_2$.

In experiments by Wells *et al.* (2007), juvenile Pacific tarpons ($\bar{x}$ = 186.36 ± 25.62 mm FL, 89 ± 38.02 g, *n* = 36) were swum in a freshwater flume at 27°C. Forced swimming speeds were 0.1 (slow) and 0.3 (fast) m/s for 30 min at normoxia (20.8 kPa) or hypoxia (6.1 kPa). Air-breathing rate (f_{ab}) in the slow-speed-hypoxia configuration was $\bar{x}$ = 0.49 (± 0.14) breaths/min and $\bar{x}$ = 0.63 (± 0.0.17) breaths/min at high-speed-hypoxia, a difference not statistically significant.

Another 24 tarpons were exercised vigorously by forcing them into burst swimming until exhaustion, determined as no response to being touched, usually by ≈5 min. This treatment was labeled strenuous exercise. After a 30-min rest (time enough for lactate to shift from muscles to blood), some fish were returned to normoxic water for 2 h ($\bar{x}$ = 216.6 ± 24.6 mm FL, 140.8 ± 45.8 g, *n* = 12) or 4 h ($\bar{x}$ = 208.9 ± 37.0 mm FL, 130.1 ± 75.1 g, *n* = 12), with or without access to air.

These findings provide the best evaluation so far of how angling affects the respiratory physiology of tarpons. Muscle lactate was significantly elevated during slow-speed-hypoxia, but the highest values were in fish denied access to air by a submerged screen. In the two fast-speed groups subjected to aquatic hypoxia, both muscle and blood lactate were higher when access to air was prevented (Fig. 8.10). The effects on glucose and the hematological variables tested were not significant. The muscle-blood lactate of the five groups (Fig. 8.11) evaluated using linear regression showed strong positive correlation (r^2 = 0.97), *indicating consistent release of muscular lactate regardless of treatment effect.*

Blood lactate more than doubled right after strenuous exercise, and blood glucose was 50% of values observed following fast-speed-hypoxia with air access denied (Fig. 8.11). Blood lactate and glucose still fell within 4 h to values near those after slow-speed-normoxia, and attenuation occurred whether or not access to air was denied. *Air-breathing offered no advantage toward clearing blood lactate, but it helped substantially in aerobic mobilization of glucose*. Hematological factors returned to resting values during the 4-h recovery period (Fig. 8.12). Hematocrit and hemoglobin decreased and pH rose, again without significant differences between fish denied or allowed access to air.

From the standpoint of angling, forcing juvenile Pacific tarpons to swim at 0.3 m/s for 30 min in hypoxic water induced elevated concentrations of muscle lactate followed by rises in blood lactate. Both metabolites rose even higher in fish denied the opportunity to breathe air. As Wells *et al.* (2007: 1655–1656) explained: "Therefore, air breathing in this species supports a higher aerobic scope for activity than that permitted by branchial gas exchange alone." Even so, blood lactate concentrations in fish denied air access were still lower than those allowed it but subjected to strenuous exercise (i.e. swum to exhaustion). Metabolic (aerobic) scope is thus narrowed in tarpons prevented from breathing air. The salient finding by these authors is that recovery of homeostasis still occurs within about 4 h, even with air access denied, meaning that water-breathing alone is sufficient to repay the oxygen debt incurred while struggling on a line. Wells and colleagues

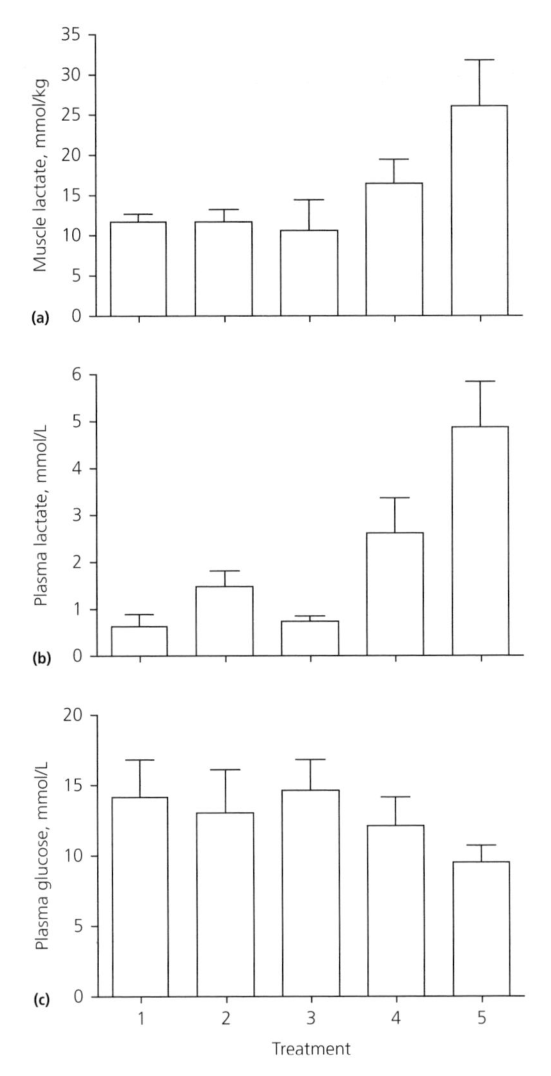

Fig. 8.10 Values for: (**a**) muscle lactate, (**b**) blood lactate, and (**c**) blood glucose in post-exercise juvenile Pacific tarpons subjected to five treatments: (1) aquatic normoxia, slow-swimming, air access; (2) aquatic normoxia, fast-swimming, air access; (3) aquatic normoxia, fast-swimming, air denial; (4) aquatic hypoxia, fast-swimming, air access; (5) aquatic hypoxia, fast-swimming, air denial. Swimming speeds were 0.1 m/s (slow) or 0.3 m/s (high). Dissolved oxygen was either 20.8 kPa (normoxia) or 6.1 kPa (hypoxia). Data represent $\overline{X}$ ±SD. Source: Wells *et al.* (2007: 1654 Fig. 1).

(p. 1657) concluded: "*Accordingly, air breathing probably permits Pacific tarpon to maintain high aerobic activity in hypoxic water through access to air, but an incurred oxygen debt appears to be repaid by aquatic breathing* [emphasis added]."

They stated that such a mechanism "makes ecological sense" by allowing an exhausted fish to avoid being killed by aerial predators while recovering, a conjecture perhaps appropriate for Pacific tarpons, but adult Atlantic tarpons,

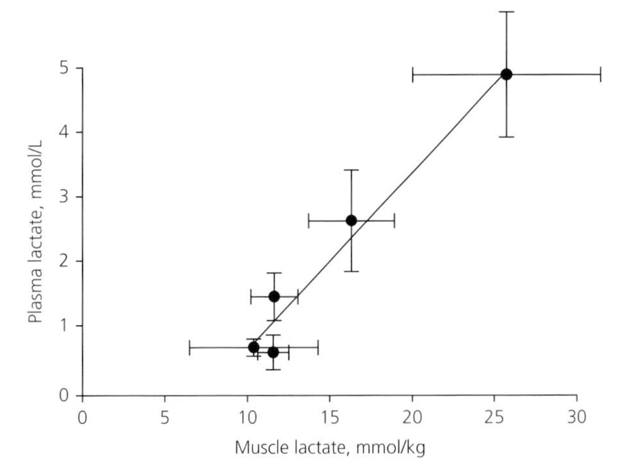

Fig. 8.11 Relationship between blood and muscle lactate for juvenile Pacific tarpons under functional and environmental normoxia and hypoxia from the five groups (Fig. 8.10) Data represent $\overline{X}$ ±SD (n = 6). Curve fit: $y = 0.27x - 2.05$. Source: Wells *et al.* (2007: 1655 Fig. 2).

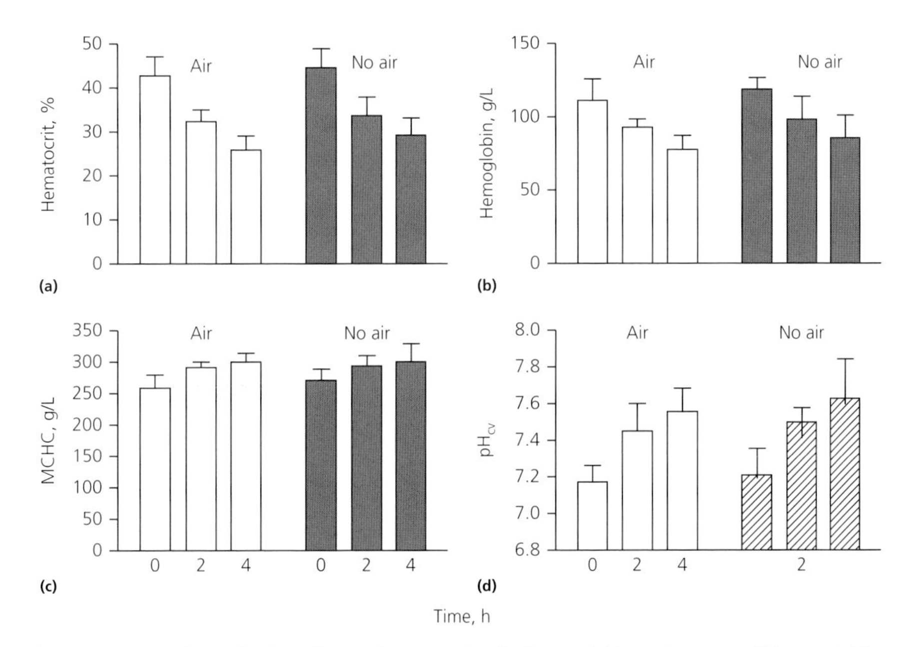

Fig. 8.12 Hematological values for (**a**) hematocrit, (**b**) hemoglobin, (**c**) mean cell hemoglobin concentration (MCHC), and (**d**) caudal venous pH (pH$_{cv}$) in juvenile Pacific tarpons swum to exhaustion and allowed to recover for 4 h, with and without access to air. Data represent $\overline{X}$ ±SD (n = 6). Source: Wells *et al.* (2007: 1656 Fig. 3).

not juveniles, are pursued by the recreational fishery, and no extant aerial predator could capture an adult and lift it out of the water. Humans are the Atlantic tarpon's only aerial predators, and sharks its principal undersea enemies, especially to those fought to exhaustion. Being hooked and played on a line has no natural equivalent for tarpons or any other aquatic creature in terms of how – or how quickly – an oxygen debt is repaid (Wells *et al.* 2007: 1659).

It can therefore be said that air-breathing enables tarpons to sustain a higher level of aerobic activity during exercise than could be achieved by water-breathing alone, but so far as I can tell this offers no advantage to a hooked tarpon, which, except for intermittent jumps, remains submerged. Add to this the fact that we have no knowledge of whether air is taken in when jumps do occur. Restoration of acid-base balance apparently is by water-breathing, meaning that its rate is restricted by several important factors, including slow rates of diffusion and ion exchange caused by water's high density and viscosity.

No account of angling for Atlantic tarpons is complete without tales of fishermen being maimed or killed in pursuing this manly sport, or in one case while innocently lighting his pipe. Randall and Moffett (1958: 142) wrote: "There are at least seven recorded instances of people killed outright, or knocked overboard and drowned, by tarpon." No sources were cited. Carson (1914: 402) offered this:

> *The sport is not unattended with danger; for when a big fish has not been properly gaffed, he is sometimes stirred into fresh activity, lashing out with his tail with a force strong enough to stave in a canoe. His cutting jaws can also inflict ugly wounds. A well-known American angler, while fishing for tarpon off the Florida coast, hooked a monster weighing considerably over a hundred pounds. During the combat the great fish made a leap which landed him with a crash on the angler's back, inflicting injuries which nearly killed the unfortunate fisherman, laying him up for nearly two years.*

Kaplan (1937: 101) referred to the experience of Richard L. Sutton, whose hooked tarpon "jumped into the boat ... and slapped the angler unconscious." Haldeman (1892: 122) had earlier described the strange experience of Uncle Charlie, a denizen of southern Florida, who was returning to port one afternoon in his sailboat. He had steadied the craft with a knee against the tiller, struck a match on his coat, and cupped both hands around his pipe to shield it from the wind. As Haldeman related,

> *he was in the midst of this interesting act when, suddenly, a mullet leaped from the water to port and darted clean across the stern of the boat, directly in front of him. He had not time to express his astonishment ere, in close pursuit of the mullet, a large tarpon rose, and came across the boat like a bolt from a catapult ....The huge fish stuck him full in the chest, and tumbled him like a log over the side of the boat. The shock of the collision threw the tarpon into the bottom of the boat, and left "Uncle" Charlie struggling in the water.*

His companion pulled him aboard, and a drowning by tarpon was averted. Uncle Charlie was subsequently bed-ridden for 3 weeks and never fully recovered. The fish, which remained aboard while Uncle Charlie took its place in the water, was examined in port. It weighed 164 lb (73.2 kg).

Commercial tarpon fishing is also not without dangers. Haldeman (1892: 112) claimed, "The Pensacola [Florida] seine-fishermen dread it [the Atlantic tarpon] while dragging their seines, for they have known of persons having been killed or severely injured by its leaping against them from the seine in which it was inclosed [*sic*]." According to Seraine (1958: 28), fishermen at Acaraú, Brazil entered the coastal weirs with some trepidation, wary of moray eels and stingrays, but also of the tarpons because of their great size and strength.

8.5 Commercial fisheries

The Atlantic tarpon's worth to the US commercial fisheries is essentially zero; elsewhere, its contribution is usually minor, either as a targeted species or part of the by-catch in more highly valued fisheries. Local abundance, ease of capture, market price, and consumer preferences influence the species compositions of regional fisheries. For example, Seraine (1958: 34) wrote, "*Entre os acarauenses o peixe mais cobiçado e sem rival no gôsto é o camorupim, que se captura em currais ....*" [Among citizens of Acaraú [northeastern Brazil] the most coveted fish and beyond compare in taste is the *tarpon*, which is caught in weirs ....] Seraine had this to say about residents of another city in Ceará State: "*Já em Paracuru, embora se associem ao curral e ao* camorupim *emoções apreciativas, não atingem elas o grau das registradas naquela outra área piscatória, sendo compartidas com outras espécies ictiológicas, como a cavala e a garoupa, e com outros tipos de pescarias ....*" [Today in Paracuru, although they appreciate the [fish] weirs and tarpons, their enthusiasm is less than at other fishery locations because of [the availability of] different ichthyological species like the mackerel and grouper, and other types of fisheries ....]

In other words, the tarpon is neither consistently preferred nor disliked even within narrow geographical ranges, being esteemed at one location but not another. Consumer preferences are obviously local at some places, but they can also be cultural. The tarpon is a delicacy in parts of western Nigeria, where it is served during special occasions, such as marriage ceremonies and festivals (Anyanwu and Kusemiju 2008).

Whether tarpons are even considered edible has a contradictory history. If we go back in time, most favorable reports by Europeans tend to be old ones, although reviews nonetheless varied by consumer, some labeling it tasty, others inedible. Many of these early writers were explorers, adventurers, diplomats, sailors, and colonists accustomed to eating rotten beef and biscuits baked with weevil-infested flour. Maybe anything reasonably fresh tasted good.

Near the end of their original species description of the Atlantic tarpon, Cuvier and Valenciennes (1846: 401) wrote: "*M. L'Herminier ... confirme qu'il vit bien dans les mares d'eau douce; que sa chair, blanche est à peu près du goût du Merlan; que les arêtes sont faciles à séparer; aussi est-il d'un avis contraire à celui de Marcgrave, en le donnant comme un poisson bon à manger et qui ne devient pas véuéneux.* [Monsieur L'Herminier ... confirms that it [the Atlantic tarpon] thrives in freshwater pools; its white flesh has a taste similar to Whiting; that the edges are easily separable; so he is of an opinion contrary to that of [Georg] Marcgrave, calling it a good fish to eat and one which does not become poisonous.] As for Marcgrave, (Cuvier and Valenciennes 1846: 400), *"cet auteur ne l'a pas trouvé agréable à manger"* [this author has not found it enjoyable to eat]. Whether the tarpons referred to at Guadeloupe by L'Herminier, Valenciennes' correspondent, were purposely cultured in freshwater pools or entered them on their own to be subsequently captured in the local fishery is not mentioned. Not becoming "poisonous" could mean that consuming them did not cause ciguatera, but this is conjecture.

As to the tarpon's edible qualities, Dampier (1909: 118) wrote of fish from México's Bay of Campeche, "'Tis good sweet wholesome Meat, and the Flesh solid and firm. In its Belly you shall find two large Scalops of Fat, weighing two or three Pound each ...." Here Dampier might have been referring to the ovaries of females in pre-spawning condition. Nietschmann (1973: 250) listed the tarpon among fishes known to the Miskito Indians of eastern Nicaragua, which they called "tarpam" and presumably ate. More recently, many who considered the fish edible appear to have been ichthyologists and other biologists who might actually have tasted it. One scientist (Röhl 1956: 455) considered tarpon roe, evidently prepared in salt in the manner of caviar, to be "exquisite."

Today, Mayan villagers in the Sian Ka'an Biosphere Reserve of central Quintana Roo, México operate a minor subsistence fishery for Atlantic tarpons in the larger karst sinkholes, catching them by hook and line during the dry season (Schmitter-Soto *et al.* 2002). Throughout México, the only legal tarpon fishery is now the recreational fishery, and tournaments take place mainly in summer at Tuxpan, Tecolutla, Veracruz, and Coatzacoalcos (Oviedo Perez *et al.* 2002). However, tarpons remain part of the incidental catch in the snook and shark fishery at Veracruz (Oviedo Perez *et al.* 2002).

According to Nichols (1912: 181), tarpons appeared occasionally in the Havana, Cuba market, and Griswold (1913: 98, 1921: 25) reported that Cubans gladly accepted the tarpons he caught on hook and line, eating them "fresh and salted." Heilner (1953: 214), after a day of angling with friends during which 105 tarpons were hooked, kept three fish, "as our Cuban friends wished to salt them down for future use." The species also has been used as food in the Bahamas (Breder 1933). Zaneveld (1962: 144 Table 2, 161 Table 9) rated it "moderately" appreciated as food and an occasional quarry of spear fishermen in the Netherlands Antilles. A commercial fishery that includes Atlantic tarpons is active in the Bigi Pan area of Suriname, at one time harvesting 6000–12 000 kg/month (Mol *et al.* 2000), and wildlife illustrator Karl E. Karalus reported having seen dried smoked tarpon for sale in Guyana,[6] adding parenthetically, "To me it smelled bad."

Tarpons in French Guiana are sold at some local markets (Anonymous 1992), and the species composes a minor part of the Colombian commercial fishery (Leal-Flórez *et al.* 2008: 368 Table 1; Rueda *et al.* 2011: 122, 123 Table 1). For example, it is listed among by-catch species of the Golfo de Morrosquillo southern pink shrimp (*Farfantepenaeus notialis*) fishery (Herazo C. *et al.* 2006), and Angel C. (1992) and

[6]http://www.karlsstudio.com/what-is-this-thing-called.pdf

Mercado Silgado (1981) included it among species fished commercially in Golfo de Morrosquillo. According to García and Solano (1995: 47), "'sábalo' soup is still one of the traditional dishes along the Caribbean coast of Colombia." Dahl (1965) observed fishermen there pursuing aggregations of tarpons and killing them by tossing sticks of dynamite. Ward *et al.* (2008) mentioned collecting scale samples from Atlantic tarpons at fish markets in México, Colombia, and Brazil, although exact locations were not specified.

Of tarpons in Nicaragua's Río San Juan, Simmons (1900: 329–330) wrote:

> *Wherever there is a shoal place in the river it [the tarpon] is to be seen breaking by the hundreds, and at the Toro Rapids, above Castillo, they are so numerous that they frequently jump into the boats ascending or descending. As many as five, measuring from four to six feet in length, have been known to jump into a boat on one trip down the rapids, which are only fifteen miles long. They are apt to bite the occupants of the boat or injure them by floundering about, and so a boatman usually stands ready, armed with a machete, to cut their heads off as soon as they strike the deck. They are not esteemed for eating, and so nobody attempts to catch them.*

Atlantic tarpons nonetheless remain part of the commercial fishery in Lake Nicaragua and some of its tributaries (Brown and Severin 2008). Meek and Hildebrand (1923: 174) reported that the tarpon in Panamá "is used as food to some extent and may be seen in the Colon market from time to time." However, its food value was thought to be generally low. Hildebrand (1937: 239) later wrote, "the natives and particularly the West Indian immigrants, however, are very fond of tarpon ...." He added that commercial tarpon fishing is done occasionally and remarked that tarpons are seen frequently in the Colon market, echoing Meek and Hildebrand (1923). According to Breder (1925a: 141), tarpons were the "favorite food fish" of the San Blas Indians of Panamá, although he never mentioned eating tarpon himself. This seems odd, considering the many years he spent studying Atlantic tarpons in Florida and elsewhere.

Perhaps the tarpon has never been popular among North Americans. Brice (1897) listed it as a valuable recreational fish to anglers fishing Indian River Lagoon in eastern Florida, although Wilcox (1897) omitted it from his list of two dozen or so commercially valuable fishes found there. However, Gill (1907: 37) gave a second-hand account of tarpons being salted and sold in the Key West market. Bean (1903: 179) quoted Dr. Smith, who claimed tarpons to once be so common in Massachusetts during late September that "an effort has been made to find a market for them in New Bedford, but the people did not like them, owing to the toughness of the flesh." W. E. Carson was an American who angled for tarpons extensively in México. In his opinion (Carson 1914: 402–403), "The only disappointing feature of tarpon fishing is that the dead fish is of no value whatever, the flesh being flavorless and rarely eaten." Kaplan (1937: 119) wrote, "The flesh of the tarpon, especially the larger ones, is coarse, dry and bony." He remarked that native Americans in southern Florida near the Everglades ate them.

Dampier (1909: 118) described how the "Moskito" Indians at México's Bay of Campeche along the Gulf of Mexico caught tarpons for food. Although they fished other species at sea, Dampier stated: "I never knew any [tarpons] taken with Hook and line; but are either with Nets, or by striking them with Harpoons, at which the Moskito-Men are very expert." Of net-fishing, Dampier wrote:

> *The Nets for this Purpose are made with strong double Twine, the Meshes five or six Inches square. For if they are too small, so that the Fish be not intangled therein, he presently draws himself a little backward, and then springs over the Net: Yet I have seen then [sic] taken in a Sain made with small Meshes in this manner. After we have inclosed a great Number, whilst the two ends of the Net were drawing ashore, ten or twelve naked Men have followed; when a fish struck against the Net, the next Man to it grasped both Net and Fish in his Arms, and held all fast till others came to his Assistance. Besides these we had three Men in a Canoa, in which they mov'd side-ways after the net; and many of the Fish in springing over the Net, would fall into the Canoa: And by these means we should take two or three at every draught.*

In the words of de Paiva Carvalho (1964: 4), "*No Maranhão, hã uns 15 anos, por ocasião da safra e em uma só maré, foram capturados 4500 quilos de "Camurupim", índice por si só capaz de server de base para se avaliar a importância dessa riqueza.*" [Some 15 years ago in Maranhão [State], during the season and on a single tide, 4500 kilos of "tarpons" were captured, in itself an index that can serve as a basis for assessing the importance of this resource.] He remarked that Brazilians living along the northeastern coast appreciated its food value, as did people in Central America, Guadeloupe, Martinique, Santo Domingo, Puerto Rico, and Haiti. According to da Silva de Almeida (2008: 240 Table 11), tarpons contribute in a minor way to the marine and estuarine fisheries of Maranhão State.

Giacometti Mai *et al.* (2012) listed the tarpon among 117 species of fishes caught in three weirs set along the coast of Piauí State, Brazil and monitored around the clock during 6 d each month for 11 months.

Tarpons in the western Atlantic have historically had few commercial uses except food. Whitehead (1891/1991: 202) stated, "The scales are two inches in breadth on the large fish, and are pure silver as a pearl, and are frequently used as ornaments." He continued: "Sometimes they are drilled and sewed on to a belt, making a conspicuous girdle of studded silver." Bean (1903: 178) wrote, "The scales are an article of commerce as curiosities." According to Röhl (1956: 456), Venezuelan artisans valued the scales for polishing tortoiseshell (actually, the carapace of the hawksbill turtle, *Eretmochelys imbricata*, one of the sea turtles). All sea turtles are endangered, and that application is of doubtful use today. According to Norman and Fraser (1938: 96), "it is said that these [Atlantic tarpon scales] can be sold at from 5 to 25 cents apiece. They are used for fancy work, being made up in various ways by the curiosity dealers to attract the fancy of the winter visitors to Florida." This information might have come from Goode (1887: 406):

> *The fish, when alive, presents a very brilliant metallic appearance, and the scales are much prized by curiosity hunters and for fancy work in the Florida curiosity shops. They are a staple article of*

the trade, selling for from ten to twenty-five cents each, the price paid to the fishermen being about fifty cents per dozen.

Gill (1907: 39) also mentioned this, using similar phrasing. Fifty cents would be the equivalent of ≈ $25 today, making collecting scales worthwhile during a time in US history when the average workman's daily wage was ≈ $1.50, often for a 15-h day.

I found little information about tarpon fisheries in the eastern Atlantic. Artisanal fishermen occasionally catch tarpons while trolling in pass areas of Aby Lagoon, Ivory Coast (Charles-Dominique 1993: 107). Bianchi (1986: 15) listed the tarpon among species fished commercially in Angola. The Ondo region is apparently the only targeted fishery for tarpons in Nigeria. There they are caught all year at depths of 5–39 m, peak abundance occurring in June–July and November–December (Anyanwu and Kusemiju 2008). Elsewhere in the country the species is fished by artisanal fishermen using weirs, beach seines, gill nets, and hook and line (Anyanwu and Kusemiju, 2008).

The Pacific tarpon has been a minor seasonal component of the Chennai (Madras), India commercial fishery for many years, appearing in the markets during August, September, October, and January (Moses 1923: 134, 138, 159).

Of the Pacific tarpon's suitability as food, Roughley (1953: 7) wrote that "as an edible fish it is of indifferent quality; its flesh, although white and of good flavour, lacks firmness, and the prevalence of fine bones renders the eating of it troublesome." Wells *et al.* (2007: 1650) mentioned its "dismal table qualities." According to Merrick and Schmida (1984: 54), the Pacific tarpon "is not valued for the table in Australia because of soft, flavourless flesh...." The Pacific tarpon is a source of protein for inland communities of Malekula and Epi, Vanuatu, Melanesia (Amos 2007: 150). Mann (2000: 66) considered the species "to be of poor eating quality ...." Pollard (1980: 54) wrote that it is "not valued as food because flesh is soft, very bony, and relatively flavourless ...."

8.6 Aquaculture

The future of Atlantic and Pacific tarpons as cultured commercial species seems bright if world fisheries continue their precipitous decline, relinquishing an increasingly greater market share to aquaculture. Tarpons tolerate poor water quality, are not fussy about what they eat, and can be easily reared in captivity. Taste aside, the composition of tarpon flesh is imminently suitable for human consumption, being high in protein and low in fat. Proximate analyses by Emmanuel *et al.* (2011) of Atlantic tarpons reared in pens and concrete ponds in Nigeria yielded favorable results for filets taken near the tails of fish of 402–421 mm TL and 0.78–0.80 kg. Mean dry-weight percentages: water (31.84 ± 0.03), protein (29.43 ± 0.01), lipid (0.97 ± 0.01), carbohydrate (22.25 ± 0.03), ash (14.97 ± 0.03), fiber (0.57 ± 0.03). Variances presumably represent standard deviations. These fish, although labeled adults by the authors, could only have been juveniles.

To my knowledge the Pacific tarpon has not yet been cultured through its life-cycle, although it has been captive-raised for food since at least the nineteenth century (Gill 1907: 39). The same is true of the Atlantic tarpon in the Western Hemisphere, although perhaps not in the eastern Atlantic, where the species putatively has been induced to spawn and the eggs hatched. Anyanwu and Kusemiju (2008: 120) made these tantalizing statements: "Fertilized eggs ... hatched into tarpon fry when stocked in nursery ponds. Leptocephalus larval stages were not encountered, probably due to the selectivity of the sampling equipment." Later in the same article (p. 125) they wrote that at Orioke, Ondo State, Nigeria large aquaculture ponds have been constructed "specifically for natural spawning of *M. atlanticus*." The authors provided no other details, and this terse description is unclear and open to skepticism. From ponds at these locations, "Large specimens were sometimes restocked in large ponds for natural spawning. Tarpon spawned naturally in the brackish water ponds in the study region." Perhaps such events are routine at aquaculture facilities in the eastern Atlantic, although no data were provided, and that eggs or larvae were not sought to confirm spawning is unfortunate. In any case this would be the first known occasion not just of Atlantic tarpons spawning in captivity, but doing so in waters more dilute than full-strength seawater. These observations certainly require confirmation.

Atanda (2007: 363 Table 7.14.1) listed the Atlantic tarpon as a "seed" fish in Nigerian freshwater aquaculture, although its contribution is minor. More important is the vigorous trade in juvenile Atlantic tarpons, which are captured and shipped throughout the country for aquaculture grow-out (Anyanwu and Kusemiju 2008). Most efforts in Nigeria and elsewhere have been limited to grow-out of juveniles netted from the wild and confined in floating pens, natural ponds and lagoons of fresh or brackish water, or excavated earthen ponds sometimes lined with concrete (e.g. Anyanwu *et al.* 2010, de Araújo Santos 2013, Emmanuel *et al.* 2011, Santhanam *et al.* 1990: 81).

The technology used in these applications is primitive or nonexistent, the water ordinarily refreshed by rainfall or tidal flow (e.g. de Araújo Santos 2013: 16, de Normandes Valadares 2013: 17, Mello Lopes 2013: 43). Floating pens are often suspended in estuaries and lagoons (Santhanam *et al.* 1990: 140). Above-ground plastic tanks have been tested, the water supply presumably controlled by pumping (Okoro *et al.* 2009). Captive tarpons during grow-out are fed bait fishes or fish scraps until harvested at ≈2 kg and sold, largely for regional consumption.

Most grow-out efforts in the Western Hemisphere have been in northern South America. The reports issued to date, like those from western Africa, have been short on details. Harvey *et al.* (1998: 217 Table 1), for example, listed the Atlantic tarpon as a commercial aquaculture species in Colombia, but gave no other information. A brief description was provided by de Araújo Santos (2013: 16) of an operation in Tutóia, Maranhão State, Brazil (translation in Appendix H).

An informal report by Farias Costa (2014)[7] states that 400 juvenile Atlantic tarpons of ≈ 80 g each were netted from coastal lagoons at Ceará during January and February, distributed into four floating net pens (*tanques-rede*) of 20 m^3 each, and fed fish scraps. After 10 months the average fish was > 3 kg, and total production ≈1200 kg. According to the author, "no issues of mortality, disease, deformity, or cannibalism during cultivation were observed."

Dahl (1971: 159–160) claimed that well-fed Atlantic tarpons in pond culture could reach 400 mm SL within 1 y and 600 mm SL by 15 months. He also noted that such ponds must be "seeded" with young tarpons because adults tarpons do not reproduce inshore, and that juveniles require live prey, which is not true.

Fishermen at Tutóia, Maranhão State who collected juvenile Atlantic tarpons for aquaculture were interviewed by de Normandes Valadares (2013). They told her, "*Há alguns anos atrás, as quantidades coltados já 250 kg de peixes por barco, atualmente esse número é de no máximo 100 kg.*" [A few years ago the amounts [of juveniles] collected reached up to 250 kg of fish per boat; now that number is a maximum of 100 kg.] These were fish 100–210 mm and 5–50 g, sizes suitable for stocking aquaculture ponds. The decline, according to de Normandes Valadares, could be attributed to increased water pollution, more fishing vessels, and lack of rainfall. Not listed as a possible reason was overfishing large juvenile and adult tarpons.

A similar trend has been reported in the eastern Atlantic. Anyanwu *et al.* (2010: 6) wrote that in Nigeria, "Presently, only very few farmers culture Megalops [*sic*] in their ponds unlike in the 80s and 90s when a booming Tarpon fingering trade existed in the coastal communities of Ondo State." Community-based cage culture of Atlantic tarpons is being investigated in Lagos State (Solarin *et al.* 2010). How this fishery affects the numbers of juvenile tarpons locally should be examined.

Stocking densities are seldom known when the space used is an irregularly-shaped natural body of water of variable depth and subject to changes in volume with rainfall and tides. Mercado Silgado (2002) briefly described results of a pilot study in which Atlantic tarpons were cultured at several locations in Colombia at ionic strengths ranging from freshwater to seawater. The sites included coastal lagoons linked with the sea, man-made earthen ponds in contact with estuaries and mangrove swamps, and several freshwater habitats. Wild-caught Stage 2 larvae and juveniles (80–350 mm TL) were held in pens at 3000–5000/m^3, the larvae until metamorphosis was completed. Densities of all age groups were

[7] http://www.panoramadaaquicultura.com.br/paginas/revistas/52/pisciculturamarinha.asp. Downloaded 06 October 2014.

then reduced to 1500–2000/m^3 and the fish held another 3–4 months prior to stocking and grow-out. Mortality through this period was <10%, most of it occurring during the capture and transport of leptocephali.

One site was 420-ha Luruaco Lagoon, which was stocked with 6700 tarpons ($\bar{x}$ = 431 mm TL). After six years, the lagoon was opened for four months to the artisanal fishery, and a six-day recreational tournament was also held. These events in sum yielded 6050 tarpons of 1000–1500 mm TL ($\bar{x}$ = 11 kg). At some sites, tarpons were stocked with other species (polyculture); at other locations, just tarpons were stocked (monoculture). Juveniles in monoculture experiments were of similar size to reduce the incidence of cannibalism.

During grow-out tarpons were kept in 100 m^2 pens or in earthen ponds of 30–3000 m^2. The ponds were stocked at starting densities of 3 fish/ha, 10 000/ha, 30 000/ha, and 50 000/ha. After nine months in ponds containing fish of ≈200 mm TL at densities of one fish/30 m^2 and fed cut trash fish, growth averaged 690 mm TL and 3.0 kg. After 15 months, these parameters had increased to $\bar{x}$ = 830 mm TL ($\bar{x}$ = 6.0 kg).

Wild tarpons show promise for cichlid control (e.g. Mozambique tilapias, Leal-Flórez *et al.* 2008), and cichlids provide a continuous source of food for Atlantic tarpons when reared with them in polyculture, as demonstrated by Mercado Silgado (2002) and Anyanwu and Kusemiju (2008). Tarpons of ≈225 mm TL and 72 g were stocked in 6000 m^2 ponds at 0.03 fish/m^2 with a resident population of redbreast tilapias, *Tilapia rendalii* ($\bar{x}$ ≈ 100 mm TL). After 13 months the tarpons averaged 825 mm TL and 6.0 kg, and the surviving tilapias were individuals big enough to have escaped predation. Santhanam *et al.* (1990: 140) recommended stocking Pacific tarpons and tilapias together in pen culture.

Anyanwu and Kusemiju (2008) reported stocking densities of 1–7 fish/m^2 in Nigerian aquaculture ponds, and 1–3 fish/m^2 in pens. In earthen ponds, the tarpon:tilapia ratio was 1:87 to 1:23. Grow-out was 1.5–3 y, at which point tarpons were >1000 mm FL and ≈5 kg. Survival was 73–86.7%. In one case, of 617 tarpons stocked in a 0.1 ha pond, 535 specimens lived to harvesting.

Captive tarpons thrive and grow on a variety of foods, including trash fishes (Mercado Silgado 2002, Okoro *et al.* 2010), bait fishes both dead and alive (de Araújo Santos 2013), commercial fish pellets (Okoro *et al.* 2010, Seymour *et al.* 2004), cattle and poultry by-products (Mercado Silgado 2002), poultry feed, fresh and smoked fish, crabs, crayfish, and bread (Anyanwu and Kusemiju 2008), and products made of cassava or cooked rice (Mercado Silgado 2002). In Mercado Silgado's (2002) extensive pilot-scale experiments Atlantic tarpon juveniles were given trash fish and chicken guts; juveniles <50 mm TL were fed floating pellets. In many cases when tarpons are reared in polyculture and the secondary species is a herbivore or omnivore (e.g. a tilapia), exactly which of them is consuming the food provided is seldom clear. For example, tarpons might not consume bread or cassava and rice products, although tilapias would undoubtedly eat them.

Bagarinao *et al.* (1986) wrote a manual with instructions for identifying, collecting, and handling fish and shrimp fry for aquaculture in the Philippines, including (superficially) the Pacific tarpon. Indonesians have cultured Pacific tarpons in ponds as a food fish (Pollard 1980: 54). Like the Atlantic tarpon they are easily reared from juveniles to marketable size. They accept commercial fish pellets and could presumably do well on feeds formulated for salmonids, which are also carnivorous. The Pacific tarpon has a limited market in the home aquarium trade. It also is fished and sold commercially in limited quantities.[8]

8.7 Populations

Any animal that is large, conspicuous, long-lived, and valuable for one reason or another is especially susceptible to population collapses caused by human interference (Collette *et al.* 2011). Add to these features extended longevity and long generation time, geographically restricted breeding sites and short breeding periods, low natural mortality except for human predation, and excessive exploitation, and the consequence is an animal in immediate and continuous need of rigorous protection. No such creature in today's rapacious world – elephant, rhinoceros, tarpon, or tuna – can hope to survive as a self-perpetuating species without legal protection and population management (Darimont *et al.* 2015).

The status of tarpon stocks everywhere is poorly known, but anecdotal accounts of how Atlantic tarpons are faring point to a downward trend (or wariness of such a trend) in at least some regions where they are exploited as recreational or commercial resources (e.g. Angel C. 1992, Anyanwu and Kusemiju 2008, Holt *et al.* 2005: 66), and perhaps to a concomitant decline in adult size (Bortone *et al.* 2007). Holt wrote that tarpon anglers had been attracted to the Texas coast since the 1880s, and from the 1920s through the 1940s Port Aransas had been the self-designated "Tarpon Capital of the World," but the bounty soon evaporated. "By the 1950s, the vast schools of tarpon that passed along the coast for hours were essentially gone."

Holt *et al.* (2005) attempted to reconstruct a picture of the Texas recreational tarpon fishery by examining scales nailed to a wall of the Tarpon Inn at Port Aransas. It was customary there and elsewhere for an angler to remove a scale from his catch and sign his name on it, also recording the capture date and size of the fish (Fig. 8.13 and Fig. 8.14). Of ≈2700 scales Holt examined, ≈2000 included length, and both length and weight had been inscribed on 270. Annual catches (implied by numbers of scales) declined gradually after 1940, becoming spotty by the late 1950s. No reduction in size was evident, but the absence of small fish in later

[8]For example, see this site offering frozen Pacific tarpons marketed out of China. Whether the product is cultured or wild-caught is not stated: http://frozenfishmarket.com.

Fig. 8.13 Wall panel of tarpon scales signed by anglers fishing out of the Tarpon Inn, Port Aransas, Texas. They reveal the lengths and weights of fish caught through the years. Source: Holt *et al.* (2005: 68 Fig. 1).

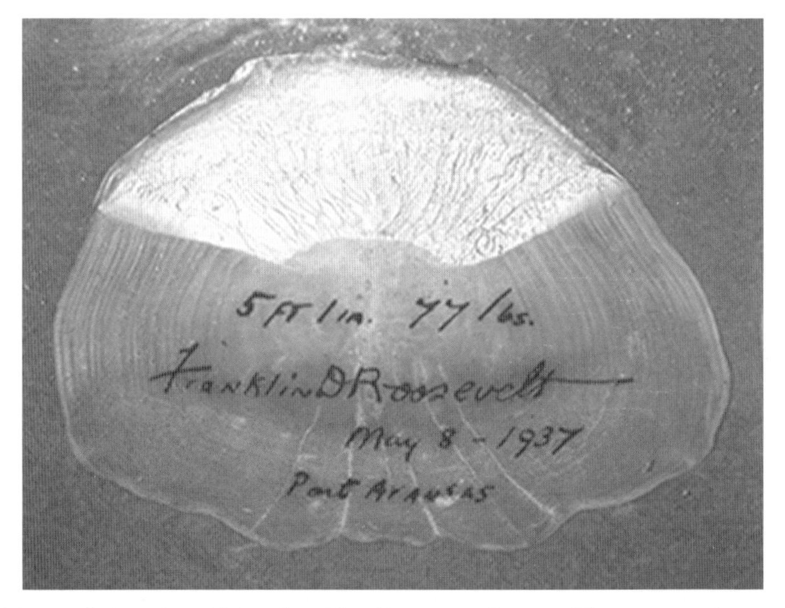

Fig. 8.14 A scale from the wall signed and dated by US President Franklin D. Roosevelt 8 May 1937. Source: Holt *et al.* (2005: 68 Fig. 1).

years was obvious. Holt suggested that a possible cause could have been recruitment failure because the recreational fishery targeted the biggest specimens.

At the population level, deriving conclusions about the direction of historical fluxes would be unproductive at the moment, considering that specific populations and their distributions have yet to be identified. Because the Atlantic tarpon has been valued solely as a recreational fish in North America the usual government catch statistics are lacking, and US landing records exist in the form of historical logbooks and personal accounts. These are scattered, incomplete, and often questionably accurate. Franks (1970: 35) wrote, "The tarpon is reportedly not as prevalent in Mississippi waters, particularly Mississippi Sound, as it was several years ago." However, "Whether this in actuality is due to its absence or decline in fishing for this species is debatable." Today's level of knowledge is little more advanced.

Bortone *et al.* (2007) came across an anglers' log titled *Useppa Inn – The Greatest Sport on Earth*, dated 1932 and published by the Izaak Walton Club. It contains a list of Atlantic tarpons hooked and landed in Boca Grande Pass from 1902–1932 (n = 13 154). There were no data for 1905 because the inn was closed. During some years, >1000 fish were recorded. They were tabulated by weight and capture date, presumably killed, and therefore not returned to the population. The inn, known today as the Useppa Island Club, is on Florida's southwestern coast.

Kaplan (1937: 106–107) reported >15 000 tarpons caught by recreational angling under auspices of the Boca Grande Tarpon Tournament Club, Gasparilla Island, Florida over the 6 y spanning 1932–1937. Bortone and colleagues compared these data with others from the recreational fishery at Port Aransas, Texas (Fig. 8.15). The pattern shows some clear spikes and sharp declines

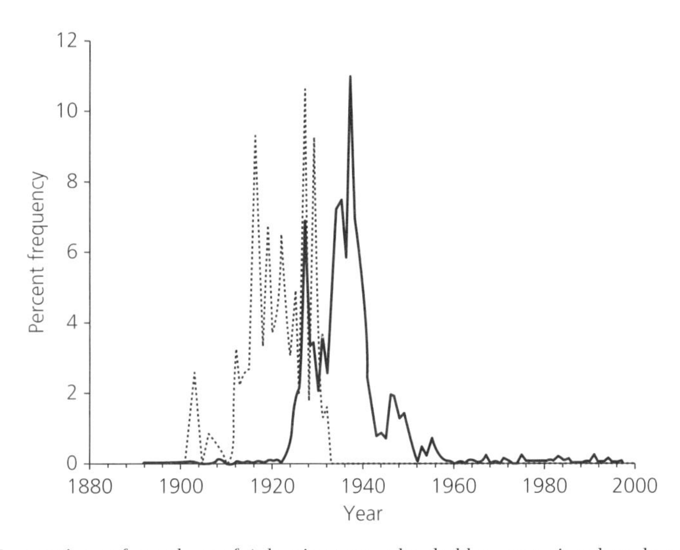

Fig. 8.15 Comparison of numbers of Atlantic tarpons landed by recreational anglers at Useppa Island, Boca Grande Pass, on Florida's southwestern coast (broken line) and Port Aransas, Texas (solid line). Source: Bortone *et al.* (2007: 34 Fig. 2).

while revealing little about any effect on the populations. The recent downward trend in the figure is obvious, but many variables confound any assumption that this is evidence of declining regional numbers. Recreational fishing was interrupted during World War II (Bortone 2008), and it must be remembered that the data refer to landings (i.e. fish actually boated, killed, and weighed), not tarpons caught and released, the current trend.

The pattern in northeastern Brazil is clearer. In the past, intense fishing pressure on tarpons has resulted in a reduction in both catch numbers and the sizes of individual fish (Ferreira de Menezes and Pinto Paiva 1966). These authors provided data on size and reproductive status of 10 052 tarpons caught commercially in wooden fish weirs set perpendicular to Almofala Beach, Acaraú County, Ceará State from 1962 through 1964. Production averaged nearly 218.5 tarpon per weir annually, and they were caught during every month of the year. Then, over the 3 y the fishery was monitored the catch fell sharply. In 1962 there were 18 weirs and 5296 tarpons caught (total weight 155 260.5 kg). In 1963 there were again 18 weirs in place and 4078 fish captured that in sum weighed 80 007.0 kg. By the end of 1964 the catch from 13 weirs was just 678 fish totaling 15 701 kg. This pattern is tentative evidence that overfishing adult Atlantic tarpons can induce a short-term regional response by the population.

The bulk of the tarpon catch obtained from Ceará State's weirs is seasonal. Carvalho Collyer and Alves Aguiar (1972: 4) provided additional catch data for commercial fishes and sea turtles taken from some of them during 1968–1970: *"Camurupim – o índice annual de captura mais expressivo registrou-se no ano de 1968. As melhores capturas desta espécie ocorreram nos meses de novembro a março."* [Tarpon – the highest annual catch was registered in 1968. The best catches of this species occurred from November to March.] Although active throughout the year the fishery recorded its highest numbers in austral spring and summer (October–January), which probably represent pre-spawning aggregations. Fonteles-Filho and Aguiar Espinola (2001) also reported seasonal trends and generally decreasing catches at seven fish weirs monitored at Ceará, indicative of continuing decline. Stamatopoulos (1993: 102, 202) classified the commercial Atlantic tarpon catch in the Caribbean and South America as "declining" from 1970–1991.

Raimundo Bastos *et al.* (1973, Appendix I) evidently thought the Ceará weir fishery sufficiently robust to help sustain a tarpon cannery and described the bacteriological organoleptic, shelf life and gustatory properties of tarpon flesh after an experimental canning process. Teixeira Alves *et al.* (1972, Appendix J) conducted a pilot study designed to ultimately produce smoked-canned and salt-dried tarpon roe on an industrial scale to replace the salted artisanal product prepared by fishermen at Ceará, which they considered crude and unhygienic.

García and Solano (1995) mentioned that aggregations of breeding tarpons had not been seen in Colombian waters in recent times, indicating a possible population decline, and pointed out that landings of Atlantic tarpons for just the port of Barranquilla from 1964–1970 exceeded those for the entire country from

1986–1993. Assuming this is true, wasteful and unregulated fishing practices have probably been contributing factors. Forty-five years ago Dahl (1971: 159) castigated his countrymen for their method of hunting tarpons in the Colombian Caribbean:

> *"Desgraciadamente, este gran recurso natural ha sido, y sigue siendo tratado de un modo estúpido y criminal. En vez de una pesca deportiva, sensata en los ríos y ciénages (con la probabilidad del influjo del turismo) y comercial en el mar, utilizando redes agalladeras de malla grande que permitan el paso de los individuos inmaduros, se está pescando a todo lo largo de la costa con dinámita. En otras palabras, cuando pasa un cardúmen de sábalos, los mal llamados "pescadores" salen en sus botes y le lanzan "tacos" (cargas de dinámita). De los peces muertos y lisiados por tales descargas, se recoge posiblemente un 15%, pues el resto se pierde totalmente convirtiéndose en alimento para los tiburones. Por esta razón no es raro que la pesca del "Sábalo" disminuya."* [Disgracefully, this great natural resource has been, and continues to be, exploited in a stupid and criminal way. Instead of sensible sport fishing in the rivers and marshes (with the probability of influencing tourism) and commercial [fishing] in the sea using gill nets of large mesh that allow immature individuals to escape, the fishery everywhere mainly uses dynamite. In other words, when encountering a school of tarpons these so-called "fishermen" remain in their boats and simply throw out "tacos" (sticks of dynamite). Of the fish killed and crippled by such discharges, possibly 15% are picked up because the rest are lost, becoming shark food. For this reason it is not surprising that tarpon fishing is declining.]

Patterns elsewhere are nebulous and impossible to interpret. During the early 1990s artisanal fishermen working in Colombia's Golfo de Morrosquillo reported declining catches of tarpons and other large species, although one perceived reason could be artifactual: those without effective motor boats were unable to reach remote sites (Angel C. 1992). Other observers have reported declines in the Atlantic tarpon fisheries both for recreational (Holt *et al.* 2005, Landry 2002) and commercial purposes (e.g. García and Solano 1995), although reliable catch data for either fishery are scarce.

The Atlantic tarpon was recently listed on the IUCN Red List as "vulnerable," a status attained by a species when its overall population has declined 30% (Adams *et al.* 2012,[9] 2013). Adams and colleagues noted that a formal stock assessment has yet to be made for any part of the species' range, adding, "multiple lines of evidence suggest that populations of Atlantic tarpon appear to have declined from historic levels throughout their range ...." Part of their assessment (Adams *et al.* 2012: 4) reads:

> *Although patchy, data on total commercial landings in Central and South America show large historical declines. Total global landings of M. atlanticus declined 84.5% between 1965 and 2007 (4600 metric tons versus 712 metric tons), particularly in Brazil, and mostly during the early years of that time period, reflecting a drop in population size, not a change in fishery effort.*

[9] http://www.iucnredlist.org/details/full/191823/0. Downloaded 22 August 2014.

Based on the FAO statistics[10] and other factors, they estimated that from 1969–2007 landings of Atlantic tarpons fell "at least 60%" and inferred that the worldwide decline in abundance was 30%.

8.8 Final note: whom should we save?

Trophy fishing, like trophy hunting, targets the largest and oldest members of a species, which happen to be the most valuable to sustaining their populations (Darimont *et al.* 2015, Shiffman *et al.* 2014, Worm 2015). Their fecundity and quality of larvae (including likelihood of survival) correlate positively with size (Birkland and Dayton 2005), and in some cases this also applies to quality of the eggs (Palumbi 2004 and references). Size matters because bigger fishes have greater volumes and can hold more eggs, often exponentially more (i.e. volume = L^3). Not surprisingly, harvesting of these fish can lead to an exponential decline in numbers of larvae produced, a shortening of the reproductive season, a lowering of the probability of larvae encountering conditions conducive to survival, and selection for earlier reproduction and slower growth (Birkeland and Dayton 2005, Conover and Munch 2002).

Large females are ultimately many times more fecund than smaller ones and their progeny likely to be more successful. Larvae of old female black rockfish (*Sebastes melanops*) were found to grow more than three times faster and survive starvation twice as long as larvae of younger females (Berkeley *et al.* 2004). Older individuals of some species are more successful spawners, even capable of extending the spawning season (Birkeland and Dayton 2005 and references). Protecting the bigger, older specimens in "no-take" marine reserves is of utmost importance. In some instances they can double the egg output of their species, even in a reserve comprising just 5% of the total habitat (Palumbi 2004 and reference).

The survivorship of fishes actually increases with size, a phenomenon that both recreational and commercial fishing turn upside-down by targeting the largest, and introducing a form of predation unknown throughout their evolutionary history. In modern times the advantages to growing old and large have become disadvantages.

The lifeboat is too small to include everyone, so when the mother ship christened *All Life* takes on water and starts to sink, who among its members should be given priority? I shall address this difficult question within the context of fitness and narrow it to assume that "who" refers to the life-history stages of a species to be rescued. This restricts the possibilities by limiting the answer to whichever stage (or stages) offers greatest value in terms of reproductive potential. For a fish like the tarpon, a potential "who" can then be a gamete (egg or sperm), a leptocephalus

[10] http://www.fao.org/fishery/statistics/global-capture-production/query/en. Downloaded 12 July 2015. The data now cover 1950–2013.

(Stage 1), a metamorphosing leptocephalus (Stage 2), a fry (Stage 3), a small juvenile (early Stage 4), a late juvenile approaching maturity (late Stage 4), or a mature adult (Stage 5).

We could, in theory, plot a survivorship curve for each of these dynamic stages in the tarpon's life-cycle and determine both the energetic cost and probability of survival to each level. The result would be a series of separate curves depicting proportional survivorship and specific contribution to the survival and fecundity of progeny. Unfortunately data for at least one or more important category is usually insufficient or lacking entirely. In the absence of real information we make mathematical models, filling the blank spaces with estimates and projections, and hoping in the end to gain a fuzzy understanding of what really matters, which is the rate of population growth (indicated by lambda, λ).

Despite the huge number of gametes generated and released the odds of a fertilized tarpon egg eventually becoming an adult tarpon are less than bleak. Although millions are discharged into the plankton during a single spawning event, many die or are gobbled by predators. Some ova, of course, become fertilized, but the larvae that hatch from them are tiny, abandoned without parental care and unable to protect themselves.

In many long-lived species it is older and larger adults of reproductive age and not their immediate offspring, that have the greatest positive effect on λ (Section 8.7). Known examples, in addition to the fishes discussed above, include certain invertebrates, reptiles, and seabirds (e.g. Doherty *et al.* 2004, Hyslop *et al.* 2011, Linares *et al.* 2007). An adult female tarpon can produce millions of eggs in a single breeding season, the males millions of sperm cells (Chapter 3.2). The odds of a larva attaining reproductive age are similar to winning a lottery; those of a juvenile lower but still extremely high. All life stages of an exploited population have value and are worth protecting (e.g. Vasilakopoulos *et al.* 2011), but there can be no denying that older juveniles at the onset of maturity and adults of reproductive age constitute the genuine survivors. Protecting one or both these stages is a better conservation strategy than releasing hatchery-reared larvae or juveniles and hoping a few survive to reproduce.

Size and survival in young fishes correlate positively (Chapter 3.2), and larger individuals have greater survivorship. Stocking red drum (*Sciaenops ocellatus*) larvae failed to produce even a minor ripple through a regional population, although positive effects were seen when juveniles of 25–50 mm were stocked instead.[11] I found no models or empirical evidence that releasing hatchery-reared tarpon larvae or early juveniles could positively affect a wild population, and the effort might be nothing but a feel-good exercise. Adams *et al.* (2012: 4)[12] listed the loss of sexually immature individuals of 0–9 y as a major threat to Atlantic tarpon

[11] http://www.dnr.sc.gov/marine/stocking/research/reddrumstockinglarvae.html. Downloaded 6 July 2015.

[12] http://www.iucnredlist.org/details/full/191823/0. Downloaded 6 July 2015.

populations, but this remains to be determined and might be valid only for the extreme upper part of the range (i.e. the oldest and therefore biggest juveniles).

Elasticity analyses and similar studies often show that λ is more sensitive to changes in survival than in fecundity. The opposite is more likely true in animals having relatively short lives, that breed at early ages, are highly fecund, reproduce more than once annually, and experience extreme natural mortality as larvae and juveniles. In the Atlantic herring (*Clupea harengus*), for example, 1% of larvae perhaps survive to the juvenile stage.[13]

Stocking of larval vs. juvenile tarpon is likely to give similar results (Winemiller and Dailey 2002). Production costs rise with each step in the life-history sequence deferred until release from the hatchery. The cheapest stage is usually the earliest. In aquaculture the costs of labor, equipment, and supplies are minimum if fertilized eggs are released into the wild. Larvae require feeding, monitoring, and other forms of care, and rearing tarpons to near-maturity would be most expensive of all, partly because they are carnivorous, and protein is the most expensive ingredient in animal feeds. This is especially true when the targeted species has no commercial market as human food and is valued only by a small percentage of recreational anglers.

Our knowledge of tarpon survivorship contains too many empty spaces to make predictions, but a useful species for comparison might be the loggerhead sea turtle (*Caretta caretta*), for which at least some life-history data are available. In the context of turtles in general, loggerheads are considered reasonably fecund, although based on sheer numbers no turtle's production can match a tarpon's. Like tarpons the loggerhead ranges widely, and its populations are difficult to tag and monitor, in part because only adult females come ashore for at most a few hours to lay eggs during the nesting season. Also like tarpons, loggerheads are iteroparous, long-lived, and the juvenile stage to first reproduction is lengthy. First breeding occurs at ≈22 years, and maximum lifespan is ≈54 years, again reminiscent of the tarpon's natural history. The model devised by Crouse *et al.* (1987) and described here assumed sexes to be present in equal numbers in a closed, or self-sustaining, population. Data were collected over several years at Little Cumberland Island, Georgia.

Current management practices focus on protecting the loggerhead's fertilized eggs buried on sandy beaches and then protecting the hatchlings until they abandon the nest and try to reach the surf. In Florida, biologists and volunteers regularly patrol beaches during the nesting season, tracking females by their nocturnal "crawls" left in the sand, and marking and monitoring nests until hatching (Fig. 8.16). Eggs are sometimes dug up and moved to safer locations (e.g. closer to the dunes to protect them from high tides). Persons living near the beach are advised to use only red outdoor bulbs and to dim inside lights. Hatchlings are phototropic; artificial illumination can lure them away from the sea into the dunes, or onto roads

[13] http://www.gma.org/herring/biology/life_cycle/default.asp. Downloaded 6 July 2015.

Fig. 8.16 Loggerhead sea turtle (*Caretta caretta*) nest on Longboat Key, Florida located and identified by Mote Marine Laboratory staff members and volunteers. Stakes mark the nest's perimeters; notes on the stakes record the date the eggs were laid and the expected hatch date. Source: Stephen Spotte.

where they die of sun exposure, dehydration, or are killed by terrestrial predators and motor vehicles. From a conservation standpoint, are these measures implemented to assist hatchlings helpful or are they employed merely because we can control them using the rationale that any assistance is better than none?

Crouse and colleagues devised a stage-based model using survivorship and reproductive potential of each of seven arbitrary stages in the loggerhead's lifecycle. These were: (1) fecundity (eggs and hatchlings, age <1 year); (2) small juveniles (1–7 years); (3) large juveniles (8–15 years); (4) subadults (16–21 years); (5) novice breeders (22 years); (6) first-year re-migrants (23 years); and (7) mature breeders (24–54 years). They estimated fecundity, or reproductive output (F_i), probability of surviving to the next stage (G_i), and probability of surviving in the same, or present, stage (P_i). The probability of transitioning (G_i and P_i) are then estimates based on stage-specific probabilities of survival (p_i) and duration of the stage evaluated (d_i). Some individuals will have remained in given stages for 1, 2, …, d_i y. If the proportion of those of stage i is set at 1, and survival probability to the next year within that cohort is set to p_i, the probability of living d years becomes p_i^d. If the population is at equilibrium and represents a stable age distribution, relative abundance of individuals by stage is then 1, p_i, p_i^2, $p_i^{d_i-1}$. Between t and $t + 1$ the oldest then shift to the next stage.

Skipping intermediate steps in the derivation and rewriting the geometric series yields:

$$P_i = \left(\frac{1 - p_i^{d_i - 1}}{1 - p_i^{d_i}}\right) p_i \quad (8.1)$$

Thus the number of survivors in any cohort of a stage diminishes with time as a function of the probability of stage-specific annual survival plus the time spent in that stage. The proportion of a population that shifts to the next stage and survives (G_i) is expressed similarly; that is, by the proportion of turtles in the oldest cohort of its stage times the annual survival for that stage. Again skipping intermediate steps of the derivation, this can be written:

$$G_i = \left(p_i^{d_i}(1 - p_i)\right) / \left(1 - p_i^{d_i}\right) \quad (8.2)$$

To simplify the terminology by omitting its mathematical connotations, if $\lambda = 1$ and r (the intrinsic rate of increase of a population) equals 0, the population is neither falling nor rising. After evaluating the resultant population matrix Crouse *et al.* (1987) performed a sensitivity analysis by manipulating data for fecundity, growth, and survival to produce a new matrix, then re-calculated values of λ and r. They simulated a 50% reduction in these parameters for each life-history stage while holding others constant then calculated P_i and G_i using equations 8.1 and 8.2. The outcome was a drop in λ for Stages 1–4, with Stages 2–4 showing the steepest decline. After 1 y the reduction in adult survival was similar in intensity to the patterns in λ and r for Stage 1.

What would be the stage-sensitivity if mortality for any stage could be eliminated? This is impossible to achieve, but interesting to test. Predictably, swiftness of the fall is reversed. *However, the population eventually reaches extinction, regardless of efforts expended to protect eggs and hatchlings on the beach.* Furthermore, diminution of the population was not reversible unless egg production was doubled, also an impossibility considering fewer adults in the model are now reproducing annually.

Crouse *et al.* (1987) also changed age at first reproduction to 16 y and 28 y (± 6 y) by adding two years to calculations of P_i and G_i for Stages 2–4. Lowering the age nearly halted the population decline. Proportional sensitivity, or elasticity, of λ is the proportional change in its value caused by a proportional change in one of the life-history parameters. Elasticity results of λ calculated for F_i, P_i, and G_i demonstrated a minor effect of F_i on λ. More important were changes in P_i and G_i. Changes in Stages 2–4 and 7 affected λ more than in Stages 1, 5, and 6. The most critical stages in terms of population stability comprised Stages 2–4. In fact, even if 100% survival could be achieved in Stages 1, 5, and 6 – Stage 1 being eggs and hatchlings – this model shows a population decline. Increasing survival of stages 2–4 and 7 one at a time showed equilibrium ($\lambda = 1$, $r = 0$) to be obtainable merely by increasing Stage 3 survival 14%, *demonstrating that preserving more large juveniles was the most efficient means of achieving population stability.*

It seems intuitively logical that mature adults should be the first priority of conservation efforts. After all, their survival to reproductive age is an impressive biological and statistical achievement. This is certainly true of albatrosses, an extreme *k*-selected group that is also long-lived, but the fecundity of which is very low. Albatrosses lay only one egg, and some species do not even reproduce every year. The gray-headed albatross (*Diomedea chrysostoma*) at Bird Island, South Georgia in the Southern Ocean is mostly a biennial breeder, <1% reproducing 2 y in succession (Prince *et al.* 1994). Most adults (68%) return to nest 2 y later, 11% the third year, and 5% not until the fourth year. Moreover, >50% of adults that failed to rear a chick reproduced the following year, but 23% put off reproduction still another year.

About the loggerhead sea turtle, Crouse *et al.* (1987: 1419) wrote: "By increasing survival of large juveniles (who have already survived some of the worst years) a much larger number of turtles are likely to reach maturity, thereby greatly magnifying the input of the increased reproductive value of the adult stages." Large juveniles, in other words, are on the cusp of reproducing, with years of potential reproduction ahead.

The above discussion has been presented merely as an exercise in conservation modeling. Tarpons are not sea turtles, and future evaluation might reveal a completely different pattern of survivorship, requiring a very different strategy. Both groups of animals face dangers in the wild even as large adults (e.g. shark predation), but the greatest risk is death by human activities, turtles as by-catch in the commercial fishery where otherwise they are not killed for food or ornaments, and tarpons directly or indirectly in the recreational fishery and in parts of the world where they are fished commercially. Lessons from the conservation of sea turtles and long-lived seabirds such as albatrosses clearly demonstrate the wisdom of protecting individuals approaching maturity and those having already arrived.

The fecundity of female Atlantic tarpons from southern Florida shows significant positive correlation with weight. The smallest mature specimen examined by Crabtree *et al.* (1997) contained an estimated 4.5 million oocytes, the largest held ≈20.7 million. Cyr (1991: 51, 63 Table III-3) reported the mean age of tarpons taken in the Florida recreational fishery to be 25.16 y. The mean age of females was 25.8 y, that of males 21.1 y. Directly killing just one fish this old, or stressing it by hooking it so that it dies anyway within days, subtracts millions of gametes from the population and numerous future late juveniles and breeding adults (Chapter 3.2) needed to sustain the population.

In the tarpon's case, the assumption must be that survivorship is lowest in the egg, larval, and early juvenile stages before leveling off and becoming nearly constant at some point thereafter. Because the tarpon is long-lived, longevity and mortality rate should demonstrate a negative association. Were their later mortality rate high, none would survive to old age. Selection theory indicates that such a life-history pattern should produce a type III survivorship curve. Hoenig (1983) devised a simple model having a strong empirical basis using unexploited stocks of several mollusks, cetaceans, and fishes (tarpons were not included specifically):

$$\frac{n_t}{n_0} = e^{-zt} \qquad (8.3)$$

in which z is the constant and instantaneous mortality rate, n_t represents the number of individuals surviving to age t, and n_0 is the starting number in the population (i.e. $\frac{n_t}{n_0}$ is the proportion surviving to age t).

Assume that estimated time of longevity (i.e. survivorship, s) of a population of tarpon or some other species is defined as the age, t_s, to which a fraction, or proportion, f, survives. Consider f to be a small arbitrary constant (e.g. < 0.01). Then:

$$f = e^{-zt_s} \qquad (8.4)$$

and

$$\ln(f) = zt_s \qquad (8.5)$$

At this point equation 8.5 describes a hyperbola that can be made linear by plotting mortality rate against $\frac{1}{t_s}$ or by plotting log z against log t_s (Hoenig 1983: 899). However, these exercises can be avoided by determining the ages of a few large specimens in the population, t_{max}, and substituting the values for t_s. Hoenig subsequently plotted mortality against maximum age (log z vs. log t_{max}) in the form:

$$\ln(z) = a + b\ln(t_{max}) \qquad (8.6)$$

for three diverse taxonomic groups (fishes, cetaceans, mollusks) of 134 stocks, containing individuals ranging in age from 1–123 y. The results displayed high coefficients of determination (r^2) indicating good predictive power. This was true of both the combined regression and individual plots by species. Such an approach might work for tarpons, unless it can be demonstrated that mortality rate increases with age. Data used in Hoenig's analysis were from unexploited stocks, which tarpons essentially are in North America, excepting the recreational fishery where the effect is unknown despite implementation of catch-release. Hoenig (1983: 899) wrote: "If age truncation is a common phenomenon among the stocks for which data were available, then the application of this technique to heavily exploited stocks may result in an underestimate of the mortality rate."

The high mortality of the tarpon's early life stages suggests that releasing fertilized eggs into the sea using captive brood stock, as proposed by MacKenzie and Chavez (2002), is an unlikely way of rebuilding regional populations. A better approach is to forge sensible international conservation programs and follow them closely. Aquaculture can certainly be one component, but emphasis could be on grow-out of wild-caught fry and juveniles, which can then be stocked in impoundments designed for commercial or recreational fisheries. Fry and juveniles represent a targeted fishery for aquaculture grow-out in Nigeria's Ondo State (Anyanwu and Kusemiju 2008). Fry are collected using mosquito netting, and the fishery commences in March or April. Its intensity led Anyanwu and Kusemiju (2008: 116) to express concern that "Owing to the high demand and exploitation of tarpon fry, there appears to be a relatively high fishing pressure on tarpon population [*sic*] in the coastal waters of western Nigeria." However, with conservation and monitoring there and elsewhere, licensing collectors to capture young tarpons for culture ought to assure the perpetuation of wild stocks. Where this has not been done and fishing pressure has increased (e.g. at Tutóia, Maranhão State, Brazil), catches of juveniles for stocking aquaculture ponds has decreased (de Normandes Valadares 2013).

The habitat requirements of tarpons are so general that concern about controlling specific abiotic and biotic factors within narrow limits can be pushed into the background. Unlike a trout, a tarpon does not require a particular kind of substratum, a narrow range of temperatures, and high concentrations of dissolved oxygen to survive and reproduce. All tarpons apparently spawn at sea where conditions are nearly uniform. Their young, having moved inshore, seem indifferent to composition of the substratum, water flow and water quality, characteristics of the submerged and emergent vegetation, and so forth (Chapter 7). After all, tarpons survive and grow in drainage ditches; they can breathe air. The point is, I seriously doubt that a metamorphosing leptocephalus could even recognize an "ideal" inshore habitat. Such places do not exist for species whose requirements are this general.

Conservation strategies should be less about identifying important characteristics of tarpon habitats and focused instead on preserving large swaths of the habitats themselves – salt marshes and tidal creeks, mangroves, lagoons, estuaries. For example, impoundments for mosquito control deny tarpons access to tropical and subtropical wetlands and lower the diversity of fish species that occupy them routinely for all or parts of their life-cycles (Brockmeyer *et al.* 1996, Gilmore *et al.* 1981, O'Bryan *et al.* 1990, Poulakis *et al.* 2002). Some tarpons composing the US recreational fishery spend part of the year in the southwestern Gulf of Mexico outside American jurisdiction (Luo *et al.* 2008b: 296). Any effective habitat preservation plan must inevitably include other countries.

APPENDIX A

Partial list of countries where Atlantic and Pacific tarpons have been reported

Atlantic tarpons

Country (state or province)	Source
Angola	Penrith (1978: 182); Richards (1969); Sérat (1981: 84–85)
Azores	Arruda (1997: 25); Costa Pereira and Saldanha (1977); Serrão Santos *et al.* (1997: 120)
Bahamas	Barton and Wilmhoff (1996: 9 Table 2); Böhlke and Chaplin (1968: 36–38); Breder (1933, 1944: 237 Table XI, 1945); Crabtree (1995); Heilner (1953: 216); Layman and Silliman (2002)
Barbados	Babcock (1936: 63); Butsch (1939); Gill (1907: 38); Goode (1876: 69)
Belize	Eigenmann (1912: 444); Fedler and Hayes (2008); Lowe (1962: 678, 693); Pritchett (2008); Smith *et al.* (2003: 7)
Benin	Lalèyè *et al.* (2004: 334 Table 1)
Bermuda	Beebe and Tee-van (1933: 33–34); Breder (1944: 250); Goode (1876: 68–69); Smith-Vaniz *et al.* (1999: 126–127)
Brazil	Aquino Menezes *et al.* (2003: 31)
Brazil (Amapá)	Camargo and Isaac (2001: 140 Table 1)
Brazil (Bahia)	de Paiva Carvalho (1964: 4); Duarte Lopes and Porto Sena (1996); Duarte Lopes *et al.* (1998)
Brazil (Ceará)	Carvalho Collyer and Alves Aguiar (1972); de Paiva Carvalho (1964: 4); Ferreira de Menezes (1967, 1968a, 1968b); Ferreira de Menezes and Pinto Paiva (1966); Fonteles-Filho and Aguiar Espínola (2001); Pinto Paiva and Ferreira de Menezes (1963); Raimundo Bastos *et al.* (1973); Teixeira Alves *et al.* (1972)
Brazil (Espírito Santo)	de Paiva Carvalho (1964: 4)

Tarpons: Biology, Ecology, Fisheries, First Edition. Stephen Spotte.

Country (state or province)	Source
Brazil (Maranhão)	de Araújo Santos (2013: 16); Camargo and Isaac (2001: 140 Table 1); de Normandes Valadares (2013: 17); Mello Lopes (2013); Rocha and Rosa (2001)
Brazil (Pará)	Barletta-Bergan *et al.* (2002: 19 Table 1); Camargo and Isaac (2001: 140 Table 1); de Paiva Carvalho (1964: 4)
Brazil (Paraíba)	Torelli *et al.* (1997: 70 Table 1)
Brazil (Piauí)	Giacometti Mai *et al.* (2012); Mello Lopes (2013); de Paiva Carvalho (1964: 4)
Brazil (Rio de Janeiro)	de Paiva Carvalho (1964: 4)
Brazil (Rio Grande do Norte)	de Paiva Carvalho (1964: 4)
Brazil (São Paulo)	Aquino Menezes (2011: 5); Sadowsky (1958)
Cameroon	Monod (1927.654)
Canada (Nova Scotia)	Halkett (1913: 45); Leim and Scott (1966: 85); Nichols and Breder (1927: 33); Vladykov and McKenzie (1935: 53)
Canada (unspecified)	McAllister (1990: 44)
Cape Verde Islands	Wirtz *et al.* (2013: 117)
Colombia	Angel C. (1992: 132–133); Álvarez-León (2003: 80, 85, and refs.); Cataño and Garzón-Ferreira (1994); Dahl (1965, 1971: 158–160); Herazo C. *et al.* (2006); Leal-Flórez *et al.* 2008; Medellín-Mora *et al.* (2013: 234, 236 Table 1); Mercado Silgado (1971, 1981, 2002); Mercado S. and Ciardelli (1972); Rueda and Defeo (2003); Rueda *et al.* (2011: 122–123); Ward *et al.* (2008: 136 Table 10.1)
Congo	Roux (1960); Sérat (1981: 84–85)
Costa Rica	Angulo *et al.* (2013: 989, 1011); Anonymous (1975); Bussing (1987: 62–63); Chacón Chaverri (1993, 1994); Chacón Chaverri and McLarney (1992); Crabtree *et al.* (1997); Dailey *et al.* (2008); Ward *et al.* (2008: 136 Table 10.1); McMillen-Jackson *et al.* (2002, 2005); Nordlie and Kelso (1975); Ward *et al.* (2008: 136 Table 10.1)
Cuba	Breder (1944: 237 Table XI); Eigenmann (1904: 222, 1921); Griswold (1913: 91–108; 1921: 23–34, 171; 1922: 8–9, 12–15); Heilner (1953: 207–208, 211–215, 218– 219); Kaplan (1937: 89) Vilaró-Diaz (1893: 12, 148–149); Viñola Valdez *et al.* (2008)
Dominican Republic	Marín Erausquin (2007: 15)
Equatorial Guinea	Blache 1962: 23); Moloney (1883: 7)
France (Basque coasts)	Quero and Delmas (1982); Quero *et al.* (1982: 1022–1024)
French Guiana	Lowe (1962: 678, 693); Seymour *et al.* (2008); Wade (1962: 589)
Gabon	Decker *et al.* (2003: 81); Mamonekene *et al.* (2006: 287 Table 2); McMillen-Jackson *et al.* (2002, 2005); Sérat (1981: 84–85)

(Continued)

(Continued)

Country (state or province)	Source
Ghana	Edwards *et al.* (2001: 28–29)
Granada	Luo *et al.* (2008b: 283 Table 18.2)
Guinea	Blache *et al.* (1970: 136); Sérat (1981: 84–85); Stiassny *et al.* (2007: 54–55)
Guadeloupe	Cuvier and Valenciennes (1846: 399, 401)
Guyana	Anonymous (1992); Planquette *et al.* (1996: 54–55)
Haiti	Beebe (1927, 1928: 67–74, 228–230); Beebe and Tee-Van (1928: 33–36); Breder (1944: 237 Table XI); Heilner (1953: 219); Nichols (1929: 199)
Honduras	Matamoros *et al.* (2009: 7, 33)
Ireland	Twomey and Byrne (1985); Wheeler (1992: 7, 17; 2004: 6)
Ivory Coast	Charles-Dominique (1993: 107); Koffi *et al.* (2014)
Jamaica	Alleng (1997: 35); Breder (1944: 237 Table XI); Cuvier and Valenciennes (1846: 400); Dampier (1906: 118)
Madeira Island	Wirtz *et al.* (2008)
Martinique	Cuvier and Valenciennes (1846: 399–401)
México	Cruz-Ayala (2002); Arce Ibarra (2002); Breder (1944); Carson (1914: 394–403); Castro-Aguirre *et al.* (1999: 88–90, 2002: 125); Dailey *et al.* (2008); García de León *et al.* (2002); Grey (1919: 1–7); Heilner (1953: 208, 216); Luo *et al.* (2008b: 280–281 Table 18.1); Oviedo Perez *et al.* (2002); Sanvicente-Añorve *et al.* (2011: 186 Table II); Schmitter-Soto *et al.* (2000: 147, 2002); Vásquez-Yeomans *et al.* (1998); Vega-Cendejas and Hernández (2002); Ward *et al.* (2004); Ward *et al.* (2008: 136 Table 10.1)
Netherlands Antilles	Personal observation, Zaneveld (1962: 144 Table 2, 161 Table 9)
Nicaragua	Astorqui (1976: 28–29); Brown and Severin (2008); Cotto S. (undated: 32–33); Nietschmann (1973: 250, unnumbered table); Simmons (1900: 329–330); Villa (1982: 77–78)
Nigeria	Ajayi and Okpanefe (1995); Anyanwu and Kusemiju (2008); Hollister (1939); Lalèyè *et al.* (2004: 334 Table 1); MacPherson (1935: 447); Pellegrin (1923: 44–45); Ward *et al.* (2008: 136 Table 10.1)
Pakistan	Bianchi (1984: 3)
Panamá	Anonymous (1975); Breder (1925a: 140–141; 1944: 237 Table XI); Griswold (1922: 27); Heilner (1953: 208, 219); Hildebrand (1938, 1939); Hollister (1939: 456); Meek and Hildebrand (1923: 173–175); Swanson (1946); Ward *et al.* (2008: 136 Table 10.1)
Portugal	Carneiro *et al.* (2014: 16); Costa Pereira and Saldanha (1977); Quero (1998)

Country (state or province)	Source
Puerto Rico	Austin and Austin (1971); Breder (1944: 237 Table XI); Cuvier and Valenciennes (1846: 399, 401); Dailey *et al.* (2008); Erdman (1960); Evermann and Marsh (1902: 79–80); Figueroa and Zerbi (2002); Jordan and Evermann (1904: 85); Mateos-Molina *et al.* (2013); Neal *et al.* (2009); Nichols (1929: 199); Rundle *et al.* (2002); Ward *et al.* (2008: 136 Table 10.1); Zerbi and Aliaume (2002); Zerbi *et al.* (1999)
Santo Domingo	Cuvier and Valenciennes (1846: 399, 401)
Sénégal	Pellegrin (1923: 44–45); Sérat (1981: 84–85)
Sierra Leone	McMillen-Jackson *et al.* (2002, 2005)
Spain	Arronte *et al.* (2004)
Suriname	Mol *et al.* (2000)
St. Vincent and the Grenadines	Beebe and Hollister (1935: 211); Luo *et al.* (2008b: 282–283 Table 18.2)
Trinidad and Tobago	Boeseman (1960: 86); Luo *et al.* (2008b: 281 Table 18.1); Phillip *et al.* (2013: 7, 12 (Table 2), 14, 27); Ramsundar (2005)
Turks and Caicos Islands	Nichols (1921), Personal observation
U.S. Virgin Islands	Randall (1967: 684?); Smith-Vaniz and Jelks (2014: 22)
United States (AL)	Boschung and Mayden (2004: 127–128); Breder (1944: 237 Table XI); Coker (1921)
United States (FL)	Bean (1903: 178); Breder (1944: 237 Table XI); Briggs (1958: 252); Caine (1935: 134–137, 233–235); Cairns (editor, 1992: 1–2, 5); Crabtree *et al.* (1992); Eldred (1967, 1968, 1972); Evermann and Bean (1897: 18, plate 5); Fedler (2009, 2011); Fritz (1963: 89); Gill (1907: 36–37); Gilmore (1977: 111); Gilmore *et al.* (1981); Griswold (1921: 170–171); Gunter (1942); Harrington (1958); Heilner (1953: 206–207, 215, 218–219); Jordan (1884: 107); Kushlan and Lodge (1974: 112, 116); Loftus and Kushlan (1987: 183, 185, 280, 284, 288, 291, 296); Luo *et al.* (2008b: 280 Table 18.1); McMillen-Jackson *et al.* (2002); Moffett and Randall (1957); Nelson *et al.* (1992: 84–87); O'Bryan *et al.* (1990); Randall and Moffett (1968); Shenker *et al.* (1995, 2002); Smith (1896: 173); Snelson (1983: 190); Storey (1937); Storey and Gudger (1936); Storey and Perry (1933); Tabb and Manning (1961: 606–607); Tagatz (1968, 1973); Ward *et al.* (2004, 2008: 136 Table 10.1)
United States (GA)	Luo *et al.* (2008b: 280 Table 18.1); Rickards (1968)
United States (LA)	Dailey *et al.* (2008: 57–68); Gunter (1938) Luo *et al.* (2008b: 283 Table 18.2); Nelson *et al.* (1992: 88–91); Randall and Moffett (1968); Shanks (2014: 19); Ward *et al.* (2004; 2008: 136 Table 10.1); Whiton and Townsend (1928)
United States (MA)	Bean (1903: 177–179); Bigelow and Schroeder (1953: 87); Gill (1907: 36); Nichols and Breder (1927: 33); Radcliffe (1916)

(Continued)

(Continued)

Country (state or province)	Source
United States (MD	Hildebrand and Schroeder (1928: 80); Lugger (1878: 121–122); Murdy *et al.* (1997: 60)
United States (MS)	Franks (1970: 35–36); Nelson *et al.* (1992: 88); Overstreet (1974); Ross (2001: 47)
United States (NC)	Hildebrand (1934); Lee *et al.* (1980: 57); Luo *et al.* (2008b: 278; 282–283 Table 18.2); Smith (1907: 114–115); Tucker and Hodson (1976); Wade (1962: 555, 591); Ward *et al.* (2008: 136 Table 10.1)
United States (NJ)	Babcock (1936: 62); Breder (1925b); Breder and Redmond (1929); Fowler (1906: 90–91, 1910: 599–600, 1925a: 3, 1925b: 43, 1928: 608); Nichols and Breder (1927: 33)
United States (NY)	Babcock (1936: 18–19); Bean (1903: 177–179); Breder and Nigrelli (1939); Girard (1858); Griswold (1922: 5); Hickey *et al.* (1976); Nichols and Breder (1927: 33)
United States (RI)	Gordon (1974: 21–22); Griswold (1922: 6)
United States (SC)	Allen *et al.* (1982: 7–22 [this is a single page number]); Hammond (2005); Luo *et al.* (2008b: 278 Table 18.1); Wiggers (2010: 42)
United States (TX)	Breder (1944: 237 Table XI); Brown and Severin (2008); Clark and Landry (2002); Dailey *et al.* (2008); García de León *et al.* (2002); Gill (1907: 36); Griswold (1921: 170); Hoese (1958: 320); Luo *et al.* (2008b: 280 Table 18.1); Marwitz (1986); Moore (1975); Nelson *et al.* (1992: 91–94); Pew (1971: 18–19); Qualia (2002: 106–107); Randall and Moffett (1958); Simpson (1954); Ward *et al.* (2004; 2008: 136 Table 10.1); Zappler (editor, 1993)
United States (VA)	Fowler (1927: 89); Hildebrand and Schroeder (1928: 80); Moseley (1877: 9); Murdy *et al.* (1997: 60)
Venezuela	Dailey *et al.* (2008); Marín (2000: 64); McMillen-Jackson *et al.* (2002, 2005); Röhl (1956: 455–456); Schultz (1952: 33); Shreves (1959)

Pacific tarpons

Country (state or province)	Source
Australia (New South Wales)	Bishop *et al.* (2001: 18); McCulloch (1929: 34); Merrick and Schmida (1984: 53); Pollard (1980: 53); Roughley (1953); Tuma (1994)
Australia (Northern Territory)	Bishop *et al.* (2001: 18–29); Larson (1999: 26); Merrick and Schmida (1984: 54); Richardson (1846: 310); Taylor (1964: 59–60)

Country (state or province)	Source
Australia (Queensland)	Burrows *et al.* (2009: 6 Table 3); Davis *et al.* (2014); Herbert and Peeters (1995: 21); Hitchcock *et al.* (2012); Kottelat (2013: 33); Ley (2008); Power and Marsden (2007: 10, 12, 27); Russell and Garrett (1983)
Australia (Western Australia)	Bishop *et al.* (2001: 18); Moore *et al.* (2014: 174); Roughley (1953: 7); Tuma (1994)
Bangladesh	Rahman (1989: 234–236)
Ceylon	Day (1878: 650–651); Losse (1968: 80–81); Munro (1955: 22–23)
China	Day (1878: 650–651)
China (Hainan)	Nichols (1943: 18); Nichols and Pope (1927: 325)
China (South China Sea	Liu *et al.* (2010: 5); Shen (1982)
China (Taiwan Strait)	Liu *et al.* (2010: 28); Shih *et al.* (2004)
Fiji	Boseto and Jenkins (undated); Ley (2008: 5); Seeto and Baldwin (2010: 60)
India	Day (1878: 651); Losse (1968: 80–81)
India (Chennai [Madras], Tamil Nadu)	Alikunhi and Rao (1951); Chidambaram and Menon (1947); Job and Chacko (1947); Kuthalingam (1958); Pandian (1968); Moses (1923: 134, 138, 159); Raj (1916: 253)
India (Ganges River estuary)	Kottelat (2013: 33)
India (Sundarban, West Bengal)	Vyas (2012: 308)
India (Thiruvananthapuram [Trivandrum], Kerala)	Gopinath (1946: 9)
India (Kundapur, Karnataka)	Alikunhi and Rao (1951)
India (Kozhikode [Calicut], Kerala)	Alikunhi and Rao (1951)
Indonesia (Ambon)	Kottelat (2013: 33)
Indonesia (Java)	Kottelat (2013: 33); Sunier (1922: 226)
Indonesia (Karawang, West Java)	Delsman (1926)
Indonesia (Sulawesi)	Kottelat (2013: 33)
Indonesia (Sumatra)	Kottelat (2013: 33)
Japan (Boso Peninsula, Chiba)	Tsukamoto and Okiyama (1997)
Japan (Ishigaki, Okinawa)	Tsukamoto and Okiyama (1993)
Japan (Iwawada, Chiba)	Tsukamoto and Okiyama (1993
Japan	Losse (1968: 80–81)
Kenya (Malindi)	Losse (1968: 80–81)

(Continued)

(Continued)

Country (state or province)	Source
Kenya (Mombasa)	Bell-Cross and Minshull (1988: 91); Losse (1968: 80–81)
Madagascar	Kottelat (2013: 33); Losse (1968: 80–81)
Malaysia	Day (1878: 651); Ley (2008: 5)
Malaysia (Sandakan, Borneo)	Herre (1933b: 6)
Mauritius	Losse (1968: 80–81)
Melanesia	Losse (1968: 80–81)
Melanesia (New Caledonia	Kulbicki *et al.* (2005)
Melanesia (Tanna Island)	Kottelat (2013: 33)
Melanesia (Vanuatu)	Amos (2007: 149–150); Kottelat (2013: 33)
Micronesia	Losse (1968: 80–81)
Mozambique (Maputo Bay)	Barnard (1925: 104–105)
Mozambique	Mann (2000: 66, 2013: 120)
Palau	Ley (2008: 5)
Papua New Guinea	Allen and Boeseman (1982: 71); Allen and Coates (1990: 33 Table 1, 43, 52–53); Coates (1987, 1993); Hitchcock (2002: 120 Table 1); Merrick and Schmida (1984: 53); Powell and Powell (1999); van der Heijden (2002); Roberts (1978)
Philippines (Dumaguete, Negros Oriental)	Herre (1933a: 2)
Philippines	Fowler (1941: 519–523); Gonzalez (2013: 27); Herre (1953: 55–56); Losse (1968: 80–81)
Society Islands (Tahiti)	Kottelat (2013: 33); Merrick and Schmida (1984: 53); Pollard (1980: 53)
Polynesia	Bagnis *et al.* (1987: 272); Day (1878: 651); Losse (1968: 80–81)
Seychelles	Losse (1968: 80–81)
Singapore	Kottelat (2013: 33)
South Africa	Harrison (2001); Losse (1968: 80–81)
South Africa (Natal, Algoa Bay)	Smith (1965.86)
South Africa (KwaZulu-Natal)	Harris and Cyrus (1995); Mann (2000: 66, 2013: 120)
South Africa (Durbin)	Mann (2013: 120)
Tanzania (Dar-es-Salaam)	Losse (1968: 80–81)
Tanzania (Tanga)	Losse (1968: 80–81)
Tanzania (Zanzibar)	Boulenger (1909: 27); Eccles (1992: 32); Losse (1968: 80–81); Playfair (1866: 122)

Country (state or province)	Source
Tanzania (Mafia Channel)	Losse (1968: 80–81)
Taiwan	Jordan and Richardson (1909: 165); Losse (1968: 80–81)
Vanuatu	Ley (2008: 5)
Zimbabwe	Bell-Cross and Minshull (1988: 42, 91–92)

APPENDIX B

Contribución a la morfología y organogenésis de los leptocéfalos del sábalo *Megalops atlanticus* (Pisces: Megalopidae)

Contribution to the morphology and ontogenesis of the leptocephalus of the tarpon *Megalops atlanticus* (Pisces: Megalopidae)

J. E. Mercado Silgado and A. Ciardelli

Bulletin of Marine Science **22**, 153–184, 1972

Stage I (Leptocephalic Growth)

Phases I and II (1.7–11.0 mm SL) (Fig. 2)

The characteristics of these phases have been described briefly by Mansueti and Hardy (1967: 34) as "Putative yolk-sac larvae" and "Putative larvae". Because certain developmental stages immediately following those described by the above-cited authors are unknown, it is not yet possible to [define] the characteristics and/or morphological changes of leptocephali ranging from the "putative yolk-sac larvae" and smallest leptocephalus seen by Wade (1962: 555 and 604) of 11.0 mm SL.

Phase III (11.0–17.5 mm SL (Fig. 3)

In this phase we include the leptocephali described by Wade (1962: 555–559) of 11.7 and 17.5 mm SL. Because we did not obtain specimens in this category, we cannot provide data on morphological changes of the leptocephali, the only descriptions available being those of the aforementioned authors.

Translation from Spanish. See source publication for figures, tables, and literature cited in this Appendix.

Tarpons: Biology, Ecology, Fisheries, First Edition. Stephen Spotte.

Phase IV (17.5–24.0 mm SL) (Fig. 4A)

In this phase it is not possible to give data on the development of the leptocephali because the smallest specimen we observed was 24.0 mm SL. We include a description of it to compare with Wade's (1962: 59) of two specimens of 21.3–23.0 mm SL, and by doing so obtain an idea of the morphological changes undergone by the leptocephali.

Although the leptocephalus of 23 mm SL described by Wade (1962: 599 [*sic*]) is classified by him as belonging in Stage II, we place it in Stage I because [our] laboratory experiments demonstrate that these larvae keep growing. We therefore include his size [length] in the description of Stage I Phase V (Table 1).

Leptocephalus of 24.0 mm SL (Fig. 4A, Tables 1 and 2)

The difference this specimen presents with Wade's (1962: 559 [*sic*]) of 23.0 mm SL is the position of the coelom, which is located between myomeres 22 and 24, and an invagination in the anterior half [of the body] at the contact point of myomeres 22 and 23 (Fig. 4A). Also the pelvic fins have advanced forward from myomere 24 to myomere 23. In specimens observed *in vivo* the heart is still not functional and located posteroventrally to the pectoral fin. In this leptocephalus we noticed the appearance of cartilaginous structures that will later form the opercula. Incipient formation of the vertebral disks was also observed. The pigmentation looked like stars of black or dark brown. Starting at the posteroventral part of the head is a star on the heart and three or four posteriorly, which change later to short lines separating the intestine and kidney where they become present once again in the form of four stars located on the kidney.

Phase V (24.0–28.0 mm SL) (Fig. 4A, B; Table 1)

During this phase the leptocephali reach maximum size, approximately 28.0 mm SL (the maximum length obtained *in vivo* in specimens kept in the laboratory was 27.9 mm SL). The form of the body is triangular, laterally compressed, and transparent (when observed *in vivo*), which allows observation of the internal structures. The triangular form disappears gradually with increasing size.

The greatest body height was at the pelvic fins, and it was taken [measured] vertically to the dorsal surface, a distance that gradually increases with growth of the leptocephalus. During this phase the height of the caudal peduncle, as well as the height in the region of the pectoral fins, increases slightly. The changes in height, as well as changes in the body, [become modified] to produce the typical hydrodynamic form characteristic of many fish[es].

The head, which has become slightly lengthened, is defined by formation of the operculum. In this phase the internal structures of the head cannot be appreciated because the brain presents as an opaque, milky-white mass, making it stand out in relation to the rest of the body, which is transparent. During this phase the otic capsules and otoliths can be seen. The eyes are oval, dorsoventral, and gradually become rounded with age. The jaws are cartilaginous, directed dorsoanteriorly, and lack teeth. The snout is round and a little blunt.

The central nervous system is clearly visible without microscopy. Also apparent is the notochord [around] which vertebral rings can be seen, which will later give rise to the vertebrae.

The nares are visible as shallow concavities, although in this phase it is not possible to see bifurcations. Those appear later during Phase VIII Stage II.

Through the transparent body it was possible to observe four branchial arches, from which the branchiostegal spines are not distinguishable. This, together with the fact that the branchial arches are still poorly defined, indicates that breathing is still by osmosis [*sic*, diffusion].

The heart is eight-shaped in a horizontal position [like the sign for infinity]. It is dull white, essentially in the form of a thick tube. Neither contractions nor blood cells were observed.

The number of myomeres of the examined specimens was between 54 and 56, which coincides with observations of Wade (1962), who reports 54 myomeres, and with Mansueti and Hardy (1967), who reported 53–57.

The digestive apparatus consists of a central tube extending approximately three-quarters of the SL to the level of myomere 44, where it [terminates at] the anus. The digestive apparatus in its terminal portion is covered by a line of melanin. The intestine, which is still not functional, terminates at an angle of $\approx 45°$.

The coelom is evaginated dorsoposteriorly at the level of myomeres 22–24 at the beginning of the phase, shifting later to myomeres 23–25 in leptocephali of 27.9 mm SL.

The kidney is dorsoposterior to the intestinal tract, beginning almost even with myomere 35 and extending gradually until slightly thickened and obtusely shaped and terminating at myomere 45 (Fig. 4B). With age this organ gradually [separates from] the intestine and becomes located dorsally to the swim bladder, concluding metamorphosis.

The dorsal fin is located at myomere 42 and shows 12 incipient rays, the last presenting a longitudinal rift when observed closely under a microscope. Although the number of rays can occasionally vary among specimens, the number in a given specimen remains constant from first appearance until complete development.

The pectoral fins are positioned low and possess a fleshy [literally pulpy] base from which the fin membrane arises. The base of the fin, being opaque, hides the incipient rays. The rays are only discernible within the membrane.

The pelvic fins are located at the level of myomere 23. The incipient rays are not easily seen and only occasionally can they be distinguished in the base of the membrane.

The anal fin, located at myomere 44, has 20 incipient rays. However, there are specimens in which 21 or 22 are visible. This fin reveals a slight concavity at the border of the membrane.

The rays of the caudal fin show as truncated lines. This fin is forked with the dorsal and ventral lobes rounded. The urostile is pointed and directed dorsoposteriorly. With the techniques used it was not possible to distinguish the hypurals clearly.

Pigmentation in this phase is limited to the ventral part of the body. The melanin is distributed in the following way: a spot in form of a star in the ventral part of the head, anterior to the heart. Another star-shaped pattern [is visible] on the heart, and later there are three or four of them. In the leptocephali of 24.0–25.0 mm is an array of short lines. These continue to the previously described patterns and extend until separation of the kidney from the intestine. On the kidney, four patterns of melanin are distinguished in leptocephali of 24.0 mm SL, which later lengthen and form lines. On the base of the incipient rays of the anal fin a series of lines can be seen that, as the leptocephalus ages, spread and merge forming a single line of black chromatophores. On the ventral half of the caudal peduncle four lines of melanin can be seen in the dorsoventral position that in specimens of 27.9 mm SL increased in number, appearing as a line in each of the last five myomeres.

In leptocephali of 26.2–27.9 mm SL, the patterns of the chromatophores located in the ventral part of the head and enveloping the heart have become lines.

Another pigment observed on the leptocephali, besides the previously described melanin, is guanine [*sic*], which appears as agglomerations of white crystals located along the body. This pigment [*sic*] was observed previously in juvenile tarpon (Harrington 1966: 868).

Leptocephalus of 26.3 mm SL (Table 2)

The main characteristics distinguishing leptocephali of this length compared with the lengths [just] described are: snout a little sharper; dorsal concavity less pronounced; pectoral fin has shifted ventrally; coelom bigger and extends between myomeres 23–25; pelvic fin now located at myomere 24; starry spots of pigmentation have become lines beginning in the ventral part of the head and also in the heart, intestine, kidney, and at the base of the anal fin. A line appears more pronounced in the myomeres of the caudal peduncle, increasing the myomere number with pigment to four.

Generally at this length, because of major differentiation, the diverse existent structures can be appreciated.

Leptocephalus of 27.9 mm SL (Fig. 4B; Table 2)

In this leptocephalus the organs are similar to those of the previous specimens, but more defined. [This larva] is distinguished from the previous one by the following characteristics: snout sharper; lower jaw tends to be longer; pectoral fins slightly more ventral; concavity of the coelom prominent, and this structure now occupies half the SL; pelvic fins still located at myomere 24, but are more pronounced; all patterns of pigmentation have become lines, and pigmentation of the five myomeres of the caudal peduncle now assumes the shape of a "J". The other characteristics are similar to those described previously.

Stage II (Shrinking Leptocephalus)

Phase VI (28.0–25.0 mm SL) (Fig. 5A, B; Table 1)

It is in this phase when decreased growth of the leptocephalus begins, and with it gradual metamorphosis to the juvenile.

In this phase the heart begins to work and has sufficient blood, although [the blood] is rather scarce and pale. Contractions of the heart are rhythmic and quick. The blood is only distinguishable in the pericardial region as a slight rosy spot (Fig. 4B). Circulation is not evident in the rest of the body, leading us to think that the animal breathes mainly by osmosis [*sic*, diffusion].

The head is slightly flattened, its [visible] delimitations caused by the appearance of cranial bones, opercula, and a more compact cephalic mass. The otic capsule with the otoliths is more evident.

The vertebrae that began forming around the notochord are more discernible, although their number still cannot be determined clearly.

The mouth, jaws, nares, and gills have not undergone appreciable changes. The eyes are rounder.

The esophagus appears as a tube connected anteriorly with the mouth and posteriorly with the intestine, which is prolonged, its latter part extending to the base of the anal fin.

There is no evidence of the stomach, nor indication of its formation. The alimentary canal reaches from the ventral border of the leptocephalus and, in its anterior part, moves upward, passing below the fleshy part of the pectoral fins until arriving at the mouth.

In the coelomic cavity the protuberance mentioned previously is observed as in Stage I (Fig. 5A, B). Inside the coelomic cavity an oval sac can be seen forming – the swim bladder – and on its dorsal surface the formation of a black cap. In this phase it is still not possible to observe a connection between the swim bladder and the esophagus, although it is possible that the swim bladder originates from an evagination of the intestine.

Appreciable changes were not observed in sizes of the pectoral and pelvic fins. The dorsal fin, which during Phase V Stage I had been located at myomere 42, has changed position and now, toward the end of Phase VI Stage II, is located at myomere 40. No apparent changes in the incipient rays have occurred. The anal fin in Phase V Stage I, located at myomere 44, has shifted at the end of this phase to myomere 42, and 20 incipient rays are visible. The caudal fin remains unchanged.

Pigmentation varies, being seen as seven black lines in the ventral part of the caudal peduncle and as black, star-shaped pigment above the central nerve cord, this now being the last characteristic of the phase.

Melanic chromatophores shaped like stars and rays have expanded to cover three-quarters of the kidney. There are no fundamental changes in existing pigmentation on the digestive system.

Leptocephalus of 27.3 mm SL (Fig. 4A; Table 2)

The snout has lost its sharp form and is blunter; the dorsal concavity has disappeared almost completely.

The coelomic cavity stands out more, and highlighted in its dorsal part is the appearance of a kind of vertical, elliptic globe with two black points in its dorsal aspect: the swim bladder, which is located between myomeres 23 and 24.

The dorsal fin has changed position to myomere 42 from myomere 41 at the end of the previous phase. No changes were seen in the number of incipient rays. The anal fin has also changed position, shifting from myomere 44 to 42 with an average of 20 incipient rays. Positions of the pectoral fin and peduncle remain unchanged.

The heart is clearly discernible, but signs of activity are still not apparent.

Bones that will form the opercular structure have become more discernible, but still not clearly visible *in vivo*.

Pigmentation begins to change, but remains in transition. Two ventral black spots appear at the level of the dorsal part of the swim bladder [mentioned above].

Leptocephalus of 25.0 mm SL (Fig. 5B; Table 2)

The loss of sharpness in the snout becomes more obvious. The body has less depth; the back is widening, assuming a more hydrodynamic shape. The coelomic cavity is slightly higher, the swim bladder completely visible with its black cap merged in the dorsal part. Whether the physostomic connection [with the esophagus] has formed is unclear. Movement of the heart (very fast and rhythmic) can be observed. Red pigmentation in the heart area indicates the existence of blood. The circulatory system is not still operative. The dorsal fin has continued its positional change, being located at the level of myomere 40. The anal fin has also changed position and is now located at myomere 41. Structural or positional changes were not seen in the pectoral, pelvic, and caudal fins. Apparent in the dorsal part of the caudal region are two disks of dark pigmentation shaped like stars. Also observed in the ventral part of the caudal peduncle are seven lines of black pigmentation, which have lost the "J-shaped" form described previously. Part of the pigmentation has again gone from lines to starry disks. A pigmented cap is evident on the dorsal [surface] of the swim bladder (Fig. 5B). The eyes are rounder, the otic capsule with otoliths more visible.

Phase VII (25.0–20.0 mm SL) (Figs. 5B; 6A, B; 7)

In this phase the heart becomes activated, with pulsations varying between 140 and 145 [beats] per minute. The circulatory system is functional and blood can be seen circulating [through] the gills and several other parts of the body, indicating branchial breathing. Also beginning in this phase is aerobic breathing, evident by frequent rising of leptocephali to the surface to take gulps of air. The head is blunter toward the end of the phase, assuming an almost rounded shape. The cartilaginous opercula distinguish the head from the rest of the body. One can also observe the formation of the nares. The otoliths are formed and positioned inside the otic capsule.

The vertebrae can be seen clearly, and the numbers of myomeres and vertebrae correlate one-to-one. Myomeres now total 54. At the end of this phase the mouth takes on a more dorsal position, and the jaws are longer.

The eyes become rounded and more pigmented.

Inside the coelomic cavity the connection of the swim bladder with the esophagus can be observed. This connection was not distinguishable in the previous phase. The esophagus extends without modification until arriving at the coelomic cavity, where its connection with the intestine is clearly evident. Posterior to this connection, one can observe a ring that later becomes the stomach and subsequently is transformed into the intestine that continues to the anus, which is still not functional. Dorsoposteriorly to the intestine and ventral to the swim bladder, and also inside the coelomic cavity, a completely red circle appears (Figs. 6A; 10C), which is apparent first as a small point clearly discernible at the end of this phase; the red pattern is seemingly the origin of the liver.

The swim bladder, located at myomeres 22–25 in the beginning of this phase, is at the end of it, located in an oblique position between myomeres 24 and 27. The pneumatic conduit is easily observable and, as possible evidence of a functional connection, one could observe the leptocephalus taking gulps of air at the surface of the water. The swim bladder assumes a metallic, mercury-like color, observed earlier by Harrington (1966: 868).

At the end of the phase one can observe, at the level of myomere 23, the formation of an elliptical spot of yellowish-green anterior and ventral to the coelom. This spot has not been identified completely because it is impossible to see its final position without making histologic sections. Although important changes were not observed in the position of the kidney, [this organ] seemed to decrease slightly in size. In the previous phase it was positioned a little beyond the anal fin; in this phase it hardly reaches to the anterior of the base [of the anal fin]. A rosy coloration was observed in the kidney, [along with] separation from the intestine caused by the swim bladder having been interposed between them.

The pectoral fins change position from anteroposterior to dorsoventral, [assuming] an angle of ≈ 45° with the border of the ventrum. They also increase in size toward the end of the phase, and more incipient rays can be seen clearly.

Structural and locational changes in the pelvic fins were not observed, although at the end of the phase there was an augmentation in size. The dorsal fin continues moving forward from myomere 40 at the beginning of the phase to past myomere 38, and by the end of it the rays are clearly discernible. The anal fin does not change in either form or position.

By the end of this phase the head has become strongly pigmented, especially the part dorsoposterior to the eyes. Pigmentation is also seen on the operculum, along the lateral line, and on the dorsal border. Pigmentation of the swim bladder increases, becoming a very visible black area connecting to a near-continuum of lines along the border of the ventrum. Pigmentation of the anal fin almost disappears proximally and is notable for its increase distally. Pigmentation also appears on the hypural bones. Pigmentation becomes heavier on the caudal fin, especially on the upper and lower lobes of the bifurcation, and pigments appear on the pectoral and pelvic fins.

This is the phase in which lipochrome pigments first appear and spread over the whole body.

Leptocephalus of 22.8 mm LS (Fig. 6A; Table 2)

The head, excepting the snout, is more rounded, the mouth points forward with slight anteroventral inclination. The circulatory system is defined, and the heart is a red ellipse positioned even with the base of the pectoral fin. In addition, circulation is observed through the principal arteries, along with four functional branchial arches. The fins are unchanged in form and position. The coelom is much bigger and slightly inclined dorsoposteriorly. Inside the coelom is a very defined swim bladder with its pneumatic conduit now functional as a respiratory system, [and] the leptocephalus ascends frequently to the surface to take in air. No changes in the esophagus are apparent, but on [examining] the coelom, and later the pneumatic conduit, one can observe a knot, which in its ventral part will give rise to the stomach. The prolonged tube reaching to the anal fin is the origin of the intestine. Dorsoposteriorly to the beginning of the knot and in contact with it can be seen a red disk, the incipient liver. Pigmentation appears now more than ever like a series of ventral lines. The black cap is still observed on the swim bladder, which looks like a bag of metallic mercury. Lipochrome pigments begin to appear over the whole body. Changes in the form or position of the kidney are not apparent.

Leptocephalus of 22.2 mm SL (Fig. 6B; Table 2)

The main differences from the previous larva [are as follows]. The head loses some of its sharpness, becoming almost round. The mouth is a little more inclined anteroventrally. The circulatory system is much clearer, and organs inside the coelom are better defined. The dorsal fin has changed position and is now located at myomere 39. Pigmentation is similar to the previous larva except for the cap on the swim bladder, which has been replaced by a series of disks. The series of lines on the anal fin has disappeared, and a long line appears on its base. Other characteristics are similar to those of the previous leptocephalus.

***Leptocephalus of 21.1 mm SL* (Fig. 7, Table 2):**

At this length the main changes observed [include those of] the head, which is totally rounded and heavily pigmented, especially dorsoposterior to the eye. The eyes are [now] almost completely round. The heart is well defined, and circulation through arteries, veins, and capillaries is clearly visible. The coelomic cavity has enlarged and extends from myomeres 19–26. The formation of a brilliant black cap appears for the first time above the swim bladder, losing [pigment density] toward the ventral borders of the coelom. The dorsal fin begins at the level of myomere 38, the anal fin at myomere 42. Changes in pigmentation are observed on the urostyles, and the caudal fin reveals little pigmentation. Definition of the pelvic fins becomes clearer. Inside the coelomic cavity the formation of the stomach is clearly visible (Fig. 10C). The kidney has diminished in length and reaches only to the beginning of the base of the anal fin. The stomach is still not functional, and neither is the anus.

Phase VIII (20.0–15.0 mm SL) (Figs. 8A, B, Table 1)

The head is already well defined and has assumed a round form, [as described] in the previous larva. The lower jaw continues its upward trajectory. Cartilages forming the head start to ossify. The maxillary has become wide by the end of the phase and possibly ossified. The first two fine, weak teeth appear on each side of the lower jaw [and] continue increasing [in number] with age. We could only count up to five teeth on each side.

The circulatory system is complete and functional; the heart beats fast and rhythmically. The gills are functional, and the leptocephalus is observed ascending regularly to the surface to obtain atmospheric air.

The vertebrae start to become ossified. The four branchial arches can be seen clearly, as can the branchial spines. The nares, observed clearly [too], are bifurcated; the otic capsules with their two otoliths are also visible.

The esophagus is well defined, as is the rest of the digestive system. The coelomic cavity is complete and by the end of the phase is located [between] myomeres 15–27. Inside the coelomic cavity the swim bladder is positioned dorsoanteriorly to the stomach (Fig. 8B). The elliptical spot of yellow observed in the previous phase anteroventral to the coelom and above the esophagus is, by the end of this phase, near the liver (Fig. 8B). As for the kidney there is diminution in length but increased size posteriorly; at the same time, it continues its separation from the interposing effects of the intestine and swim bladder.

The pectoral fins maintain the same position as in the previous phase, and at their fleshy bases the incipient rays are clearly visible. The dorsal fin continues moving forward from myomere 38 at the beginning of the phase to past myomere 37 by the end. The anal fin at the end of the phase is located at myomere 40.

The pelvic fins continue to occupy the same position, appearing bigger with the rays discernible. The location of the pigments continues evenly, except that in this phase the pigmentation is denser over the whole body. Lipochrome pigments are more discernible, being increasingly evident on the head and caudal peduncle.

In this phase the leptocephalus begins to take food. Because of its transparency food can be seen in the stomach and intestine. Voracity increases with age. The food offered consisted mainly of *Artemia salina* nauplii, cladocerans, and mosquito larvae (genus *Culex*). The anus becomes functional, and feces are observed on the bottoms of the aquariums.

Leptocephalus of 19.0 mm SL (Fig. 8A, Table 2)

The notable changes observed at this length: ossification of principal bones of the head and dilation of the coelomic cavity caused by growth of its enclosed organs. The liver becomes more visible and assumes an elliptical shape; the stomach is defined, and the swim bladder lengthens. The leptocephali start to accept food at this size, and their voracity increases with age. The stomach and the anus begin to function. Pigmentation increases in intensity and extends to the whole body. Each myomere contains a chromatophore along what will become the lateral line. The dorsal fin is located at myomere 38, the anal fin at myomere 40. Definition of the pelvic fins becomes clearer. The pectoral fins do not shift position, but become larger.

Leptocephalus of 16.4 mm SL (Fig. 8B, Table 2)

Quite similar to the previous larva except for the dorsal fin, which is now located at myomere 37, and the anal fin, now at myomere 39. The pelvic fins, although retaining the same location, become better defined. The protuberance located under the coelom begins to diminish, and the esophagus spreads to take the form of a "S" instead of its previous arc-shape. The coelomic cavity has grown and is now located between myomeres 15 and 27. The liver is in contact with the green spot, which seems to be a biliary vesicle.

Phase IX (15.0–13.0 mm SL) (Fig. 9, to B)

The head is well defined, flattened, and looks ossified. The upper and lower jaws are well formed with a tendency observed of the lower jaw to be more prominent. The maxillary is wide and occupies a third of the lower jaw. The small [villiform?] teeth can be seen clearly. Eyes are black and round. The urostyles and hypurals, now ossified, can be seen; the gular plate is forming.

In this phase, although the leptocephalus is losing transparency, one can still observe the circulatory system and blood circulating through it. Contractions of the heart are already apparent without the aid of a microscope.

The vertebrae are not clearly observable because of the body's opacity. The dorsal and lateral part of the body is silvery. The branchial arches are covered by the opercular bones. The nares are clearly visible, but not the otoliths because of advanced ossification.

The coelomic cavity has increased in size, extending from myomeres 14–27. The swim bladder, now silvery, is horizontal. The esophagus, because of forward displacement of the liver and stomach, has shortened. The stomach has increased in size, is now completely formed, and usually filled with food. The liver continues its forward displacement and is in direct contact with the yellowish-green spot, seemingly a biliary vesicle. The spot is united with the liver and esophagus by a membrane and requires sufficient illumination to be seen. The anterior part of the kidney continues to rise above the swim bladder, becoming tape-shaped and wider posteriorly.

The pectoral and pelvic fins are larger, their rays more visible. The dorsal and anal fins have not changed [lateral] position, but are higher, their rays clearly apparent. The caudal fin is already forked, its lobes symmetrical.

Pigmentation is concentrated on the back of the head, extending along the dorsal surface of the body. On the ventral part of the body the most highly pigmented zone covers the coelomic cavity. This pigmentation extends along the ventral border where it joins the dorsal edge by a row of chromatophores subsequently bordering the tail bones.

In this phase the formation of scales is not noticed in the leptocephalus. These leptocephali prefer *Culex* mosquito larvae as food, although because of their voracity and growth they also ingest other spineless aquatic invertebrates introduced into the aquariums.

It is in this phase when the leptocephalus begins to resemble a small fish, not just because of its pigmentation but also as a result of the forward displacement of the coelomic cavity and thinning of the back, which begins to reveal a characteristic torpedo shape. Also contributing is the fact that the head is already practically formed.

At the end of this phase the size [length] of the animals grown in the laboratory was never less than 13.0 mm SL (Fig. 9B), these having the aspects of small tarpons, which can begin to be classified within Stage III of Wade (1962).

Leptocephalus of 14.0 mm SL (Fig. 9A; Table 2)

The head has lengthened and is almost completely ossified. The lower jaw seems to have a slight tendency to be longer than the upper; the eyes are round, and the gills already completely covered by the opercular bones. The coelom now extends to myomere 14, and inside it can be seen the completely formed stomach and liver. The latter is already in contact with the green spot, the one that, as mentioned previously, is presumably a biliary vesicle. Because of the forward displacement of the coelom the esophagus is now shorter. Pigmentation becomes denser and it is now found mainly on the dorsal surface of the head from where it continues in a line along the dorsal border. The line of pigmentation on the coelomic cavity continues along the ventral border until uniting with dorsal pigmentation skirting the posterior part of the hypurals and hemoneurals[?]. The dorsal fin has shifted to myomere 37, the anal fin to myomere 39. The other fins remain in the same position but have increased of size and are more defined. The kidney, gradually assuming a continuous, tape-like shape, occupies a position dorsal to the swim bladder.

Leptocephalus of 13.0 mm SL (Fig. 9B, Table 2)

This is the minimum size of the Stage II leptocephalus *in vivo* [reared] by us in the laboratory. The head is shorter, the body well defined. Dentition of the upper jaw stands out. The body begins to take the juvenile form. The forward part anterior of the coelomic cavity extends to behind the head, the posterior part reaches to the anus. The body is opaque, the internal structures difficult to observe, although the swim bladder, kidney, heart, gills, and part of the circulatory system are clearly distinguishable. The pectoral fins have lengthened considerably and shifted slightly downward to their final locations. The other fins have also become larger. Pigmentation is similar to the specimen of 14.0 mm described previously except that more pigmentation appears on the dorsal part of the dorsal fin and on the lobes of the caudal fin. Some star-shaped spots also appear, scattered on the anal fin. The leptocephali in this phase are extraordinarily voracious, and their main food consisted of mosquito larvae. At this length obtained in the laboratory, metamorphosis of the tarpon *Megalops atlanticus* Valenciennes is complete.

Stage III (Juvenile Growth)

Phase X (13.0 mm and larger) (Figs. 9B; 10A, B)

We include tarpons in this phase in Stage IIIA of Wade (1962). These have completed metamorphosis.

Because the body is opaque there can be no descriptions of the organs. In our work a specimen of 15.9 mm SL described by Wade (1962: 564) as belonging to Stage IIIA is classified as Stage II Phase VIII.

Juvenile of 16.9 mm SL (Fig. 10A, Table 1)

Head almost completely ossified. Lower jaw projected slightly forward. Opercular bones still projected posteriorly and ellipsoid in shape. These bones take their definitive form when [the fish] attains 41.1 mm SL (Fig. 10B). The dorsal part of the body has a gentle sigmoid shape, which is conserved in the adult. The ventral part of the body presents to the level of the pectoral fin at a concavity at which the coelom is located. Immediately posterior is a ventral convexity that extends to the anus, where it finishes in a point. Between the anus and the tail is another dorsal concavity. The convexities and the concavity disappear later, leaving only a gentle convexity. Also, the posterior concavity diminishes, becoming barely visible in an adult tarpon. The body is completely opaque, allowing observation of only the kidney and part of the posterior circulatory system. The coelomic cavity is completely covered with silver pigment under the epidermis. Signs of scale formation were not observed in this specimen. All fins are in their final positions. The last ray of the dorsal fin has not begun elongation, which commences at ≈ 71 mm SL.

APPENDIX C

Desarrollo temprano del sábalo, *Megalops atlanticus* (Pisces: Megalopidae)

Early development of the tarpon, *Megalops atlanticus* (Pisces: Megalopidae)

Didiher Chacón Chaverri and William O. McLarney

Revista de Biología Tropical **40**, 171–177, 1992

According to Wade (1962), Mercado and Ciardelli (1972), and Jones *et al.* (1978), the larval development of the tarpon (*Megalops atlanticus*) Valenciennes, 1846, comprises three stages up to the juvenile. Stage I: period of initial growth until maximum size as a leptocephalus larva. Stage II: beginning of the feeding leptocephalus larva, size reduction (TL), and initial formation of the juvenile stage. Stage III: formation of the juvenile stage and its [subsequent] total transformation.

Rickards (1968), Tagatz (1973), Jones *et al.* (1978), and Smith (1980) established that this species commonly spawns in bodies of brackish water or dilute coastal areas where currents can carry the eggs to estuarine waters of rivers and coastal lagoons.

On the other hand, Wade (1969), Wade and Robins (1973), Crabtree *et al.* (1991), and Cyr (1991) established spawning areas to be oceanic blue waters far from the coast, based on the absence of Stage I leptocephalus larvae in coastal ecosystems. From analysis of some previous publications it seems that the spawning season can include the months of July to November (Harrington 1966, Mercado and Ciardelli 1972, Tagatz 1973, Tucker and Hodson 1976, Smith 1980), although some authors, such as Mansueti and Hardy (1967) and Fischer (1978), established the spawning season to be May and July, and Wade and Robins (1973) determined it to be June to August. In the case of the Caribbean coast of Costa Rica, Nordlie and Kelso (1975) found that in the dry season

Translation from Spanish. See source publication for figures, tables, and literature cited in this Appendix.

Tarpons: Biology, Ecology, Fisheries, First Edition. Stephen Spotte.

(April–May and September–October) there is typically an increase in the density of larval marine fishes. For Costa Rica there is no investigation that verifies the observations about tarpons postulated to occur in other places of the tropical and subtropical Atlantic Ocean. It therefore is necessity to clarify the biology and ecology of *Megalops atlanticus* in the Costa Rican Caribbean. For this, Gandoca Lagoon was chosen (a coastal lagoon on the Caribbean coast of Costa Rica), located inside the limits of the National Living Forest Refuge Gandoca/Manzanillo, Limón (Fig. 1). The present investigation was carried out because early stages of the tarpon had been collected here in 1973 and 1987. The present work confirms the presence of leptocephalus larvae and juvenile stages of *M. atlanticus* in Gandoca Lagoon and bordering sectors and offers new information about ecology and larval development.

Material and methods

Samplings of the lagoon were carried out in marine, estuarine, and freshwater areas. A protocol was developed in each area of the lagoon, three times per week between the fourth week of July and first week of November 1988. To collect larvae of *Megalops atlanticus*, a channel net was used having a mesh of 300 microns [μm], and a General Oceanics 20030R flow meter (Bozeman and Dean 1980). This method was applied by Tucker and Hodson (1976). Sampling of the creeks (Middle Creek, Black Creek, Don Nati Creek) was carried out by means of a hand net of 2.0 mm mesh and stationary nets (gill net of elastic mesh, double-entrance traps, and a dragnet of 2.0 mm mesh [measuring] 0.92 m high by 5.0 m long).

The fish were fixed in a solution 33.3% formaldehyde [followed by preservation in] 66.6–80% alcohol.[1]

The study of larval development was based on the works of Mercado and Ciardelli (1972), Jones *et al.* (1978), and Smith (1979).

The identification of each stage was carried out making use of biometric characters such as length of the head and predorsal and preanal lengths, in addition to meristic characters (Tables 2 and 3).

Leptocephalus size was compared among *M. atlanticus*, *Elops saurus*, and *Albula vulpes* because the larvae of the first-mentioned species are the smallest in their early stages. Another aspect taken into account was the presence of melanophores in the ventral part of the larvae and in each myoseptum, a

[1]Assuming 37% commercial formaldehyde to represent 100% for dilution purposes, pouring 33.3 mL into a graduated cylinder and bringing the volume to 100 mL with water would produce a 33.3% formalin solution. A 5% solution using filtered water from the collection site as the diluent seems more appropriate for delicate larvae to prevent their shrinking and deformation.

typical character used to differentiate Elopiformes from Anguilliformes and Notacanthiformes (Smith 1979).

In each sampling area the salinity was determined directly by means of a refractometer. Also taken were surface temperature of the water and its transparency using a Secchi disk. The concentration of dissolved oxygen was measured according to Sánchez *et al.* (1984), based on the Winkler method. The pH was measured by paper colorimetry.

For collecting juvenile specimens in the lagoon and creeks a net gill net of 1.5 cm of mesh was used [hung] in vertical position for 1 h, three times a day.

Results and discussion

Morphometry appears in Tables 2–3 and development in Fig. 2.

In the case of estuaries such as the Gandoca Lagoon the circulation of water depends on the tide, flow of the Gandoca River, marine currents, wind, and topography, in addition to external masses of water [such] as those introduced by the Sixaola River and its countercurrent, which are responsible in great measure for distribution and dispersion of the leptocephalus larvae of *Megalops atlanticus*.

Station 3 (Don Nati Creek) lacks a permanent connection with the sea, and the arrival of larvae is a result of strong rains and periods of high tide.

In Station 2 (Middle Creek), tidal influence introduces larvae, but not many, given the necessity of precipitation and gradually sloping outlet topography. The fact of having collected only one larva could be caused by flaws in the collection method. These include occlusion of pores of the mesh and the effect of the river current on emptying of the net, or on dispersion processes of larvae [causing them to avoid] this sector; the case for Black Creek [was similar?] in its two sectors.

In Gandoca Lagoon are two sectors where larvae were found: the outlet and Lizard Pipe. In the case of the first sector the tide and the marine current carry organisms to this place of contact between lagoon and sea with a salinity of 34.5 ppm [*sic*, ppt], explaining the presence of 41 leptocephalus larvae collected in July; to a depth of 3.0 m with a Secchi disk transparency of 2.5 m. At Lizard Pipe [it is] a combination of wind (NE-SO) perpendicular to the coast, tidal flow, and weak fluvial flow; factors that carry water rich in planktonic organisms, including tarpon leptocephalus larvae.

According to Beebe (1939), Wade (1962), Rickards (1968), and Tucker and Hodson (1976) it is possible to check the influence of abiotic factors such as turbidity and H_2S on the presence and distribution of larvae and juveniles. In all the collections except at Station 2 the organisms were distributed in waters of brown color and high turbidity, and in the presence of H_2S and variable salinity and temperature where concentrations of dissolved oxygen are generally low.

This situation facilitates the survival of larvae, in addition to which this species can breathe atmospheric air (Shlaifer 1941). In the case of leptocephalus larvae ([Stages] I, II), all were in salinities greater than 29.0 ppm [*sic*, ppt], in agreement with Smith (1979), who reported on the distribution of larvae both before and after metamorphosis in areas of high, stable salinities. Juveniles were collected in waters having salinities of 0.0 ppm [*sic*, ppt], demonstrating the tendency of the tarpons to move from marine to freshwaters in the course of their metamorphosis and development into juveniles.

As to one contribution of this study in the larval development of *Megalops atlanticus* it should be noted that measurements of some individuals collected had never before been recorded. This is the case at Station 1 where the maximum standard length (SL) was 29.0 mm, whereas the previous maximum was 27.9 mm (Mercado and Ciardelli 1972). The difference of 1.1 mm in SL might be because these other authors raised their larvae in the laboratory. Conditions could not be the same as the natural environment, and the effect of handling abrasion during collection of the larvae and their [subsequent] handling during rearing could account for the reduction in size (Thielacker 1980). Another factor that could influence this difference is the possible load energetics given by adults to the eggs, a fact that can markedly affect the "quality" of larvae, although Smith (1979, 1980) reports a possible maximum growth of 30.0 mm and 28.0 mm, respectively, in larvae of this type.

This same stage received new contributions in the sense that the range of proportional variation with respect to head length, dorsal [fin] length, and preanal [fin] length [as percentage of] standard length increased; a Student's t-test at a confidence interval of 95% showed no significant difference. These new values are greater than those reported by Jones *et al.* (1978).

The average standard length at this stage was 25.05 mm with laterally compressed transparent body and with melanophores present in the ventral area of the myomeres and some in the myosepta and in the upper part of the head, fin-ray bifurcation, the tract or alimentary canal stuck to the ventral region of the myomeres, and the swim bladder located on average at the 22nd myomere. There were no teeth, indicating that these leptocephalus larvae [Stage] I have begun the shrinking phase and metamorphosis to enter Stage II; average total myomere count was 55.

For Stage II, two important facts stand out. First, in the final phases of this stage the individual possibly shrinks by reabsorption of tissues (Mercado and Ciardelli 1972, Jones *et al.* 1978), and the dorsal and anal fins begin migrating forward of the body, lessening their position in two myomeres while the head grows in length and increases as a percentage of SL. Second, transformation into the juvenile begins, at which [time] some vital organs are generated and melanophores partly cover the body.

Stages II and III reveal differences in the proportions of prepelvic, predorsal and preanal [fin] lengths, following Jones *et al.* (1978), although here the low number of larvae does not allow us to make a more extensive analysis.

The transport and presence of larvae in waters of the Gandoca Lagoon are a result of the reproductive behavior of adults as described by Dando (1984). This species uses marine currents so that eggs and larvae are carried to areas of high productivity, such as estuaries and coastal areas.

The organisms collected in the study confirm [?] that this area (or at least the east sector of the refuge) is used as a nursery for larval and juvenile stages. It is possible that after having passed through these stages they find a means to arrive in oceanic waters and become part of the "stock" of tarpons on the Caribbean Costa Rican coast.

The early leptocephali are typical of pelagic waters where they are carried toward the coast by currents, winds, and tides while developing; having arrived, metamorphosed larvae or leptocephali in Stages II and III inhabit coastal ecosystems of low salinity and high turbidity, whereas the juveniles are common in freshwater environments.

APPENDIX D

Aspectos biométricos de una población de sábalo, *Megalops atlanticus* (Pisces: Megalopidae)

Biometric aspects of a tarpon population, *Megalops atlanticus* (Pisces: Megalopidae)

Didiher Chacón Chaverri

Revista de Biología Tropical **41**(Suppl. 1), 13–18, 1993

The tarpon (*Megalops atlanticus*) Valenciennes 1846 is among the more sought-after species by the sport fishing world, and maintaining the growing recreational fishery is important economically. Although the tarpon is fished from Virginia through the Gulf of Mexico and south to Brazil (Hildebrand 1963, Fischer 1978), its [sport] fishery is more developed in Costa Rica and Florida. Despite the tarpon's importance in these regions the effects of human activities on its abundance are poorly understood (Robins 1977).

Although the tarpon is distributed through the tropical and subtropical Atlantic, most investigations have been conducted in southern Florida. The mature tarpon possesses great mobility and has the capacity to travel considerable distances in open waters. The larvae also have the potential to be transported from the Caribbean to Florida waters (Crabtree *et al.* 1992).

Although mature tarpons are very common along the Caribbean coasts of Costa Rica, Nicaragua, and Panamá, larvae and juvenile are rare or absent. Chacón and McLarney (1992), however, documented the presence of larvae and juvenile tarpons in the Gandoca Lagoon of the Gandoca-Manzanillo Wildlife Refuge located in the Talamanca district near the frontier with Panamá. The ultimate objective of this work is to describe the morphometric structure and some other biological aspects of a mature population of *M. atlanticus*; for the larval and

Translation from Spanish. See source publication for figures, tables, and literature cited in this Appendix.

Tarpons: Biology, Ecology, Fisheries, First Edition. Stephen Spotte.

juvenile phase the objectives are to describe the extension of the habitat of the early stages in Costa Rica and to estimate the temporal patterns of recruitment.

Material and methods

During every month from September 1990 to August 1992 we visited sport-fishing locations on the Colorado River (Fig. 1) where tarpons killed in the fishery were taken for analysis. The length and weight of the fish and a variety of morphometric and meristic measurements were recorded according to the method set down by Robins and Ray (1986) (Fig. 2). The fish were sexed, and their gonads removed and weighed individually; a 1 g sample was fixed [*sic*] in 95% ethyl alcohol and stored in 10 mL vials. The ethanol was changed 24 h after the first fixation [*sic*]. These gonad samples were taken to determine state of maturity.

The gonadosomatic index (GSI) was calculated according to the procedure of Kaiser (1973):

$$\text{GIS} = \text{total gonad wet weight} / \text{weight of the fish} \times 100$$

Having known the location of a habitat for larvae and juveniles in the Gandoca area (Fig. 1), we sampled it monthly from July 1988 to August 1992, following the method described by Chacón and McLarney (1992) to study recruitment patterns and size frequencies. It [the method] could be extended to habitats of the larvae and juveniles in Costa Rica to explore other areas that could be potential areas of tarpon year classes. Morphometrics were analyzed using techniques described in Cyr (1991).

Rainfall patterns for the study areas were obtained from the National Meteorological Institute (1988).

Results and discussion

We intended to collect mature tarpons monthly over 23 months at fishing locations along the Colorado River, Costa Rica. Collections ranged from none (July and August 1990) to 23 (February 1991) fish per month. Collecting tarpons was not possible in every month for several reasons. The fishing areas were closed several times during the year, and at times other factors interfered with the fishing. The earthquake of 1991 also disrupted the normal sampling procedure and created numerous logistical problems. To date, 113 mature tarpons constitutes the examined sample (Table 1). Weights of the fish examined ranged from 7.00 kg to 74.00 kg with a mean of 29.68 kg. The relationship of fish weight (W) to fork length (FL) for tarpons in the sample was:

$$\text{Log}_{10} W = -8.413 + 3.144(\log_{10} FL) \tag{D.1}$$

Male tarpons weighed, on average, 21.36 kg (range 7.00–29.90 kg) and were significantly smaller that the females, which had a mean weight of 36.90 kg (range 19.09–74.00 kg, *t*-test, $p < 0.001$). A female of 19.09 kg presented with vitellogenic oocytes, indicating attainment of sexual maturity. A male of 10.00 kg. presented with gonads containing mature sperm. These two fish are, at present, the smallest [known] mature individuals of their respective sexes. Both are smaller than those seen in Florida. Nonetheless, more Florida fish have been examined (Cyr 1991). Minimum sizes reported here are similar to those of a Brazilian population of tarpons examined by Ferreira and Pinto (1966).

The fish showed a range in GSI from 0.19%–7.17%; the value of the cumulative average of this index in females was generally bigger than that of males (Fig. 3). Gonadal weight (GW) correlated positively with weight of the fish of both sexes (males: $GW = 0.2063 + 0.463W$, $p < 0.001$, $R^2 = 0.26$, $n = 46$; females: $GW = 0.9525 + 0.67W$, $p < 0.001$, $R^2 = 0.30$, $n = 63$).

A tendency was clearly apparent in the monthly GSI values for both sexes, suggesting a bimodal pattern with two peaks where a correlation exists between the cumulative average GSI values of males and females. During every month the indices were higher, corresponding with mature individuals and demonstrating year-round reproduction. However, the sample size is too small to draw a definitive conclusion.

Fishing in the Bar region of the Colorado takes place in the ocean, usually six or eight kilometers off the coast, and in the river, often 70 or more kilometers upstream. The size of the fish captured in the ocean (mean $L = 1384$ mm SL, $n = 67$) was not significantly different from those captured in the river (mean $L = 1387$ mm LS, $n = 46$, *t*-test [value of *t* omitted in original source publication], $p > 0.05$. Total size of the fish varied over time, with the biggest captured at the end of the rainy season (November, December, and January) and at the beginning of the dry season (February and March) coincident with findings presented by Ferreira and Pinto (1966). These similarities follow seasonal changes in the percentage of sexes of the tarpons captured. The percentage of females was greater during November, December, and January and declined in June, July, and August. Females were larger than males, the average total weight reflecting the percentage of females captured (42.2% males vs. 57.8% females). The significance of this tendency is unclear and could be due simply to the small sample size, although it clearly reflects results of Ferreira and Pinto (1966) and Cyr (1991).

Juvenile tarpons were concentrated at Don Nati Creek in the Gandoca-Manzanillo Wildlife Refuge and in other areas along the Costa Rican coast where the presence of early stages was demonstrated. According to Chacón and McLarney (1992), this creek flows intermittently during storms and floods during extreme high tides when pre-metamorphic tarpon larvae enter the creek and undergo metamorphosis. In sum, 240 juveniles were collected, ranging from 19.00 to 260.00 mm SL, together with 107 pre-metamorphic leptocephali (Table 1). Recruitment occurs seasonally from July to October (leptocephali) and

December to February (juveniles) (Fig. 4a and 4b). Standard lengths and relative frequencies were: 30.00–45.00 mm (17%) and 45.10 mm–60.00 mm (23%) (Fig. 5). Recruitment of larvae is consistent with Crabtree *et al.* (1992), considering that the leptocephalus stages collected for the most part represent Stages II and III, as classified according to Wade (1962).

The preliminary data suggest that Costa Rican tarpons reach sexual maturity at smaller sizes than Florida fish; however, size data at sexual maturity are inadequate. Duration of reproductive status is poorly defined. Tarpons in several stages of maturity have been found during every month, possibly pointing to year around reproduction with a bimodal model having two peaks, one in February–May, the other August–November.

The significance of the juvenile habitats in Costa Rica remains unclear. On several occasions dead tarpons have been found next to sandy banks during periods when the creeks experience very low water levels. Whether several or all the tarpons entering the creeks survive to adulthood is unknown. Examination of a large stretch of the Caribbean coast of Costa Rica did not reveal many juvenile habitats. This can be related to a number of factors, including large quantities of freshwater that limit the extent of mangrove swamp habitats. In places like the Colorado River, salinities at the mouth are 0.00 ppm from surface to bottom. Such large quantities of flowing freshwater can impede recruitment of larval tarpons. The extent of mangrove swamp habitat is also unknown in this region in contrast with areas such as southern Florida, which sustain substantial juvenile populations. It seems that juvenile habitats are more disseminated in other Central and South American countries like Belize and Venezuela.

APPENDIX E

Partial list of food items consumed by Atlantic tarpons

Common name	Scientific name	Location	Source
Reptiles			
Loggerhead turtle hatchlings	*Caretta caretta*	Florida	Stewart and Wyneken (2004)
Fishes			
Atherinidae	–	Ceará, Brazil	Ferreira de Menezes (1968b)
Atlantic bumper	*Chloroscombrus chrysurus*	Ceará, Brazil	Ferreira de Menezes (1968b)
Atlantic cutlassfish	*Trichurus lepturus*	Louisiana (USA)	Whiton and Townsend (1928)
Atlantic Spanish mackerel	*Scomberomorus maculatus*	Ceará, Brazil	Ferreira de Menezes (1968b)
Atlantic tarpon	*Megalops atlanticus*	Florida, Georgia	Moffett and Randall (1957), Randall and Moffett (1958), Rickards (1968)
Atlantic thread herring	*Opisthonema oglinum*	Ceará, Brazil	Ferreira de Menezes (1968b)
Blue tilapia	*Oreochromis aureus*	Texas	Howells and Garrett (1992)
Cichlidae	–	Ceará, Brazil	Ferreira de Menezes (1968b)
Cyprinodonts	–	Florida	Tabb and Manning (1961)
Eleotridae	–	Ceará, Brazil	Ferreira de Menezes (1968b)
Fish eggs	–	Georgia	Rickards (1968)
Fish larvae	–	Ceará, Brazil	Ferreira de Menezes (1968b)
Fish remains	–	Ceará, Brazil; Georgia	Rickards (1968), Ferreira de Menezes (1968b)
Fish scales	–	Bahamas	Breder (1933)
Gerreidae	–	Isla Contoy, México, Ceará, Brazil	Ferreira de Menezes (1968b), Vega-Cendejas and Hernández (2002)

Tarpons: Biology, Ecology, Fisheries, First Edition. Stephen Spotte.

Common name	Scientific name	Location	Source
Gobiidae	–	Ceará, Brazil	Ferreira de Menezes (1968b)
Gulf menhaden	*Brevoortia patronus*	Louisiana	Dailey *et al.* (2008)
Marsh killifish	*Fundulus confluentus*	Florida	Harrington (1966)
Mullets	*Mugil* spp.	Colombia	Cataño & and Garzón-Ferreira (1994)
Mosquito fish	*Gambusia affinis*	Georgia	Rickards (1968)
Mummichog	*Fundulus heteroclitus*	Georgia	Rickards (1968)
Nile tilapia	*Oreochromis niloticus*	Colombia	Leal-Flórez *et al.* (2008)
Parassi mullet	*Mugil incilis*	Colombia	Cataño and Garzón-Ferreira (1994)
Poeciliidae	–	Ceará, Brazil; Colombia	Cataño & and Garzón-Ferreira (1994), Ferreira de Menezes (1968b)
Rough silverside	*Membras martinica [=vagrans]*	Texas	Simpson (1954)
Sailfin molly	*Poecilia* (= *Mollienesia*) *latipinna*	Georgia, Florida	Harrington (1966), Moffett and Randall (1957), Rickards (1968)
Scup	*Stenotomus chrysops*	New York	Hickey *et al.* (1976)
Sheepshead minnow	*Cyprinodon variegatus*	Bahamas	Breder (1933), Moffett and Randall (1957)
Silver mullet	*Mugil curema*	Florida	Griswold (1913: 101, 1922: 11)
Striped mullet	*Mugil cephalus*	Georgia, Florida	Harrington (1966), Rickards (1968)
Wild molly	*Poecilia gilli*	Colombia	Cataño and Garzón-Ferreira (1994)
Annelid worms			
Polychaetes	*Eunice* (= *Leodice*) *fucata*	Florida	Ellis (1956)
Crustaceans			
Blue crab	*Callinectes sapidus*	Florida	Norman and Fraser (1938: 97), Whiton and Townsend (1928)
Brine shrimp	*Artemia* sp.	Aquaculture	Mercado Silgado (2002)
Class Ostracoda	–	Georgia	Rickards (1968)
Crab larvae	–	Florida	Moffett and Randall (1957)
Crustacean larvae	–	Ceará, Brazil	Ferreira de Menezes (1968b)
Crustacean remains	–	Ceará, Brazil	Ferreira de Menezes (1968b)

(*Continued*)

(Continued)

Common name	Scientific name	Location	Source
Family Portunidae	*Callinectes* spp.	Ceará, Brazil	Ferreira de Menezes (1968b)
Grass shrimp	*Palaemonetes intermedius*	Florida	Tabb and Manning (1961)
Grass shrimps	*Palaemonetes* spp.	Georgia	Rickards (1968)
Order Amphipoda	*Corophium* spp.	Isla Contoy, México	Vega-Cendejas and Hernández (2002)
Order Anostraca	–	Ceará, Brazil	Ferreira de Menezes (1968b)
Order Cladocera	–	Ceará, Brazil	Ferreira de Menezes (1968b)
Order Copepoda	*Cyclops* spp.	Florida	Moffett and Randall (1957)
Order Copepoda	–	Aquaculture	Mercado Silgado (2002)
Order Copepoda (Family Cyclopoidae)	–	Ceará, Brazil	Ferreira de Menezes 91968b)
Order Harpacticoidea	–	Ceará, Brazil; Isla Contoy, México	Ferreira de Menezes (1968b), Vega-Cendejas and Hernández (2002)
Order Isopoda	*Aega psora*	Georgia	Rickards (1968)
Penaeid shrimp	*Penaeus* sp.	Colombia, Puerto Rico	Austin and Austin (1971), Cataño & and Garzón-Ferreira (1994)
Insects			
Corixidae	–	Ceará, Brazil; Florida	Ferreira de Menezes (1968b), Moffett and Randall (1957)
Diving beetle	*Dystiscus* sp.?	Bahamas	Breder (1933)
Family Chironomidae larvae	–	Ceará, Brazil	Ferreira de Menezes (1968b)
Family Chloropidae	–	Ceará, Brazil	Ferreira de Menezes (1968b)
Family Culicidae larvae	–	Ceará, Brazil	Ferreira de Menezes (1968b)
Family Dolichopodidae	–	Ceará, Brazil	Ferreira de Menezes (1968b)
Family Formicidae	–	Ceará, Brazil	Ferreira de Menezes (1968b)
Family Lygaeidae	–	Ceará, Brazil	Ferreira de Menezes (1968b)
Family Notonectidae	–	Ceará, Brazil	Ferreira de Menezes (1968b)
Insect fragments	–	Georgia	Rickards (1968)
Insect larvae	–	Georgia	Rickards (1968)
Insect larvae	–	Ceará, Brazil	Ferreira de Menezes (1968b)
Insect remains	–	Ceará, Brazil; Colombia	Ferreira de Menezes (1968b), Cataño & and Garzón-Ferreira (1994)

Common name	Scientific name	Location	Source
Mosquito larvae	*Culex* sp.	Laboratory	Mercado Silgado and Ciardelli (1972)
Mosquito larvae	–	Aquaculture	Mercado Silgado (2002)
Order Coleoptera Order Odonata (larvae and adults)	–	Ceará, Brazil	Ferreira de Menezes (1968b)
Planthopper	Fulgoridae	Georgia	Rickards (1968)
Water boatman	*Trichocorixa reticulata*	Haiti	Beebe (1927), Beebe and Tee-Van (1928)
Miscellaneous			
Order Ctenophora	–	West Indies	Randall (1967: 684?)
Digested residue	–	Georgia	Rickards (1968)

Partial list of food items consumed by Pacific tarpons

Common name	Scientific name	Location	Source
Fishes			
Ambassids	Family Ambassidae	Australia	Merrick and Schmida (1984: 53)
Bagred catfish (adults)	*Mystus* sp.	India	Pandian (1968)
Barbs (adults)	*Barbus* sp.	India	Pandian (1968)
Cichlids	*Etroplus* spp.	India	Kuthalingam (1958, larvae)
Delicate blue-eye	*Pseudomugil tenellus*	Australia	Bishop *et al.* (2001: 25, 27 Table 5)
Eeltail catfishes	Family Plotosidae	Australia	Merrick and Schmida (1984: 53)
Eeltail catfishes	*Neosilurus* spp.	Australia	Bishop *et al.* (2001: 25, 27 Table 5)
Empire gudgeon	*Hypseleotris compressa*	Australia	Bishop *et al.* (2001: 25, 27 Table 5)
Glassfish	*Ambassis interrupta*	Papua New Guinea	Allen and Coates (1990: 53)
Glassfishes	*Ambassis* spp.	Australia, India	Bishop *et al.* (2001: 25, 27 Table 5), Kuthalingam (1958, larvae)
Grunter (fry)	*Therapon* sp.	India	Pandian (1968)
Grunters	Family Terapontidae	Australia	Merrick and Schmida (1984: 53)
Grunters	*Therapon* spp.	India	Kuthalingam (1958, larvae)
Gudgeons	Family Eleotridae	Australia	Merrick and Schmida (1984: 53)

(Continued)

(Continued)

Common name	Scientific name	Location	Source
Ladyfish (fry)	*Elops* sp.	India	Pandian (1968)
Milkfish (fry)	*Chanos chanos*	India	Pandian (1968)
Milkfish (fry)	*Chanos chanos*	Australia	Merrick and Schmida (1984: 53)
Mosquitofish	*Gambusia* sp.	Australia	Merrick and Schmida (1984: 53)
Mosquitofish (adults)	*Gambusia* sp.	India	Pandian (1968)
Mozambique tilapia	*Oreochromis mossambicus*	Papua New Guinea	Allen and Coates (1990: 53)
Mullets (juveniles)	*Mugil* sp.	India	Pandian (1968)
Pacific tarpon	*Megalops cyprinoides*	India	Chacko (1949), Job and Chacko (1947)
Pennyfish	*Denariusa bandata*	Australia	Bishop *et al.* (2001: 25, 27 Table 5), Merrick and Schmida (1984: 53)
Rainbowfish	*Melanotaenia splendida inornata*	Australia	Bishop *et al.* (2001: 25, 27 Table 5)
Sentani gudgeon	*Oxyeleotris heterodon*	Papua New Guinea	Allen and Coates (1990: 53)
Sleeper goby	*Oxyeleotris* sp.	Papua New Guinea	Coates (1987: 533)
Small fishes	–	Papua New Guinea	Coates (1987: 531 Table 1)
Snakehead gudgeon	*Ophieleotris aporos*	Papua New Guinea	Allen and Coates (1990: 53), Coates (1987: 533)
Toothed river herring	*Clupeoides papuensis*	Papua New Guinea	Roberts (1978: 13)
Crustaceans			
		Australia	Bishop *et al.* (2001: 25, 26 Table 1)
Atyid shrimps	*Caridina* spp.	India, Papua New Guinea	Allen and Coates (1990: 53), Chacko (1949),
Coates (1987: 53 Table 1)			
Branchiopods	*Evadne* spp.	India	Chacko (1949)
Ciliophora	*Cyttarocylis* (= *Favella*) *ehrenbergi*	India	Noble (1973)
Cladocerans	*Diaphanosoma* spp., –	Papua New Guinea	Bishop *et al.* (2001: 26 Table 1), Coates (1987: 531 Table 1)
Cladoceran	*Daphnia* sp.	India	Pandian (1968)
Cladocerans	*Daphnia* spp.	India	Chacko (1949)
Clam shrimps	*Cyzicus* spp.	Australia	Bishop *et al.* (2001: 25, 26 Table 1)
Copepods	–	Australia, India	Kuthalingam (1958, larvae), Merrick and Schmida (1984: 53)

Common name	Scientific name	Location	Source
Copepod	*Mesocyclops* sp.	India	Kuthalingam (1958)
Copepod	*Acartia bifilosa*	India	Kuthalingam (1958)
Copepods	*Microcyclops* spp.	India	Kuthalingam (1958)
Copepod	*Acartia erythraea*	India	Noble (1973)
Copepod	*Acrocalanus longicornis*	India	Noble (1973)
Copepod	*Paracalanus aculeatus*	India	Noble (1973)
Copepod	*Oithona plumifera*	India	Noble (1973)
Copepod	*Acrocalanus gibbor*	India	Noble (1973)
Copepod	*Eucalanus* sp.	India	Noble (1973)
Copepod	*Centropages furcatus*	India	Noble (1973)
Copepod	*Centropages orsinii*	India	Noble (1973)
Copepod	*Lucifer hanseni*	India	Noble (1973)
Copepod	*Cyclops bicolor*	India	Pandian (1968)
Copepods	*Cyclops* spp.	India	Chacko (1949)
Copepods	*Diaptomus* spp.	India	Chacko (1949)
Malacostracans	*Cirolina* sp., *Grandierrella* sp.	India	(Pandian (1968)
Mantis shrimp	*Squilla* sp.	India	Kuthalingam (1958)
Mysid shrimps	*Mysis* spp.	India	Chacko (1949)
Ostracods	–	Papua New Guinea	Coates (1987: 531 Table 1)
Ostracod	*Cypris* spp.	India	Chacko (1949)
Prawns	–	Papua New Guinea	Merrick and Schmida 1984: 53), Roberts (1978: 13)
Prawns	*Penaeus* spp.	India	Kuthalingam (1958)
Prawn	*Acetes* sp.	India	Kuthalingam (1958), Chacko (1949)
River prawns	*Macrobrachium* spp.	Papua New Guinea	Bishop *et al.* (2001: 25, 26 Table 1, Coates (1987: 531 Table 1)
Speckled shrimp	*Metapenaeus monoceros*	India	Pandian (1968)
		Australia	Merrick and Schmida (1984: 53)
Insects			
Coleoptera	Fragments	Australia	Bishop *et al.* (2001: 25, 26 Table 1)
Diptera	Chaoborinae	Australia	Bishop *et al.* (2001: 25, 26 Table 1)
Diptera	Chironomidae larvae	Australia	Bishop *et al.* (2001: 25, 26 Table 1)

(Continued)

(Continued)

Common name	Scientific name	Location	Source
Diptera	Chironomidae pupae	Australia	Bishop *et al.* (2001: 25, 26 Table 1)
Diving water beetle	*Dysticus* sp.	India	Pandian (1968)
Ephemeroptera	Baetidae	Australia	Bishop *et al.* (2001: 25, 26 Table 1)
Hemiptera	Naucoridae	Australia	Bishop *et al.* (2001: 25, 26 Table 1)
Hemiptera	*Anisops* sp.	Australia	Bishop *et al.* (2001: 25, 26 Table 1)
Hemiptera	Corixidae	Australia	Bishop *et al.* (2001: 25, 26 Table 1)
Larvae and nymphs	–	Papua New Guinea	Allen and Coates (1990: 53), Coates (1987: 531 Table 1 533)
Midge larvae/ pupae	–	Australia	Merrick and Schmida (1984: 53)
Midge pupae	*Chironomus* sp.	India	Pandian (1968)
Odonata	Libelluidae	Australia	Bishop *et al.* (2001: 25, 26 Table 1)
Orthoptera	Fragments	Australia	Bishop *et al.* (2001: 25, 26 Table 1)
Terrestrial insects	–	Papua New Guinea	Allen and Coates (1990: 53), Coates (1987: 533)
Water boatman	–	Australia	Merrick and Schmida (1984: 53)
Water boatman	*Micronecta scutellaris*, *Notonecta* sp.	India	Pandian (1968)
Water boatman	*Micronecta* sp.	India	Pandian (1968)
Water boatman	*Notonecta* sp.	India	Pandian (1968)
Winged termites	–	Papua New Guinea	Merrick and Schmida (1984: 53), Roberts (1978: 13)
Miscellaneous			
Brachyuran crab larvae	–	India	Pandian (1968)
Crustacean eggs	–	India	Pandian (1968)
Organic matter	–	Australia	Bishop *et al.* (2001: 25, 26 Table 1)
Rotifers	*Brachionus* spp.	India	Pandian (1968)
Spiders	–	Papua New Guinea	Roberts (1978: 13)
various nauplii	–	India	Chacko (1949)

APPENDIX F

Sôbre a alimentação do camurupim, Tarpon atlanticus (Valenciennes), no Estado do Ceará

On food of the tarpon, *Tarpon atlanticus* (Valenciennes), in the State of Ceará

Mariana Ferreira de Menezes
Archives of the Marine Biological Station of the Federal University of Ceará, Fortaleza **8**, 145–149, 1968

Continuing our studies on the biology of the tarpon, *Tarpon atlanticus* (Valenciennes), we present in this work information on its feeding in littoral and coastal waters of the State of Ceará (Brazil).

The first news report on feeding of adult tarpons in waters of Ceará was presented by Menezes and Menezes (1965). The authors reported finding only fish in the stomachs examined, notably the species *Opisthonema oglinum* (Le Sueur) [Atlantic thread herring] and *Chloroscombrus chrysurus* (Linnaeus) [Atlantic bumper].

The tarpon's remarkable capacity for adaptation to different natural environments, its resistance to wide variations in salinity, and its migrations are determining factors influencing its feeding regime.

Material and method

We analyzed the stomach contents of 178 juveniles, of which 118 were captured in brackish areas within the municipal districts of Fortaleza and Acaraú, and 60 others from mangroves in the municipal district of Araraú [Acaraú?].

Translation from Brazilian Portuguese. See source publication for figures, tables, and literature cited in this Appendix.

Tarpons: Biology, Ecology, Fisheries, First Edition. Stephen Spotte.

We also studied the stomach content of 78 adults captured in coastal waters of the municipal district of Acaraú.

Samples were collected during the years 1962–1967, capture of the adults being limited to times when tarpons are fished (Menezes and Paiva 1966).

To capture juveniles we used two types of fishing gear, a hand-operated seine net and a gill net, both with small mesh. Adults were captured in the weir fishery.

For each individual studied we recorded the length (fork length) and the place and date of capture. We also recorded the sex of each adult.

Juveniles varied in length from 7–47 cm, adults from 51–190 cm. Maximum length reached by juveniles corresponds approximately to the [end] of the first year when they abandon estuarine waters and migrate to the sea (Menezes and Paiva 1966).

After capture the adults were eviscerated and the viscera washed and preserved [*sic*] in 10% formalin. Juveniles were preserved [*sic*] whole in 10% formalin for subsequent evisceration

For identifying stomach contents we used a magnifying glass and, when necessary, a stereoscopic microscope. Identifications were usually made to the family level, and to species when possible. When the material had already undergone digestion we attempted identification by examining skeletal parts with respect to arthropods, mollusks, and fishes. Foods in a nearly complete state of digestion making any identification impossible were considered digested remains.

In the analysis of the juvenile stomach contents we used frequency of occurrence because separating different foods for volumetric determination was impractical. Regarding the adults, in addition to frequency of occurrence we used a volumetric method based on displacement in a column of water, considered more precise in determining the volume of foods (Pillay 1952).

Discussion and conclusions

The foods found in stomachs of the juveniles, by frequency of occurrence and in order of decreasing importance, can be classified as follows: *primary foods* – insects, fishes, and crustaceans; *secondary foods* – algae and planktonic eggs; *occasional foods* – higher plants (Table 1, Figure 1).

Among the foods mentioned above, these are especially prominent: *insects* – hemipterids (mainly of the family Notonectidae) and dipteran larvae; *fishes* – the family Poeciliidae and diverse larvae; *crustaceans* – cladocerans, copepods, (mainly of the family Cyclopoidae), and decapods, although species of the last group, because of their size, represented a lot the food in terms of volume; algae – diatoms and chlorophytes. Standing out among the chlorophytes were filamentous [forms] (with an abundance in the genus *Spyrogira*) and the unicellular chlorophytes (with an abundance in the genus *Chlorella* Beverinck); higher plants – exclusively the species *Halodule* (= *Diplanthera*) cf. *wrightii* (Ascherson).

Information on feeding by juvenile tarpons presented by Harrington and Harrington (1960, 1961) and Rickards (1968) is, in general, concordant with our findings.

The salt flats and the mangroves are ecologically different. To juveniles feeding on the salt flats, these foods stand out in order of decreasing importance: crustaceans, insects, fishes, and algae; in the mangroves and in the same order of importance, these are prominent: fishes, insects, and crustaceans. These facts lead us to conclude that selectivity does not exist in feeding by juveniles, but depends on relative abundance of foods in the different ecological habitats.

Adult tarpons can be classified ecologically as coastal pelagic, accomplishing defined migrations with trophic and genetic objectives (Menezes and Paiva 1966). In the data on feeding by adults we did not find any clear relationship between the sexes, perhaps because we failed to consider food found in the stomachs of males and females separately.

Fishes and crustaceans stand out based on frequencies of occurrence and in order of decreasing importance; however, based on volume, only the fishes were or real importance (Tables II and III, Figure 1). As before, we can adopt the following classification in descending order: *primary foods* – fishes; *secondary foods* – crustaceans; *occasional foods* – mollusks and plants.

Among the items mentioned above, based on frequency of occurrence, the following fishes are prominent: species of the family Carangidae, mainly *Chloroscombrus chrysurus* (Linnaeus); species of the family Clupeidae, mainly *Opisthonema oglinum* (Le Sueur); species of the family Scombridae, mainly *Scomberomorus maculatus* (Mitchill) [Spanish mackerel]. Among crustaceans, decapods of the family Portunidae stand out, with an abundance of the genus *Callinectes* Stimpson.

APPENDIX G

Ecología básica y alimentación del sábalo *Megalops atlanticus* (Pisces: Megalopidae)

Basic ecology and feeding of the tarpon *Megalops atlanticus* (Pisces: Megalopidae)

Didiher Chacón Chaverri
Revista de Biología Tropical **42**, 225–232, 1994

Early stages of Phase III (according to Wade's 1962 classification) and the juveniles of *Megalops atlanticus* Valenciennes 1846 have the same ecological distribution in estuarine waters, coastal lagoons, canals, and swamps to the berm of the beach. This is especially true in aquatic ecosystems having low oxygen content, high turbidity, and little current flow, along with stagnation and the consequent high levels of hydrogen sulfide, H_2S (Harrington 1958, Wade 1962, Harrington 1966, Rickards 1968, Jones *et al.* 1978).

Leptocephalus larvae arrive at swamps and coastal lagoons with high tides and surf where they remain until reaching the juvenile stage at approximately 490.0 mm standard length (SL), at which point they migrate toward marine waters (Jones *et al.* 1978). Cyr (1991) maintained that 150.0 mm SL constituted the length at migration of juveniles toward the ocean, but other authors, like Ferreira and Pinto (1966), claim they enter coastal waters after previously living in the sea when climate and marine conditions allow it. They stay in those [inshore] areas feeding and growing until attaining 40.0–50.0 cm SL, then return to the ocean. On the other hand, Chacón and McLarney (1992) did not find individuals longer than 274.0 mm SL in coastal areas they studied.

The factors primarily influencing distribution of this species are both biotic and abiotic; these last are physical and chemical. According to Rickards (1968), Jones *et al.* (1978), and Chacón and McLarney (1992) the tolerable range of salinity is 0.0–47.0 ppm; temperature, 12.0–40.0°C; and pH, 4.0–7.0. Tarpons

---

Translation from Spanish. See source publication for figures, tables, and literature cited in this Appendix.

Tarpons: Biology, Ecology, Fisheries, First Edition. Stephen Spotte.

also tolerate high elevations of hydrogen sulfide and concentrations of dissolved oxygen from 0.26–8.00 mg/L. It is remarkable that the physiological characteristics and adaptability of this species enable it to be distributed in areas where severe environmental conditions are fatal to other species.

Other important factors in [the tarpon's] distribution are the feeding habits of juveniles and their place in the food chain. To Simpson (1954), Harrington (1958), Wade (1962), Ferreira and Pinto (1966), Harrington and Harrington (1966), Menezes (1968), and Zale and Merrifield (1989), [tarpons] are fully carnivorous organisms that prey on insects, fishes, crustaceans, and coelenterates. To understand the selective feeding of this species and some facts about its ecology were the general objectives of this study.

Observations were carried out in July 1988 and August 1992 in two sections inside the Gandoca/Manzanillo Live Tree Refuge [Gandoca-Manzanillo National Wildlife Refuge] (Fig. 1). These were divided into 12 stations, the first including all coastal aquatic ecosystems from the Sixaola River to Punta Mona; the second covered the area of Manzanillo Creek to Black Creek (Kèöldi, or Coclés, River). A list was made of all the coastal stations visited inside the refuge.

Sampling was carried out with a gill net 0.92 m high and 5.0 m long of light mesh (0.013 m), a cast net of 2.0 m diameter and light mesh (0.01 m), and a trawl net 5.0 m long and 1.0 m high, with a light mesh of (0.001 m). Collecting equipment was then selected to meet conditions of the sampling locations. In the case of the Lagoon of Gandoca, trawls were made with a plankton net of light mesh (300 μm) similar to that shown by Lewis *et al.* (1970), following the method of Chacón and McLarney (1992).

Specimens from the collections were placed in 95% ethanol and glycerine (66.6% and 33.3%, respectively) to fix [*sic*] them and to preserve the tissues. A milliliter of this solution was introduced into the digestive system to preserve stomach contents using a 1.0 mL pipette (±0.1 mL). Contents of the stomach and intestine were analyzed according to the method of Harrington and Harrington (1966), Rickards (1968), Odum and Heald (1972), and Nordlie (1981). Stomach contents were identified to the lowest taxon possible. Taxonomic groups from each stomach were measured by volumetric displacement (±0.1 mL). Average volumetric proportion was measured according to Ankenbrandt (1985). All the organisms were identified by appearance of external characteristics with those described in Needham and Needham (1982), Barnes (1984), and Bussing (1987).

Foods in advanced stages of digestion, and detritus (e.g. remains in microbial decomposition), were combined under the term "waste." Foods were classified according to Menezes (1968) as primary, secondary, and occasional. Biometric proportions of the fish were also measured using callipers (±0.05 mm). Developmental stages correspond to the keys of Wade (1962), Mansueti and Hardy (1967), Mercado and Ciardelli (1972), and Jones *et al.* (1978).

Temperature was measured with a regular mercury thermometer, dissolved oxygen using the Winkler test, salinity with a refractometer (±0.002

ppm [ppt?] with temperature compensation), pH with universal colorimetric paper (1–10, Merck), and observations [noted] of sediment type and presence of H_2S for qualitative analysis.

Results and discussion

The Glandoca/Manzanillo Tree of Life Refuge possesses ≈ 400 ha of swamps; some are part of the five micro-basins that encompass the region. These are characterized by areas of flat topography, poor drainage, are dendritic, and dominated by rivers of low flow (Alfaro and Rooms 1992). The soils are andosols, mainly E-1, dominated by Typic Hydaquent and Typic Tropaquent (Alfaro and Salas 1992).

The predominant vegetation at the sampling stations consisted of yolillo (*Raphia taedigera*), orey (*Campnosperma panamensis*), red mangrove (*Rhizophora mangle*) and negra forra (*Acrostichum aureum*). In these zones of coastal life with possible flooding we collected 261 juveniles in six of the 12 stations visited along the coast. This reflects the ecological importance of coastal ecosystems to this species for protecting its early growth stages, which applies as well to other fish species of economic value.

The data show that *M. atlanticus* does not have specific preferences for temperature and salinity (Table 1), which confirms the euryhaline character of the tarpon.

In the estuaries and swamps of the Costa Rican Caribbean, tidal variations of fluctuate <1 m, and changes in salinity values can be attributed to the effect of precipitation. Precipitation presents in a bimodal pattern (December–January and July–August). The driest periods are September–October and March–April (Nordlie and Kelso 1975; IMN 1988); rains tend to lower temperature and salinity.

The easy adaptation of juvenile [tarpons] to ecosystems of varied physical and chemical properties is partly attributable to the physiological capability of breathing atmospheric air (Shlaifer 1941), which allows them to inhabit waters that are nearly anoxic and characterized by low pH values and the presence of hydrogen sulfide. This anoxia and acidity of the water typified the stations during months when the most specimens were collected.

As Chacón and McLarney established (1992), tarpons are found in fluvial sources of dark coloration caused by plant decay, the tannins of the swamp, and natural sedimentation.

The highest recruitment and smallest specimens were caught during (or soon after) the rainy periods. Owing to runoff, links were created with the sea, allowing leptocephali to move toward freshwaters. In this manner, their invasion of coastal ecosystems is tied to physical and chemical aspects of the environment and is a characteristic of the life cycle.

In general the tarpon is [a] typical [inhabitant] of those marshy ecosystems of the refuge. The species can occupy ecosystems having drastic characteristics, thus avoiding intraspecific [interspecific?] competition over feeding levels and habitat and having a very special niche in the coastal swamps of southern Talamanca.

Other families of fishes found in the same environments as the tarpon were: Eleotridae: fat sleeper (*Dormitator maculatus*), bigmouth sleeper (*Gobiomorus dormitory*), spinycheek sleeper (*Eleotris pisonis*); Mugilidae: white mullet (*Mugil curema*); Poecilidae: Gill's molly (*Poecilia gillii*); Centropomidae: common snook (*Centropomus undecimalis*), swordspine snook (*C. ensiferus*), tarpon snook (*C. pectinatus*); Carangidae: jacks (*Caranx* spp.); Atherinopsidae: silverside (*Atherinella* [= *Melaniris*] *milleri*); Gerreidae: mojarras (*Eucinostomus* spp.); Characidae: banded astyanax (*Astyanax fasciatus*); Soleidae: lined sole (*Achirus lineatus*); Lutjanidae: gray snapper (*Lutjanus griseus*), lane snapper (*L. synagris*); Sygnathidae: freshwater pipefish (*Pseudophallus mindii*); Pimelodidae: three-barbeled catfishes (*Rhamdia* spp.); Haemulidae: burro grunt (*Pomadasys crocro*); Belonidae: needlefishes; Tetradontidae: bullseye puffer (*Sphoeroides anulatus*). There was a maximum of five coincident families with the tarpon in the samples.

The vertebrates observed preying on juveniles of *M. atlanticus* were mainly the *garza* [heron] and *martín peña* [bare-throated tiger heron] (Ardeidae).

Stomach contents of the 261 juveniles reflected a minimal stomach for the class up to 14.9 mm of SL (Fig. 2). There was an average volumetric proportion of 32% copepods (Cyclopoidae) and 12% notonectids (hemipterids) and 25% waste (Fig. 3). These were derived mainly from digestion of the two aforementioned groups of five invertebrate animals, although more groups were consumed in smaller quantities; all are called staples, being derived mainly before digestion of the two groups of invertebrates mentioned and five animal groups consumed in lesser quantities, all are labeled primary foods according to Menezes (1968).

Therefore the tarpon is a typical carnivorous species, corroborating findings of Ferreira and Pinto (1966), Rickards (1968), Dahl (1971), Fischer (1978), and Villa (1982).

The average volumetric proportions tend toward invertebrates because 70.1% of the sample comprised fish <90.0 mm SL that consume [mainly] invertebrates like crustaceans and insects.

These fish [tarpons] rise in the food chain with their ontogeny (Kruskal-Wallis, $p < 0.005$). There is a clear difference between the group in which 90.00 mm is the maximum length and the one in which 90.10 mm is the minimum length. Stomachs and intestines of the first are dominated by copepods (45%) and then 13% notonectids and 25% waste (Figs. 4A and 4B). These observations fully coincide with Wade's (1962) findings for [tarpons] having fork lengths (FL) of 70.0–83.0 mm, except that the samples here did not recover any fishes preyed on by this group. Beebe (1927) and Dahl (1971) also mention hemipterans as food.

Zale and Merriefield [*sic*] (1989) confirmed that stages II and III and juveniles (<125 mm LS) feed of zooplankton and secondarily on insects and small fishes, coinciding with findings here.

In the longest individuals to 90.1 mm SL, fishes predominated (30%) and [then] notonectids (20%); here waste was combined with other elements in the digestion state. Feeding was selective with preference for the biggest available particles. Harrington and Harrington (1966) and Zale and Merriefield [*sic*] (1989) found this distinction clear, demonstrating that the individual [a tarpon] takes larger foods as it grows. It stops consuming copepods and shifts progressively to eating insects and fishes. These authors also found many groups of animals that were not in the contents discovered here.

With growth, the volume of fish in the stomach contents increased at the expense of the volumes of other items. According to Harrington and Harrington (1966) the percentage of invertebrates and vertebrates depends on the size of the consumer, behavior of the prey, microhabitat, and fluctuations in the number of items consumed. To Caddie and Sharp (1988) this can often mean established size preferences and specific organisms. The voracious behavior of *M. atlanticus* causes it to move throughout the water column, often preying on organisms that take refuge in and among plants on the substratum, and [plants] later appear in the stomach.

Early stages, in general, are at lower trophic levels than adults of the same species. Their survival is more uncertain and varies with time. This is one of the reasons why variations in recruitment are so common in the sea, even in the case of high-level consumers like *M. atlanticus*.

The variation in spatial distribution, food availability, and type and size of the environment are key causes of a population's fluctuations in marine ecosystems despite the fact that such variations have their biggest impact on the early life-stages.

This type of information on trophic relationships of early-stage tarpons can improve the possibilities to manage [the species] in semi-extensive or intensive cultivation systems. It can modify the recruitment of juveniles to increase the levels of survival because of their high economic value [the species has] for use in the coastal sport fishing of the American Caribbean.

APPENDIX H

Ensaio preliminary sobre o cultivo camurupim (*megalops* [*sic*] *atlanticus*) em viveiros escavados e sua ocorrência em lagoas marginais no litoral de Tutóia – Ma

Preliminary report on tarpon culture (*megalops* [sic] *atlanticus* in excavated ponds and its [the tarpon's] occurrence in lagoons on the coast of Tutóia – Ma [Maranhão State]

G. A. de Araújo Santos

In XXIV SEMIC Seminário de Iniciação Científica: Livro de Resumos, Universidade Federal do Maranhão, Pro-Reitoria de Pesquisa e Pós-Graduação, São Luís, 405 pp. [p. 16], 2013.

The aim of this study was to monitor the culture of [Atlantic] tarpons in an estuary of Tutóia-MA [Maranhão] during the period August 2011–August 2012. Juveniles of this species are caught in stagnant ponds temporarily disconnected from the sea, and estuarine nurseries of different sizes are created where water exchange depends on the seas or the quantity of rainfall during the period. The rearing of the tarpon is now mainly done by the villages of Porto de Areia (78%) and Arpoador (22%) in ponds dug directly into the soil or (in excessively sandy soil) coated with concrete. The fish are raised in ponds of 15 m^2 to 430 m^2 with a depth of about 1 m. The stocking densities in excavated ponds are 1.3 fish/m^2 and in concrete nurseries 10 fish/m^2. Dietary management varies widely among farmers: some feed the fish every day; others, for lack of time, leave the bait fish up to three days without feeding it. The average amount given varies from 0.05–30 kg of fish/d regardless of stocking densities, becoming more a function of food availability. Because live food (small fish) is not always available, [the captive tarpons] end up eating fresh or frozen bait. [Harvested] biomass computed for the period (eight months) was 300 kg in excavated ponds and 256 kg in concrete ponds. Water quality

---

Translation from Brazilian Portuguese. See source publication for figures, tables, and literature cited in this Appendix.

Tarpons: Biology, Ecology, Fisheries, First Edition. Stephen Spotte.

was monitored in the fresh and salt water nurseries, using physical-chemical analyses with kits and portable equipment (temperature, salinity, ammonia, dissolved oxygen, pH, and water transparency). These parameters generally remained suitable for the development of tarpons, considering that this species is quite resistant to low oxygen levels and large variations in salinity. Most breeders rear tarpons only for consumption with few commercial purposes; however, the activity is informal and without concern for the environmental impact that this crop can cause in the long term. To identify these families of fish farmers of Tutóia it was important to analyze their municipal activities [and] identify the organization and its potential shortcomings.

APPENDIX I

Sobre a elaboração de conservas de pescado em leite de côco em óleos de algodão e de babaçu

On the preparation of fish preserved in coconut milk and oils of cotton and babassu[1]

José Raimundo Bastos, Tarcísio Teixeira Alves, Carlos Antônio Esteves Araripe and Francisco José Siqueira Telles
Arquivos de Ciências do Mar **13**(1), 25–29, 1973

Despite technological advances, almost nothing [new] has been developed for the preparation of fish products. However, interest exists in diversifying the traditional line of the canned fish, along with the formulation of types of preservation and/or improvement of the processing techniques (Langraf Jr. 1963).

The objective of this study is the preservation of foods made from [the] king mackerel, *Scomberomorus cavalla* (Cuvier), Spanish mackerel, *Scomberomorus maculatus* (Mitchill), and [Atlantic] tarpon, *Tarpon atlanticus* (Valenciennes), in coconut milk and in oils of cotton and babassu, as well as to observe, during a period of six months of shelving, the characteristics of the preserved products in terms of their bacteriology, organoleptic properties, and chemistry.

Translation from Brazilian Portuguese. See source publication for figures, tables, and literature cited in this Appendix.

[1] The babassu (*Attalea speciosa*) – also called babaçu and cusi – is a palm native to the Amazon basin of South America. Its seeds produce an edible oil.

Tarpons: Biology, Ecology, Fisheries, First Edition. Stephen Spotte.

Material and methods

We worked with fish caught along the coast of the State of Ceará (Brazil), acquired 2–3 d after capture, gutted, and conserved in ice. All individuals presented before processing were in an excellent sanitary state, [as] evidenced by organoleptic testing.

The coconut milk, cotton and babassu oils, and the seasonings used were commercial products.

The cans used were 500g capacity and coated internally with an insulating layer of epoxy varnish on aluminum.

For preparation of the preserves we processed 54.0kg of king mackerel, 54.0kg of Spanish mackerel, and 43.0kg of tarpon; 340 cans [in total] were used to contain the preserves formulated in coconut milk and the oils of cotton and babassu. The protocol for the preparation was as follows.

Preparation of the preserves was based on conventional methods of canning modified for use with the covering liquids.

For subsequent calculations, fish were sorted by species. Heads and fins were removed, after which each individual was cut into transverse pieces. These were washed in running water to eliminate blood, mucus, and extraneous matter.

The clean pieces were immersed for 1h in a 25% brine solution adjusted to pH 4.0 ± 2 with vinegar. They were then immersed again, this time in 2% brine for 15min; finally, portions of 320g were distributed to the individual cans washed previously and sterilized.

The meat contained in the cans was pre-baked at 100°C for 10min. This quickly eliminated the water component in the fish, which was liberated by muscular retraction.

The cans contained the dehydrated pieces plus 90 mL of coconut milk or the same volume of the oils, all seasoned and homogenized with tomato extract and salt in proportions of 1 and 3%, respectively.

Cans containing the slices [of fish] were covered [with nets?] and submitted to exhaust ventilation at 100°C for 10 min then sealed immediately.

Sterilization was at 112°C for 100 min in a vertical autoclave, after which the cans were cooled in running water until reaching room temperature. Then they were shelved.

The preserved products were left for six months. During this time samples were taken randomly each month for bacteriological, organoleptic, and chemical analyses.

Bacteriological analyses were initiated in samples preserved after 30, 60, 90, 120, 150 and 180 d of shelving, being incubated at 37°C for 15 d. In the cans sampled, total aerobic and anaerobic bacteria and the most probable number of coliforms (*Streptococcus faecalis*, Andrews and Holder) were assessed per 100g of meat according to methods described by Sharf (1973).

For organoleptic analysis, samples from the lots were collected and the products characterized according to odor, flavor, and texture. We used the following criteria for evaluation of these characteristics: odor – 1 = very pleasant, 2 = pleasant, 3 = a little pleasant, and 4 = unpleasant; flavor – 1 = excellent, 2 = good, 3 = unremarkable, 4 = bad; texture – 1 = uniform, 2 = compact, 3 = moderate, and 4 = flaccid.

Chemical analyses were made monthly from samples of the lots during the whole shelving period and included determinations of the meat and covering liquids after separation by draining.

For the meat, humidity [water content] determinations were made by desiccation at 105°C based on even-weigh constants; protein by the Kjeldahl method with 6.25 as a conversion factor[2]; fat by Soxhlet extraction using acetone as the solvent; ash by incineration at 575°C; acidity by titration; and pH using a potentiometer (Metrohn Herissau AND 350-B) in proportions of 10 g of sample homogenized in 100 mL distilled water.

The coconut milk was of known acidity, and its pH was measured using the potentiometer mentioned above; readings were obtained directly by immersion of the electrodes in the covering liquid.

Acidity of the cotton and babassu oils was determined by titration.

All analyses followed procedures of the Association of Official Agricultural Chemists (1965).

The yield was calculated by the weight difference between the raw material and the weight of the canned meat.

Results and discussion

Average yields of canned king mackerel, Spanish mackerel, and tarpon in different covering liquids were approximately 66.5, 66.6, 64.0% respectively (Table I).

According to Kalantarova (1958), the amount of water lost during cooking and before canning depends not only on baking but also on the species being processed.

In agreement with norms of the Division of Inspection of Products of Animal Origin (1952), all canned foods should be submitted to a sterilization test that consists of the incubation of the product at 37°C for a minimum of 10 d before its [commercial?] release.

To verify the efficiency of sterilization of the king mackerel, Spanish mackerel, and tarpon in coconut milk and in cotton and babassu oils, we incubated samples of the preserved products at 37°C for 15 d, verifying that no can produced a bulge. In Portugal an entire lot is declared useless if a certain percentage of cans bulge (Costa 1963).

[2] The standard used for the percentage nitrogen of protein.

Bacteriological analysis did not reveal the presence of anaerobes or coliforms (*Streptococcus faecalis*) in any of the lots during the shelving period. In the total bacterial counts we did notice the presence of coconut saprophytes, but not regularly, and considered them occasional contamination. Morais (1971) affirmed that coconuts provide the predominant organisms of the microbial flora of canned seafood products, and that they are easily destroyed by heat.

Organoleptic examination of the canned king mackerel, Spanish mackerel, and tarpon in coconut milk and oils of cotton and babassu for all the lots during the shelving period had a very pleasant odor, excellent flavor, and uniform texture. The products preserved in coconut milk had better appearance and palatability when compared with those preserved in cotton oil or babassu oil.

As to the containers, sulfur production and corrosion were not in evidence, demonstrating the efficiency of the protective internal layer of epoxy varnish.

Data from chemical analyses of king mackerel, Spanish mackerel. and tarpon meats, as well as the liquids covering them in the cans, are presented in Table II.

With regard to species, covering liquids, and length of shelving, values for water, protein, fat, and ash did not differ greatly. As for acidity, a stable balance was observed for products canned in coconut milk. This was not verified to the same extent in the other products, probably because of the low acidity of the commercial oils used.

The pH of the meat [preserved] in coconut milk varied little, being always lower than products preserved in oils of cotton and babassu, although in these, too, important variations were not observed. According to Farber (1965), the pH has little or no significance as an indicator of fish freshness.

Conclusions

1 In experimental laboratory conditions, king mackerel, Spanish mackerel, and tarpon preserved in coconut milk and in oils of cotton and babassu presented average yields of 66.5, 66.6, and 64.0%, respectively.
2 No bulging was observed in any of the preserved cans after incubation of the samples at 37°C for 15 d.
3 Bacteriological tests of incubated samples did not show evidence anaerobes or coliforms (*Streptococcus faecalis*).
4 As to the organoleptic aspect, all products presented with a very pleasant odor, excellent flavor, and uniform texture.
5 Sulfur was not detected in the cans after removal of their contents.
6 Substantial chemical variations were not observed during the shelving period.
7 Products preserved in coconut milk looked the best and were the most palatable.

APPENDIX J

Industrialização da ova do camurupim, *Tarpon atlanticus* (Valenciennes)

Industrialization of tarpon eggs, *Tarpon atlanticus* (Valenciennes)

Tarcísio Teixeira Alves, Masayoshi Ogawa, Maria da Conceição Caland-Noronha and C. A. Esteves Araripe

Arquivos de Ciências do Mar, Fortaleza **12**(2), 151–154, 1972

The tarpon, *Tarpon atlanticus* (Valenciennes), occurs along the tropical Atlantic coast of America from the state of North Carolina (USA) to the state of São Paulo (Brazil) – (Sadowsky 1958, Hildebrand 1963).

In northeastern Brazil the tarpon is commonly captured in the states of Maranhão, Piauí, and Ceará, mainly in fishing weirs. During the months of October and November to January, great shoals of this species approach the coast, gathering for intense reproduction (Menezes and Paiva 1966; Paiva *et al.* 1971).

The ovaries are quite [well] developed in females of advanced reproductive status, reaching 50 cm in length, both [ovaries] weighing up to 5 kg. According to Cervigón (1966), a female tarpon in its reproductive state releases 800 000 ova. The ripe ova have a diameter of 1.8 mm (Breder 1944).

The ripe ovaries, commonly called a pair of roes by fishermen of the Brazilian northeast, have great commercial value and wide acceptance in the regional market. For their preservation, they [the fishermen] dry them with salt, resulting in a terrible product that is processed and stocked under quite precarious hygienic conditions.

Faced with such problems we sought in the present work to define techniques for industrial processing of tarpon roe within the most rigid precepts of hygiene to obtain dry-salted and smoked-canned products.

Translation from Brazilian Portuguese. See source publication for figures, tables, and literature cited in this Appendix.

Tarpons: Biology, Ecology, Fisheries, First Edition. Stephen Spotte.

Material and methods

The material consisted of pairs of fresh ovaries collected in the municipal districts of Acaraú and Camocim (State of Ceará, Brazil). Both mature females and their eggs were weighed to obtain yields at different stages of processing.

Soon afterward the ovaries were washed in 2–3% brine after making small holes in the surrounding membrane to facilitate penetration of the brine and release coagulated blood.

The processing was aimed at preserving a canned-smoked roe [product] and a dry-salted roe [product]. Figure 1 shows a flow chart of the process with its steps and yields by weight.

In the curing step, sodium glutamate was added at 1 g per kg of refined salt. Seasonings were applied with the best flavor of the products in mind.

Chemical composition of the ovaries (mean values) before, during, and after processing is presented in Table I.

Methods used for determining the chemical composition of the ovaries were: humidity [water content] – desiccation at 105°C to constant weight; protein – Kjeldahl method [for total nitrogen] with the conversion factor 6.25; fat – Soxhlet extraction with acetone as the solvent; ash – incineration at 550°C; chlorides – Mohr's method. All follow the Association of Official Agricultural Chemists (1965).

Several analyses were performed on the preserved products; results are presented in Table II. Methods used: refractive index – read with a refractometer (Bausch & Lomb) at 40°C; saponification index – Koettstorfer's method; iodine index – Hannus' [*sic*, Hanus'] method; TBA (thiobarbituric acid) value – Yu and Sinnhuber's (1957) [method]; barium peroxide – Lea's method (in Anonymous 1965/1966). In the determination of saponification indices and of iodine we followed the Association of Official Agricultural Chemists (1965) where these methods are described. In the smoked-canned product, analyses were made after 40 d of storage; in the salt-dried product, soon after drying.

During and after processing of the smoked-canned roe, counts of total bacteria/g were made according to Sharf (1972). Results of these analyses, [expressed as] mean values, are [shown] in Table III.

Cans used in the processing of the smoked-canned roe were coated internally with aluminum epoxy. They were filled three-quarters full, and each had a gross weight of 215 g, a net weight (roe plus oil) of 154.9 g, and a solid content weight of 101.5 g under positive vacuum [?].

Conclusions and recommendations

Observations at the processing locations and commercialization of the salt-dried roe, as well as the quality of the product preserved by the fishermen, showed evidence that their methods are quite inefficient. This is caused by the terrible quality of the salt, long drying time in the sun, and permanently bad hygienic conditions.

In preserving salt-dried roe it is necessary that the salt be refined and of good quality; that the roe is treated with an anti-oxidant to guard against the fat becoming rancid; that it [the roe] be pressed to facilitate penetration of the salt; that drying takes place in shade to reduce oxidation; and that the product is packed in plastic bags under partial vacuum and stored at 3–10°C.

Because the market is good, smoking and then canning is recommended to produce a product of better quality and easier preservation.

Summary

In northeastern Brazil the ripe gonads of female tarpons, *Tarpon atlanticus* (Valenciennes), have considerable commercial value and are mostly consumed regionally. To preserve the gonads the fishermen use the salt-drying method, producing an inferior product processed and stocked under deplorable hygienic conditions.

This paper tries to describes techniques suitable for industrial processing of tarpon roe to obtain finished dry-salted and canned-smoked products. The latter proved of superior quality and were better preserved.

References[1]

Adams, A., K. Guindon, A. Horodysky, T. MacDonald, R. McBride, J. Shenker and R. Ward (2012). *Megalops atlanticus*. The IUCN Red List of Threatened Species 2012: http://dx.doi.org/10.2305/IUCN.UK.2012.RLTS.T191823A2006676.en.

Adams, A. J., A. Z. Horodysky, R. S. McBride, T. C. MacDonald, J. Shenker, K. Guindon *et al.* (2013). Global conservation status and research needs for tarpons (Megalopidae), ladyfishes (Elopidae), and bonefishes (Albulidae). *Fish and Fisheries* **15**, 280–311.

Ajayi, T. O. and M. O. Okpanefe (1995). Etudes (socio-economiques) recentes relatives aux coûts et revenus en pêche artisanale côtiere au Nigeria. In *Rapport de la Première Réunion du Groupe de Travail du DIPA sur Coûts et Revenus en Pêche Artisanale en Afrique de l'Ouest* (Dakar, Sénégal 12–13 juin 1995), A. M. Jallow (editor). Rapport Technique No. 72, Programme du DIPA. FAO, Cotonou, Bénin, pp. 53–58.

Albaret, J-J. (1987). Les peuplements de poissons de la Casamance (Sénégal) en période de sécheresse. *Revue d'Hydrobiologie Tropicale* **20**(2–3), 291–310.

Alderdice, D. F. (1988). Osmotic and ionic regulation in teleost eggs and larvae. In *Fish Physiology*, Vol. XIA, W. S. Hoar and D. J. Randall (editors). Academic Press, San Diego, pp. 163–251.

Alexander, R. M. (1993). Buoyancy. In *The Physiology of Fishes*, D. H. Evans (editor). CRC Press, Boca Raton, pp. 75–97.

Alikunhi, K. H. and S. N. Rao (1951). Notes on the metamorphosis of *Elops saurus* Linn. and *Megalops cyprinoides* (Broussonet) with observations on their growth. *Journal of the Zoological Society of India* **3**, 99–109.

Allen, D. M., S. E. Stancyk and W. K. Michener (1982). *Ecology of Winyah Bay, South Carolina, and Potential Impacts of Energy Development*. Special Publication No. 82–1, Belle W. Baruch Institute for *Marine Biology* and Coastal Research. University of South Carolina, Columbia, xvi + 259 pp. [p. 7–22; this is a single page number]

Allen, G. R. and M. Boeseman (1982). A collection of freshwater fishes from western New Guinea with descriptions of two new species (Gobiidae and Eleotridae). *Records of the Western Australian Museum* **10**(2), 67–193. [p. 71]

Allen, G. R. and D. Coates (1990). An ichthyological survey of the Sepik River, Papua New Guinea. *Records of the Western Australian Museum* (Supplement 34), 31–116. [pp. 33 Table 1, 43, 52–53]

Alleng, G. P. (1997). The fauna of the Port Royal Mangal, Kingston, Jamaica, 1997. *Studies on the Natural History of the Caribbean Region* **73**, 25–42. [p. 35]

Álvarez-León, R. (2003). Ictiofauna del complejo fluvio-lagunar-estuarino de la Ciénaga Grande de Santa Marta (Colombia), antes de los obras civiles de recuperación de los caños. *Dahlia* (**6**), 79–90. [pp. 80, 85]

[1] Pages from longer publications mentioning tarpons specifically are given in brackets.

Tarpons: Biology, Ecology, Fisheries, First Edition. Stephen Spotte.

Amos, M. J. (2007). *Vanuatu Fishery Resource Profiles.* IWP-Pacific Technical Report (International Waters Project) No. 49. Apia, Samoa, viii + 200 pp. [pp. 149–150]

Andrews, A. H., E. J. Burton, K. H. Coale, G. M. Cailliet and R. E. Crabtree (2001). Radiometric age validation of Atlantic tarpon, *Megalops atlanticus*. *Fishery Bulletin (NOAA)* **99**, 389–398.

Angel C., J. L. (1992). La pesca artesanal en el Golfo de Morrosquillo una caracterizacion general. *Ensayos de Economía* **3**(5), 127–152. [pp. 132–133]

Angulo, A., C. A. Garita-Alvarado, W. A. Bussing and M. I. López (2013). Annotated checklist of the freshwater fishes of continental and insular Costa Rica: additions and nomenclatural revisions. *Check List* **9**(5), 987–1019. [pp. 989, 1011]

Anonymous (1975). Tarpon caught in Pacific Ocean. *International Marine Angler* **37**(4), 2.

Anonymous (1978). *Background Papers and Supporting Data on the Practical Salinity Scale 1978.* Technical Papers in Marine *Science* 37. UNESCO, Paris, 144 pp.

Anonymous (1992). *Les Poissons de Guyane.* IFREMER (Institut français Recherche Exploitation de la Mer). Plouzané, 26 pp. (unpaginated)

Anyanwu, P. E. and K. Kusemiju (2008). The Nigerian tarpon: resource ecology and fishery. In *Biology and Management of the World Tarpon and Bonefish Fisheries*, J. S. Ault (editor). CRC Press, Boca Raton, pp. 115–128.

Anyanwu, P. E., M. M. Akinwale, M. A. Matanmi and M. R. Ajijo (2010). Hatching of eggs of *Megalops atlanticus*. In *2010 Annual Report*. Nigerian Institute for Oceanography and Marine Research, Victoria Island, Lagos, pp. 6–8.

Aquino Menezes, N. (2011). Checklist dos peixes marinhos do Estado de São Paulo, Brasil. *Biota Neotropica* **11**(1a), 15 pp. [p. 5]

Aquino Menezes, A., P. A. Buckup, J. Lima de Figueiredo and R. Leão de Moura (editors). (2003). *Catálogo das Espécies de Peixes Marinhos do Brasil.* Museu de Zoologia de Universidade de São Paulo, 159 pp. [p. 31]

Arce Ibarra, M. C. A. M. (2002). *Ictiofauna en cenotes del ejido maya "Xhazil Sur y Anexos" y de la reserve de Sian Ka'an, Q. Roo, México.* Informe final SNIB-CONABIO Proyecto No. S173. El Colegio de la Frontera Sur, Unidad Chetumal, Quintana Roo, México, 23 pp. (unpaginated). [p. 11 Table 6]

Armbrust, E. V. and S. R. Palumbi (2015). Uncovering hidden worlds of ocean biodiversity. *Science* **348**, 865–867.

Arronte, J. C. J. A. Pis-Millán, M. P. Fernández and L. García (2004). First records of the subtropical fish *Megalops atlanticus* (Osteichthyes: Megalopidae) in the Cantabrian Sea, northern Spain. *Journal of the Marine Biological Association of the United Kingdom* **84**, 1091–1092.

Arruda, L. M. (1997). Checklist of the marine fishes of the Azores. *Arquivos do Museu Bocage (Nova Série)* **3**(2), 13–162. [p. 25]

Astorqui, I. (1976). Peces de la cuenca de los grandes lagos de Nicaragua. *Revista de Biología Tropical* **19**(1–2), 7–57. [pp. 28–29]

Atanda, A. N. (2007). Freshwater seed resources in Nigeria. In *Assessment of Freshwater Fish Seed Resources for Sustainable Aquaculture*, M. G. Bondad-Reantaso (editor). FAO Fisheries Technical Paper No. 501. FAO, Rome, pp. 361–380. [p. 363 Table 7.14.1]

Atema, J., M. J. Kingsford and G. Gerlach (2002). Larval reef fish could use odour for detection, retention and orientation to reefs. *Marine Ecology Progress Series* **241**, 151–160.

Ault, J. S. and J. Luo (2013). A reliable game fish weight estimation model for Atlantic tarpon (*Megalops atlanticus*). *Fisheries Research* **139**, 110–117.

Austin, H. and S. Austin (1971). The feeding habits of some juvenile marine fishes from the mangroves in western Puerto Rico. *Caribbean Journal of Science* **11**(3–4), 171–178.

Babcock, L. L. (1936). *The Tarpon: A Description of the Fish Together with Some Hints on Its Capture*, 4th edition. Privately printed, Buffalo, 175 + viii pp.

Babcock, L. L. (1951). *The Tarpon: A Description of the Fish with Some Hints on its Capture*, 5th edition. Privately Printed, Buffalo, 157 pp. + index, addenda.

Bagarinao, T. U., N. B. Solis, W. R. Villaver and A. C. Villaluz (1986). *Important Fish and Shrimp Fry in Philippine Coastal Waters: Identification, Collection and Handling.* Aquaculture Extension Manual No. 10. Aquaculture Department, Southeast Asian Fisheries Development Center (SEAFDC), Tigbauan Iloilo, 52 pp. [pp. 14–15]

Bagnis, R., P. Mazellier, J. Bennett and E. Christian (1987). *Fishes of Polynesia*, 2nd edition. Times Editions, Singapore, 368 pp. [p. 272]

Bailey, K. M. and R. S. Batty (1984). Laboratory study of predation by *Aurelia aurita* on larvae of cod, flounder, plaice and herring: development and vulnerability to capture. *Marine Biology* **83**, 287–291.

Bailey, R. M. (1951). The authorship of names proposed by Cuvier and Valenciennes' "*Histoire Naturelle des Poissons.*" *Copeia* **1951**, 249–251.

Barletta, M. and A. Barletta-Bergan (2009). Endogenous activity rhythms of larval fish assemblages in a mangrove-fringed estuary in north Brazil. *Open Fish Science Journal* **2**, 15–24.

Barletta-Bergan, A., M. Barletta and U. Saint-Paul (2002). Structure and seasonal dynamics of larval fish in the Caeté River estuary in north Brazil. *Estuaries, Coastal and Shelf Science* **54**, 193–206.

Barnard, K. H. (1925). *A Monograph on the Marine Fishes of South Africa (*Part I*)*. Annals of the South African Museum **21**, 1–418. [pp. 104–105]

Barrell, J. (1916). Influence of Silurian-Devonian climates on the rise of air-breathing vertebrates. *Proceedings of the National Academy of Sciences* **2**(8), 499–504.

Bartholomew, A. and J. A. Bohnsack (2005). A review of catch-and-release angling mortality with implications for no-take reserves. *Reviews in Fish Biology and Fisheries* **15**, 129–154.

Barton, B. A. (2002). Stress in fishes: a diversity of responses with particular references to changes in circulating corticosteroids. *Integrative and Comparative Biology* **42**, 517–525.

Barton, M. and C. Wilmhoff (1996). Inland fishes of the Bahamas – new distribution records for exotic and native species from New Providence Island. *Bahamas Journal of Science* **3**(2), 7–11.

Beamish, R. J. and G. A. McFarlane (1983). Validation of age determination estimates: the forgotten requirement. In *Proceedings of the International Workshop on Age Determination of Oceanic Pelagic Fishes: Tunas, Billfishes, and Sharks*, E. D. Prince and L. M. Pulos. NOAA, NMFS, Southeast Fisheries Center, Miami, pp. 29–33.

Bean, T. H. (1903). *Catalogue of the Fishes of New York.* New York State Museum, Bulletin 60 (Zoology 9), 784 pp. [pp. 177–179]

Beebe, W. (1927). A tarpon nursery in Haiti. New York Zoological Society Bulletin 30, 141–145. [Reprinted in Beebe 1928, 67–74.]

Beebe, W. (1928). *Beneath Tropic Seas: A Record of Diving Among the Coral Reefs of Haiti.* Blue Ribbon Books, New York, viii + 234 pp. [pp. 67–74, 228–230]

Beebe, W. and G. Hollister (1935). *The Fishes of Union Island, Grenadines and British West Indies, with a Description of a New Species of Star-gazer. Zoologica (NY)* **19**(6), 1–300. [p. 211]

Beebe, W. and J. Tee-Van (1928). The Fishes of Port-au-Prince Bay, Haiti. *Zoologica (NY)* **10**(1):1–279. [pp. 33–36]

Beebe, W. and J. Tee-Van (1933). *Field Book of the Shore Fishes of Bermuda and the West Indies.* G. P. Putnam's Sons, New York, xiv + 337 pp. [pp. 33–34]

Bell-Cross, G. and J. L. Minshull (1988). *The Fishes of Zimbabwe.* National Museums and Monuments of Zimbabwe, Harare, 294 pp. [pp. 42, 91–92]

Bellow, S. (1944). *Dangling Man.* Vanguard Press, New York. [1971 Penguin Books edition, New York, xii + 143 pp.]

Berkeley, S. A., C. Chapman and S. M. Sogard (2004). Maternal age as a determinant of larval growth and survival in a marine fish, *Sebastes melanops. Ecology* **85**, 1258–1264.

Berrien, P. L., M. P. Fahay, A. W. Kendall Jr. and W. G. Smith (1978). *Ichthyoplankton from the RV* Dolphin *Survey of the Continental Shelf Waters Between Martha's Vineyard, Massachusetts and Cape Lookout, North Carolina, 1965–66.* Technical Series Reports 15. NOAA, NMFS, Sandy Hook Laboratory, Highlands, 152 pp.

Bianchi, G. (1984). *Field Guide: Commercial Marine and Brackish Water Species of Pakistan*. FAO Species Identification Sheets for Fishery Purposes, Project UNDP/FAO Pak/77/033. FAO, Rome, xi + 200 pp., 24 plates. [p. 3]

Bianchi, G. (1986). *Fichas FAO de Identificação de Especies para Actividades de Pesca: Guia de Campo Espécies Comerciais Marinhas e de Águas Salobras de Angola*. FAO, Rome, xv + 184 pp., 14 plates. [p. 15]

Bidwell, J. P. and S. Spotte (1985). *Artificial Seawaters: Formulas and Methods*. Jones and Bartlett, Boston, 349 pp.

Bigelow, H. B. and W. C. Schroeder (1953). *Fishes of the Gulf of Maine*, first revision. Fishery Bulletin 74, U.S. Fish and Wildlife Service. U.S. Government Printing Office, Washington, viii + 577 pp. [p. 87]

Birkland, C. and P. K. Dayton (2005). The importance in fishery management of leaving the big ones. *Trends in Ecology and Evolution* **20**, 356–358.

Bishop, K. A. (1980). Fish kills in relation to physical and chemical changes in Magela Creek (East Alligator River system, Northern Territory) at the beginning of the tropical wet season. *Australian Zoologist* **20**, 485–500.

Bishop, K. A., S. A. Allen, D. A. Pollard and M. G. Cook (2001). *Ecological Studies on the Freshwater Fishes of the Alligator Rivers Region, Northern Territory: Autecology*. Supervising Scientist, Report 145, Environment Australia, Darwin, xii + 570 pp. [pp. 18–29]

Bishop, R. E. and J. J. Torres (1999). Leptocephalus energetics: metabolism and excretion. *Journal of Experimental Biology* **202**, 2485–2493.

Blaber, S. J. M. (1973). Population size and mortality of juveniles of the marine teleost *Rhabdosargus holubi* (Pisces: Sparidae) in a closed estuary. *Marine Biology* **21**, 219–225.

Blache, J. (1962). II. Liste des poissons signalés dans l'Atlantique tropic-oriental sud du Cap des Palmes (4° lat. N) à Mossamedes (15° lat. S) (Province Guineo-Equatoriale). Cahiers ORSTOM, Sér Partie Noire 2, 13–102. [p. 23]

Blache, J., J. Cadenat and A. Stauch (1970). *Clés de détermination des poissons de mer signalés dans l'Atlantique oriental entre le 20° parallel nord et le 15° parallel sud. I. Clés de determination des familles, genres, espèces de poissons sélaciens signalés dans le golfe de Guinée*. Faune Tropicale Vol. 18, OSTROM, Paris, 479 pp. [p. 136]

Blackwell, B. G., M. L. Brown and D. W. Willis (2000). Relative weight (W_r) status and current use in fisheries assessment and management. *Reviews in Fisheries Science* **8**, 1–44.

Blanco, G. B. (1955). Postlarval forms of marine fishes of Siokun Bay, Zamboanga del Sur province. *Philippine Journal of Fisheries* **3**(2), 97–114. [pp. 97–98]

Blandon, I. R., R. Ward, F. J. García de León, A. M. Landry, A. Zerbi and M. Figuerola *et al.* (2002). Studies in conservation genetics of tarpon (*Megalops atlanticus*). I. Variation in restriction length polymorphisms of mitochrondrial DNA across the distribution of the species. *Contributions in Marine Science* **35**, 1–17.

Boehlert, G. W. and B. C. Mundy (1988). Roles of behavioral and physical factors in larval and juvenile fish recruitment to estuarine nursery areas. *American Fisheries Society Symposium* **3**(5), 51–67.

Boeseman, M. (1960). The fresh-water fishes of the island of Trinidad. *Studies of the Fauna of Curaçao and Other Caribbean Islands* **10**(48), 72–153. [p. 86]

Böhlke, J. E. and C. C. G. Chaplin (1968). *Fishes of the Bahamas and Adjacent Tropical Waters*. Livingston, Wynnewood, xxiii + 771 pp. [pp. 36–38]

Bolger, T. and P. L. Connolly (1989). The selection of suitable indices for the measurement and analysis of fish condition. *Journal of Fish Biology* **34**, 171–182.

Bond, C. E. (1979). *Biology of Fishes*. Saunders College Publishing, Philadelphia, vii + 514 pp.

Bortone, S. A. (2008). Insight into the historical status and trends of tarpon in southwest Florida through recreational catch data recorded on scales. In *Biology and Management of the World Tarpon and Bonefish Fisheries*, J. S. Ault (editor). CRC Press, Boca Raton, pp. 69–77.

Bortone, S. A., J. G. Holt and D. Engle (2007). Perspectives on tarpon, based on the historical recreational fishery in the Gulf of Mexico. 59th Annual Gulf and Caribbean Fisheries Institute, Belize City, pp. 31–36.

Boschung, H. T. Jr. and R. L. Mayden (2004). *Fishes of Alabama*. Smithsonian Books, Washington, xviii + 736 pp. [pp. 127–128]

Boseto, D. and A. P. Jenkins (Undated). *A checklist of freshwater and brackish water fishes of the Fiji Islands*. Institute of Applied Sciences, University of the South Pacific, Suva, Fiji, 9 pp.

Boulenger, G. A. (1909). *Catalogue of the Fresh-water Fishes of Africa in the British Museum (Natural History)*, Vol. 1, xi + 373 pp. [p. 27]

Bouyat, A. (1911). *La Pêche maritime en Uruguay*. Tip. de l'Ecole nationale des Arts et Métiers, Montevideo, 92 pp.

Brady, P. C., A. A. Gilerson, G. W. Kattawar, J. M. Sullivan, M. S. Twardowski, and H. M. Dierssen *et al.* (2015). Open-ocean fish reveal an omnidirectional solution to camouflage in polarized environments. *Science*: **350**, 965–969.

Brainerd, E. L. (1994). The evolution of lung-gill bimodal breathing and the homology of vertebrate respiratory pumps. *American Zoologist* **34**, 289–299.

Brauner, C. J. and P. J. Rombough (2012). Ontogeny and paleophysiology of the gill: new insights from larval and air-breathing fish. *Respiratory Physiology and Neurobiology* **184**, 293–300.

Breder, C. M. Jr. (1925a). Notes of fishes from three Panama localities. *Zoologica (NY)* **4**(4), 137–158. [pp. 140–141]

Breder, C. M. Jr. (1925b). Fish notes for 1924 from Sandy Hook Bay. *Copeia* **1925**(138), 1–4.

Breder, C. M. Jr. (1929). *Field Book of Marine Fishes of the Atlantic Coast*. G. P. Putnam's Sons, New York, xxxvii + 332 pp., 16 photographs. [pp. 59–60]

Breder, C. M. Jr. (1933). Young tarpon on Andros Island. *Bulletin of the New York Zoological Society* **36**, 65–67.

Breder, C. M. Jr. (1944). Materials for the study of the life history of *Tarpon atlanticus*. *Zoologica (NY)* **29**(19), 217–252.

Breder, C. M. Jr. (1945). On the relationship of social behavior to pigmentation in tropical shore fishes. *Bulletin of the American Museum of Natural History* **94**(2), 87–106.

Breder, C. M. Jr. (1959). Studies on social groupings in fishes. *Bulletin of the American Museum of Natural History* **117**(6), 387–481, Plates 70–80. [p. 107]

Breder, C. M. Jr. and R. F. Nigrelli (1939). *Helioperca* in New York harbor. *Copeia* **1939**, 49.

Breder, C. M. Jr. and A. C. Redmond (1929). Fish notes for 1928 from Sandy Hook Bay. *Copeia* **1929**(170), 43–45.

Brice, J. J. (editor) (1897). *Report on the fisheries of Indian River, Florida*. U.S. Government Printing Office, Washington, 40 pp., 37 plates. [p. 18, Plate 5]

Briggs, J. C. (1958). A list of Florida fishes and their distribution. *Bulletin of the Florida State Museum (Biological Sciences)* **2**(8), 221–318. [p. 252]

Brittain, T. (1985). A kinetic and equilibrium study of ligand binding to a Root-effect haemoglobin. *Biochemical Journal* **228**, 409–414.

Brittain, T. (1987). The Root effect. *Comparative Biochemistry and Physiology* **86B**, 473–481.

Brockmeyer Jr., R. E., J. R. Rey, R. W. Virnstein, R. G. Gilmore and L. Earnest (1996). Rehabilitation of impounded estuarine wetlands by hydrologic reconnection to the Indian River Lagoon, Florida (USA). *Wetlands Ecology and Management* **4**, 93–109.

Broussonet, P. M. A. (1782). *Ichthyologia Sistens Piscium Descriptiones et Icones*. Peter Elmsly, London, 80 pp., unnumbered. [pp. 62–65]

Brown, R. J. and K. P. Severin (2008). A preliminary otolith microchemical examination of the diadromous migrations of Atlantic tarpon *Megalops atlanticus*. In *Biology and Management of the World Tarpon and Bonefish Fisheries*, J. S. Ault (editor). CRC Press, Boca Raton, pp. 259–274.

Buchheister, A., M. T. Wilson, R. J. Foy and D. A. Beauchamp (2006). Seasonal and geographic variation in condition of juvenile walleye pollock in the western Gulf of Alaska. *Transactions of the American Fisheries Society* **135**, 897–907.

Bui, P. and S. P. Kelly (2014). Claudin-6, -10d and -10e contribute to seawater acclimation in the euryhaline puffer fish *Tetraodon nigroviridis*. *Journal of Experimental Biology* **217**, 1758–1767.

Bui, P., M. Bagherie-Lachidan and S. P. Kelly (2010). Cortisol differentially alters claudin isoforms in cultured puffer fish gill epithelia. *Molecular and Cellular Endocrinology* **317**, 120–126.

Burrows, D., C. Perna and B. Pusey (2009). A brief freshwater fish survey of northern Cape York Peninsula. Australian Centre for Tropical Fresh*Water Research*, Townsville, 16 pp. [p. 6 Table 3]

Bussing, W. A. (1987). *Peces de las Aguas Continentales de Costa Rica*. Universidad de Costa Rica, San José, 271 pp. [pp. 62–63]

Butsch, R. S. (1939). A list of Barbadian fishes. *Journal of the Barbados Museum and Historical Society* **7**(1), 17–31.

Caine, L. S. (1935). *Game Fish of the South and How to Catch Them*. Houghton Mifflin, New York, viii + 259 pp. [pp. 134–137, 233–235]

Cairns, S. (editor) (1992). Distribution of selected fish species in Tampa Bay: Final Report. Tillamook Bay National Estuary Project (TBNEP). Linthicum, viii + 50 pp.

Camargo, M. and V. Isaac (2001). Os peixes estuarinos da região norte do Brasil: lista de espécies e considerações sobre sua distribuição geográfica. *Boletim do Museu Paraense Emilio Goeldi, Zoologia (Nova Serie)* **17**(2), 133–157. [p. 140 Table 1]

Campana, S. E. (1999). Chemistry and composition of fish otoliths: pathways, mechanisms and applications. *Marine Ecology Progress Series* **188**, 263–297.

Carneiro, M., R. Martins, M. Landi and F. O. Costa (2014). Updated checklist of marine fishes (Chordata: Craniata) from Portugal and the proposed extension of the Portuguese continental shelf. *European Journal of Taxonomy* (**73**), 73 pp. [p. 16]

Carson, W. E. (1914). *Mexico: The Wonderland of the South*, revised edition. Macmillan, New York, xiii + 449 pp. [pp. 394–403]

Carvalho Collyer, E. and D. Alves Aguiar (1972). Sobre a Produção pesqueira de alguns currais-de-pescado Ceará – dados de 1968 a 1970. *Boletim de Ciências do Marinha, Fortaleza* **24**, 1–9.

Casselman, J. M. (1983). Age and growth assessment of fish from their calcified structures – techniques and tools. In: *Proceedings of the International Workshop on Age Determination of Oceanic Pelagic Fishes: Tunas, Billfishes, and Sharks*, E. D. Prince and L. M. Pulos (eds). NOAA, NMFS, Southeast Fisheries Center, Miami, pp. 1–17.

Castro-Aguirre, J. L., H. Espinosa Pérez and J. J. Schmitter-Soto (1999). *Itiofauna Estuarino-Lagunar y Vicaria de México*. Colección Textos Politécnicos, Serie Biotechnologías. Limusa Noriega Editores, Balderas, 711 pp. [pp. 88–90]

Castro-Aguirre, J. L., H. Espinosa-Pérez and J. J. Schmitter-Soto (2002). Lista sistemática, biogeográfica y ecológica de la ictiofauna. Estuárino lagunar y vicaria de México. In *Libro Jubilar en Honor al Dr. Salvador Contreras Balderas*, M. de Lourdes Lozano-Vilano (editor). Facultad de Ciencias Biológicas, Universidad Autónoma de Nuevo León, Monterrey, pp. 117–142 [p. 125]

Cataño, S. and J. Garzón-Ferreira (1994). Ecología trófica del sábalo *Megalops atlanticus* (Pisces: Megalopidae) en el área de Ciénaga Grande de Santa Marta, Caribe colombiano. *Revista de Biología Tropical* **42**, 673–684.

Chacko, P. I. (1948). Growth-rate of twenty-one species of fishes of Madras. *Proceedings of the 35th Indian Science Congress, Patna*, p. 210. [Abstract]

Chacko, P. I. (1949). Nutrition of the young stages of estuarine fishes of Madras. *Science and Culture* **15**, 32–33.

Chacón Chaverri, D. (1993). Aspectos biométricos de una población de sábalo, *Megalops atlanticus* (Pisces: Megalopidae). *Revista de Biología Tropical* **41**(Supplement 1), 13–18.

Chacón Chaverri, D. (1994). Ecología básica y alimentación del sábalo *Megalops atlanticus* (Pisces: Megalopidae). *Revista de Biología Tropical* **42**, 225–232.

Chacón Chaverri, D. and W. O. McLarney (1992). Desarrollo temprano del sábalo, *Megalops atlanticus* (Pisces: Megalopidae). *Revista de Biología Tropical* **40**, 171–177.

Chapman, L. J., L. Kaufman and C. A. Chapman (1994). Why swim upside down?: A comparative study of two mochokid catfishes. *Copeia* **1994**, 130–135.

Chapman, L. J., C. A. Chapman, F. G. Nordlie and A. E. Rosenberger (2002). Physiological refugia: swamps, hypoxia tolerance and maintenance of fish diversity in the Lake Victoria region. *Comparative Biochemistry and Physiology* **133A**, 421–437.

Charles-Dominique, E. (1993). *L'exploitation de la lagune Aby (Côte d'Ivoire) par la pêche artisanale. Dynamique des ressources, de l'exploitation et des pêcheries.* Thèse de Doctorat. Université de Montpellier, 407 pp. [p. 107]

Chen, C-N., L-Y. Lin and T-H. Lee (2004). Ionocyte distribution in gills of the euryhaline milkfish, *Chanos chanos* (Forsskål, 1775). *Zoological Studies* **43**, 772–777.

Chen, H. L. and W. N. Tzeng (2006). Daily growth increment formation in otoliths of Pacific tarpon *Megalops cyprinoides* during metamorphosis. *Marine Ecology Progress Series* 255–263.

Chen, H-L., K-N. Shen, C-W. Chang, Y. Iizuka and W-N. Tzeng (2008). Effects of water temperature, salinity and feeding regimes on metamorphosis, growth and otolith Sr:Ca ratios of *Megalops cyprinoides* leptocephali. *Aquatic Biology* **3**, 41–50.

Chidambaram, K. and M. D. Menon (1947). Notes on the development of *Megalops cyprinoides* and *Hemiramphus georgii*. *Proceedings of the Zoological Society of London* **117**, 756–763.

Choe, K. P., J. Havird, R. Rose, K. Hyndman, P. Piermarina and D. H. Evans (2006). COX2 in a euryhaline teleost, *Fundulus heteroclitus*: primary sequence, distribution, localization, and potential function in gills during salinity acclimation. *Journal of Experimental Biology* **209**, 1696–1708.

Clack, J. A. (2007). Devonian climate change, breathing, and the origin of the tetrapod stem group. *Integrative and Comparative Biology* **47**, 510–523.

Claiborne, J. B., C. R. Blackston, K. P. Choe, D. C. Dawson, S. P. Harris and L. A. MacKenzie *et al.* (1999). A mechanism for branchial acid excretion in marine fish: identification of multiple Na^+/H^+ antiporter (NHE) isoforms in gills of two seawater teleosts. *Journal of Experimental Biology* **202**, 315–324.

Clark, D. L., J. M. Leis, A. C. Hay and T. Trnski (2005). Swimming ontogeny of larvae of four temperate marine fishes. *Marine Ecology Progress Series* **292**, 287–300.

Clark, E. and K. von Schmidt (1965). Sharks of the central Gulf coast of Florida. *Bulletin of Marine Science* **15**, 13–83.

Clark, R. D. and A. Landry (2002). Early life history of the Atlantic tarpon, *Megalops atlanticus*, along the upper Texas coast. *Contributions in Marine Science* **35**, 100–101. [Abstract]

Clark, T. D., R. S. Seymour, K. Christian, R. M. G. Wells, J. Baldwin and A. P. Farrell (2007). Changes in cardiac output during swimming and aquatic hypoxia in the air-breathing Pacific tarpon. *Comparative Biochemistry and Physiology* **148A**, 562–571.

Coates, D. (1987). Observations on the biology of tarpon, *Megalops cyprinoides* (Broussonet) (Pisces: Megalopidae), in the Sepik River, northern Papua New Guinea. *Australian Journal of Marine and Freshwater Research* **38**, 529–535.

Coates, D. (1993). Fish ecology and management of the Sepik-Ramu, New Guinea, a large contemporary tropical river basin. *Environmental Biology of Fishes* **38**, 345–368.

Coker, R. E. (1921). A record of young tarpon. *Copeia* **93**, 25–26.

Collette, B. B., K. E. Carpenter, B. A. Polidoro, M. J. Juan-Jordá, A. Boustany and D. J. Die *et al.* (2011). High value and long life: double jeopardy for tunas and billfishes. *Science* **333**, 291–292.

Cone, R. S. (1989). The need to reconsider the use of condition indices in fishery science. *Transactions of the American Fisheries Society* **118**, 510–514.

Conover, D. O. and S. B. Munch (2002). Sustaining fisheries yields over evolutionary time scales. *Science* **297**, 94–96.

Cooke, S. J. and I. G. Cowx (2004). The role of recreational fishing in global fish crises. *Bioscience* **54**, 857–859.

Cooke, S. J. and C. D. Suski (2004). Are circle hooks an effective tool for conserving marine and freshwater recreational catch-and-release fisheries? *Aquatic Conservation: Marine and Freshwater Ecosystems* **14**, 299–326.

Cooke, S. J. and C. D. Suski (2005). Do we need species-specific guidelines for catch-and-release recreational angling to conserve diverse fishery resources? *Biodiversity and Conservation* **14**, 1195–1209.

Cooke, S. J., D. P. Philipp, K. M. Dunmall and J. F. Schreer (2001). The influence of terminal tackle on injury, handling time, and cardiac disturbance of rock bass. *North American Journal of Fisheries Management* **21**, 333–342.

Cooke, S. J., J. Steinmetz, J. F. Degner, E. C. Grant and D. P. Philipp (2003). Metabolic fright responses of different-sized largemouth bass (*Micropterus salmoides*) to two avian predators show variations in nonlethal energetic costs. *Canadian Journal of Zoology* **81**, 699–709.

Coolidge, E., M. S. Hedrick and W. K. Milsom (2007). Ventilatory systems. *Fish Physiology* **26**, 181–211.

Costa Pereira, N. and L. Saldanha (1977). Sur la distribution de *Tarpon atlanticus* (Val., 1847) (Pisces, Megalopidae) dans l'Atlantique oriental. *Memórias do Museu do Mar – Cascais, Portugal, Série Zoológica* **1**(1), 1–15.

Cotto S., A. (undated). *Guía de Identificacíon de Peces Marinos de Nicaragua.* Place and publisher unknown, 58 pp. [pp. 32–33]

Cowley, P. D., A. K. Whitfield and K. N. I. Bell (2001). The surf zone Ichthyoplankton adjacent to an intermittently open estuary, with evidence of recruitment during marine overwash events. *Estuarine, Coastal and Shelf Science* **52**, 339–348.

Crabtree, R. E. (1995). Relationship between lunar phase and spawning activity of tarpon, *Megalops atlanticus*, with notes on the distribution of larvae. *Bulletin of Marine Science* **56**, 895–899.

Crabtree, R. E., E. C. Cyr, R. E. Bishop, L. M. Falkenstein and J. M. Dean (1992). Age and growth of tarpon, *Megalops atlanticus*, larvae in the eastern Gulf of Mexico, with notes on relative abundance and probable spawning areas. *Environmental Biology of Fishes* **35**, 361–370.

Crabtree, R. E., E. C. Cyr and J. M. Dean (1995). Age and growth of tarpon, *Megalops atlanticus*, from south Florida waters. *Fishery Bulletin (NOAA)* **93**, 619–628.

Crabtree, R. E., E. C. Cyr, D. Chacón Chaverri, W. O. McLarney and J. M. Dean (1997). Reproduction of tarpon, *Megalops atlanticus*, from Florida and Costa Rican waters and notes on their age and growth. *Bulletin of Marine Science* **61**, 271–285.

Crouse, D. T., L. B. Crowder and H. Caswell (1987). A stage-based population model for loggerhead sea turtles and implications for conservation. *Ecology* **68**, 1412–1423.

Cruz-Ayala, L. M. A. (2002). Tarpon: tourism resource of the Gulf of Mexico. *Contributions in Marine Science* **35**, 95–96. [Abstract]

Cuvier, M. Le B. and M. A. Valenciennes (1846). *Histoire Naturelle des Poissons*, Vol. 19. Libraire de la Société Géologique de France, Paris, xix + 544 pp. [pp. 398–401]

Cyr, E. C. (1991). *Aspects of the Life History of the Tarpon,* Megalops atlanticus, *from south Florida.* Doctoral dissertation. University of South Carolina, Columbia, xvii + 138 pp.

Cyrus, D. P. and S. J. M. Blaber (1992). Turbidity and salinity in a tropical northern Australian estuary and their influence on fish distribution. *Estuarine, Coastal and Shelf Science* **35**, 545–563.

Dahl, G. (1965). La metamorphosis desde leptocephalus hasta estado post-larval en el sábalo *Tarpon atlanticus* (Cuv. et Val.). Corporación Autónoma Regional de los Valles del Magdalena y del Sinú, Bogotá, 20 pp.

Dahl, G. (1971). *Los Peces del Norte de Colombia*. Ministerio de Agricultura. INDERENA, Bogotá, xvii + 391 pp. [pp.158–160]

Dahlberg, M. D. (1979). A review of survival rates of fish eggs and larvae in relation to impact assessments. *Marine Fisheries Review* **41**(3), 1–12.

Dailey, W., A. M. Landry Jr. and F. L. Kenyon II. (2008). The Louisiana recreational tarpon fishery. In *Biology and Management of the World Tarpon and Bonefish Fisheries*, J. S. Ault (editor). CRC Press, Boca Raton, pp. 57–68.

Dampier, W. (1906). *Dampier's Voyages*, Vol. II, J. Masefield (editor). E. Grant Richards, London, vii + 624 pp. [pp. 117–118, 171]

Daniels, C. B., S. Orgeig, L. C. Sullivan, N. Ling, M. B. Bennett, and S. Schürch *et al.* (2004). The origin and evolution of the surfactant system in fish: insights into the evolution of lungs and swim bladders. *Physiological and Biochemical Zoology* **77**, 732–749.

Danylchuk, S. E., A. J. Danylchuk, S. J. Cook, T. L. Goldberg, J. Koppelman and D. P. Philipp (2007). Effects of recreational angling on the post-release behavior and predation of bonefish (*Albula vulpes*): the role of equilibrium status at the time of release. *Journal of Experimental Marine Biology and Ecology* **346**, 127–133.

Darimont, C. T., C. H. Fox, H. M. Bryan and T. E. Reimchen (2015). The unique ecology of human predators. *Science* **349**, 858–860.

da Silva de Almeida, Z. (2008). *Os Recursos Pesqueiros Marinhos e Estuarinos do Maranhão: Biologia, Tecnologia, Socioeconomia, Estado da Arte e Manejo*. Programa de Pós-Graduação em Zoologia, Universidade Federal do Pará e Museu Paraense Emílio Goeldi, Belém, 286 pp. [pp. 58, 149, 240]

Davis, B., R. Baker and M. Sheaves (2014). Seascape and metacommunity processes regulate fish assemblage structure in coastal wetlands. *Marine Ecology Progress Series* **500**, 187–202.

Davis, T. L. O. (1988). Temporal changes in the fish fauna entering a tidal swamp system in tropical Australia. *Environmental Biology of Fishes* **21**, 161–172.

Day, F. (1878). *The Fishes of India; Being a Natural History of the Fishes Known to Inhabit the Seas and Fresh Waters of India, Burma, and Ceylon*, Vol. I. Bernard Qualitch, London, xx + 778 pp. [pp. 650–651]

de Araújo Santos, G. A. (2013). Ensaio preliminary sobre o cultivo camurupim (*megalops* [*sic*] *atlanticus*) em viveiros escavados e sua ocorrência em lagoas marginais no litoral de Tutóia – Ma. In *XXIV SEMIC Seminário de Iniciação Científica: Livro de Resumos*. Pro-Reitoria de Pesquisa e Pós-Graduação, Universidade Federal do Maranhão, São Luís, p. 16.

de Beaufort, L. F. (1909). Die Schwimmblase der Malacopterygii. *Gegenbaurs Morphologisches Jahrbuch* **39**, 526–532, 1 plate. [pp. 526–532, 643, Plate XV]

Decker, C., C. Griffiths, K. Prochazka, C. Ras and A. Whitfield (editors) (2003). *Proceedings of the Marine Biodiversity in Sub-Saharan Africa: The Known and the Unknown*, Cape Town, 310 pp. [p. 81]

Deibel, D., C. C. Parrish, P. Grønkjaer, P. Munk and T. Gissel Nielsen (2012). Lipid class and fatty acid content of the leptocephalus larva of tropical eels. *Lipids* doi 10.1007/s11745-012-3670-5, 12 pp.

de Normandes Valadares, P. (2013). Ocorrência do camurupim *(Megalops atlanticus)* em lagoas marginais no litoral de Tutóia – MA. In *XXIV SEMIC Seminário de Iniciação Científica: Livro de Resumos*. Pro-Reitoria de Pesquisa e Pós-Graduação, Universidade Federal do Maranhão, São Luís, p. 17.

Delsman, H. C. (1926). Fish eggs and larvae from the Java Sea. *Treubia* **8**, 389–412.

de Paiva Carvalho, J. P. (1964). Comentários sôbre os peixes mencionados na obra "História dos animais e árvores do Maranhão" de Frei Cristóvão de Lisboa. *Arquivos da Estaçao de Biologia Marinha da Universidade Federal do Ceará, Fortaleza* **4**(1), 1–39.

Diouf, P. S. (1996). *Les peuplements de poissons des milieu estuariens de l'Afrique de l'Quest: L'exemple de l'estuaire hyperhalin su Sine-Saloum*. l'Université de Montpellier II. Thèses et Documents Microfiches No. 156. ORSTOM, Paris, 267 pp. [p. 130]

Dittman, A. H. and T. P. Quinn (1996). Homing in Pacific salmon: mechanisms and ecological basis. *Journal of Experimental Biology* **199**, 83–91.

Dixson, D. L., D. Abrego and M. E. Hay (2014). Chemically mediated behavior of recruiting corals and fishes: a tipping point that may limit reef recovery. *Science* **345**, 892–897.

Doherty, P. F. Jr., E. A. Schreiber, J. D. Nichols, J. E. Hines, W. A Link and G. A. Schenk *et al.* (2004). Testing life history predictions in a long-lived seabird: a population matrix approach with improved parameter estimation. *Oikos* **105**, 606–618.

Domenici, P., C. Lefrançois and A. Shingles (2007). Hypoxia and the antipredator behaviours of fishes. *Philosophical Transactions of the Royal Society* **262B**, 2105–2121.

Driedzic, W. R. and P. W. Hochachka (1978). Metabolism in fish during exercise. In *Fish Physiology*, Vol. VII, W. S. Hoar and D. J. Randall (editors). Academic Press, New York, pp. 503–543.

Duarte Lopes, P. R. and M. Porto Sena (1996). Ocorrência de *Tarpon atlanticus* (Valenciennes, 1846) (Pisces: Megalopidae) na Baía de Todo os Santos (Estado da Bahia, Brasil). *Sitientibus* (**14**), 69–77.

Duarte Lopes, P. R., J. Tavares de Oliveira-Silva and A. S. Alves Ferreira-Melo (1998). Contribuição ao conhecimento da ictiofauna do Manguezal de Cacha Pregos, Ilha de Itaparica, Baía de Totos os Santos, Bahia. *Revista Brasileira de Zoologia* **15**, 315–325.

Dunnette, D. A., D. P. Chynoweth and K. H. Mancy (1985). The source of hydrogen sulfide in anoxic sediment. *Water Research* **19**, 875–884.

Duthie, G. G. and G. M. Hughes (1987). The effects of reduced gill area and hyperoxia on the oxygen consumption and swimming speed of rainbow trout. *Journal of Experimental Biology* **127**, 349–354.

Eccles, D. H. (1992). *Field Guide to the Freshwater Fishes of Tanzania*. Project URT/87/016, United Nations Development Programme. FAO, Rome, v + 145 pp. [p. 32]

Edwards, A. J., A. C. Gill and P. O. Abohweyere (2001). *A Revision of Irvine's Marine Fishes of Tropical West Africa*. Darwin's Initiative. Place and publisher unknown, vi + 178 pp. [pp. 28–29]

Edwards, R. E. (1998). Survival and movement patterns of released tarpon (*Megalops atlanticus*). *Gulf of Mexico Science* **16**, 1–7.

Eigenmann, C. H. (1904). The fresh-water fishes of western Cuba. *Bulletin of the U.S. Fish Commission* **XXII**, 211–236. [p. 222]

Eigenmann, C. H. (1912). *The Freshwater Fishes of British Guiana, Including a Study of the Ecological Grouping of Species, and the Relation of the Fauna of the Plateau to that of the Lowlands. Memoirs of the Carnegie Museum* V(Serial No. 67), xxii + 578 pp., 103 plates. [p. 444]

Eigenmann, C. H. (1921). Small tarpon. *Copeia* **95**, 33.

Eiras-Stofella, D. R. and S. M. Fank-de-Carvalho (2002). Morphology of gills of the seawater fish *Cathorops spixii* (Agassiz) (Ariidae) by scanning and transmission electron microscopy. *Revista Brasileira Zoologica* **19**, 1215–1220.

Eldred, B. (1967). Larval tarpon, *Megalops atlanticus* Valenciennes, (Megalopidae) in Florida waters. Florida Board of Conservation, Marine Laboratory Leaflet Series IV, Part 1 (Pisces), No. 4, 1–9.

Eldred, B. (1968). First record of a larval tarpon, *Megalops atlanticus* Valenciennes, from the Gulf of Mexico. Florida Board of Conservation, Marine Laboratory Leaflet Series IV, Part 1 (Pisces), No. 7, 1–2.

Eldred, B. (1972). Note on larval tarpon, *Megalops atlanticus* (Megalopidae), in the Florida Straits. Florida Department of Natural Resources, Marine Research Laboratory, IV, Part 1 (Pisces), No. 22, 1–6.

Ellis, R. W. (1956). Tarpon cooperative research and program progress report 56–20. The Marine Laboratory, University of Miami Marine Laboratory, Coral Gables, 13 pp.

Emmanuel, B. E., C. Oshionebo and N. F. Aladetohun (2011). Comparative analysis of the proximate compositions of *Tarpon atlanticus* and *Clarias gariepinus* from culture systems in south-western Nigeria. *African Journal of Food, Agriculture, Nutrition and Development* **11**, 5344–5359.

Erdman, D. S. (1960). Larvae of tarpon, *Megalops atlantica*, from the Añasco River, Puerto Rico. *Copeia* **1960**, 146.

Evans, D. H. (1999). Ionic transport in the fish gill epithelium. *Journal of Experimental Zoology* **283**, 641–652.

Evans, D. H. (2010a). Freshwater fish gill ion transport: August Krogh to morpholinos and microprobes. *Acta Physiologica*. doi:1111/j.1748-1716.2010.02186.x, 11 pp.

Evans, D. H. (2010b). Co-ordination of osmotic stress responses through osmosensing and signal transduction. *Journal of Fish Biology* **76**, 1903–1925.

Evans, D. H., P. M. Piermarini and K. P. Choe (2005). The multifunctional fish gill: dominant site of gas exchange, osmoregulation, acid-base regulation, and excretion of nitrogenous waste. *Physiological Reviews* **85**, 97–177.

Evermann, B. W. and B. A. Bean (1897). Indian River and its fishes. In *Report on the Fisheries of Indian River, Florida*, J. J. Brice (Commissioner). U.S. Government Printing Office, Washington, pp. 5–26. [p. 18, Plate 5]

Evermann, B. W. and M. C. Marsh (1902). *The Fishes of Porto Rico. Bulletin of the U.S. Fish Commission* (1900) **20**(1) for 1900, pp. 51–350. [pp. 79–80]

Ezekiel, E. N. and J. F. N. Abowei (2013). Length-weight relationship and condition factor of *Heterotis niloticus* from Amassoma flood plain, Niger Delta, Nigeria. *Applied Science Reports* **4**, 164–172.

Fahay, M. P. (2007). *Early Stages of Fishes in the Western Atlantic Ocean (Davis Strait, Southern Greenland and Flemish Cap to Cape Hatteras), Vol. 1: Acipenseriformes through Sygnathiformes*. Northwest Atlantic Fisheries Organization, Dartmouth, liii + 931 pp. [pp. 12–13]

Farmer, C. (1997). Did lungs and the intracardiac shunt evolve to oxygenate the heart in vertebrates? *Paleobiology* **23**, 358–372.

Farrell, A. P. (1993). Cardiovascular system. In *The Physiology of Fishes*, D. H. Evans (editor). CRC Press, Boca Raton, pp. 219–250.

Farrell, A. P. (2007). Tribute to P. L. Lutz: a message from the heart – why hypoxic bradycardia in fishes? *Journal of Experimental Biology* **210**, 1715–1725.

Farrell, A. P. and D. R. Jones (1992). The heart. In *Fish Physiology*, Vol. XII (Part A), W. S. Hoar, D. J. Randall and A. P. Farrell (editors). Academic Press, San Diego, CA, pp. 1–88.

Farrell, A. P. and J. F. Steffensen (1987). An analysis of the energetic cost of the branchial and cardiac pumps during sustained swimming in trout. *Fish Physiology and Biochemistry* **4**, 73–79.

Farrell, A. P., D. L. Simonot, R. S. Seymour and T. D. Clark (2007). A novel technique for estimating the compact myocardium in fishes reveals surprising results for an athletic air-breathing fish, the Pacific tarpon. *Journal of Fish Biology* **71**, 389–398.

Fedler, T. (2009). The economic impact of recreational fishing in the Everglades region. Report for the Everglades Foundation, Palmetto Bay, FL, iv + 13 pp.

Fedler, A. (2011). The economic impact of recreational tarpon fishing in the Caloosahatchee River and Charlotte Harbor region of Florida. Report for the Everglades Foundation, Palmetto Bay, FL, iv + 20 pp.

Fedler, A. J. and C. Hayes (2008). Economic impact of recreational fishing for bonefish, permit and tarpon in Belize for 2007. Report prepared for Turneffe Atoll Trust, Belize City, iii + 26 pp.

Ferguson, R. A. and B. L. Tufts (1992). Physiological effects of brief air exposure in exhaustively exercised rainbow trout (*Oncorhynchus mykiss*): implications for "catch and release" fisheries. *Canadian Journal of Fisheries and Aquatic Sciences* **49**, 1157–1162.

Ferreira de Menezes, M. (1967). Relação comprimento-pêso do camurupim, *Tarpon atlanticus* (Valenciennes), no nordeste Brasiléiro. *Arquivos de Ciências do Mar, Fortaleza* **7**, 101–102.

Ferreira de Menezes, M. (1968a). Relação comprimento-pêso de jovens do camurupim, *Tarpon atlanticus* (Valenciennes), no nordeste Brasiléiro. *Arquivos de Ciências do Mar, Fortaleza* **8**, (2 pp., unpaginated).

Ferreira de Menezes, M. (1968b). Sôbre a alimentação do camurupim, *Tarpon atlanticus* (Valenciennes), no estado do Ceará. *Arquivos da Estação de Biologia Marinha da Universidade Federal do Ceará, Fortaleza* **8**, 145–149.

Ferreira de Menezes, M. and M. Pinto Paiva (1966). Notes on the biology of tarpon, *Tarpon atlanticus* (Cuvier & Valenciennes), from coastal waters of Ceará State, Brazil. *Arquivos da Estação de Biologia Marinha da Universidade Federal do Ceará, Fortaleza* **6**, 83–98.

Figueroa, M. and A. Zerbi (2002). Age, growth and reproduction of tarpon in Puerto Rico. *Contributions in Marine Science* **35**, 102–103. [Abstract]

Fiksen, Ø., C. Jørgensen, T. Kristiansen, F. Vikebø and G. Huse (2007). Linking behavioural ecology and oceanography: larval behaviour determines growth, mortality and dispersal. *Marine Ecology Progress Series* **347**, 195–205.

Fiol, D. F. and D. Kültz (2005). Rapid hyperosmotic coinduction of two tilapia (*Oreochromis mossambicus*) transcription factors in gill cells. *Proceedings of the National Academy of Sciences* **102**, 927–932.

Fiol, D. F. and D. Kültz (2006). Use of suppression subtractive hybridization to identify osmotic stress transcription factors in Tilapia *(Oreochromis mossambicus)*. *Clontechniques*, April issue, 2 pp.

Fisher, R., J. M. Leis, D. L. Clark and S. K. Wilson (2005). Critical swimming speeds of late-stage coral reef fish larvae: variation within species, among species and between locations. *Marine Biology* **147**, 1201–1212.

Florindo, L. H., C. A. C. Leite, A. L. Kalinin, S. G. Reid, W. K. Milsom and F. T. Rantin (2006). The role of branchial and orobranchial O_2 chemoreceptors in the control of aquatic surface respiration in the neotropical fish tambaqui (*Colossoma macropomum*): progressive responses to prolonged hypoxia. *Journal of Experimental Biology* **209**, 1709–1715.

Fonteles-Filho, A. A. and M. F. Aguiar Espínola (2001). Produção de pescado e relações inter-specifícas na biocenose capturada por currais-de-pesca, no Estado do Ceará. *Boletim Técnico-Científico do CEPNOR, Belém* **1**, 111–124.

Foskett, J. K. and C. Scheffey (1982). The chloride cell: definitive identification as the salt-secretory cell in teleosts. *Science* **215**, 164–166.

Fowler, H. W. (1906). *The Fishes of New Jersey*. In *Annual Report of the New Jersey State Museum, Part II*. MacCrellish and Quigley, Trenton, pp. 35–477. [pp. 90–91]

Fowler, H. W. (1910). Little known New Jersey fishes. *Proceedings of the Academy of Natural Sciences of Philadelphia* **62**(3), 599–602. [pp. 599–600]

Fowler, H. W. (1925a). Fish notes for 1924 from Sandy Hook Bay. *Copeia* **138**, 1–4.

Fowler, H. W. (1925b). Records of fishes in New Jersey 1924. *Copeia* **143**, 42–46.

Fowler, H. W. (1927). Notes on fishes from Chincoteague, Virginia, 1926. *Copeia* **165**, 88–89.

Fowler, H. W. (1928). Notes on New Jersey fishes. *Proceedings of the Academy of Natural Sciences of Philadelphia* **80**, 607–614, Plate 31.

Fowler, H. W. (1936). *The Marine Fishes of West Africa: Based on the Collection of the American Museum Congo Expedition, 1909–1915. Bulletin of the American Museum of Natural History* **70**(Part 1), vii + 605 pp. [pp. 153–156]

Fowler, H. W. (1941). *Contributions to the Biology of the Philippine Archipelago and Adjacent Regions: The Fishes of the Groups Elasmobranchii, Holocephali, Isospondyli, and Ostarophysi Obtained by the United States Bureau of Fisheries Steamer "Albatross" in 1907 to 1910, Chiefly in the Philippine Islands and Adjacent Seas*. U.S. National Museum Bulletin 100, Vol. 13, x + 879 pp. [pp. 519–523]

Franks, J. S. (1970). An investigation of the fish population within the inland waters of Horn Island, Mississippi, a barrier island in the northern Gulf of Mexico. *Gulf Research Reports* **3**, 3–104. [pp. 35–36]

Fritz, F. (1963). *Unknown Florida*. University of Miami Press, Coral Gables, 213 pp. [pp. 89, 99, 193]

Froese, R. (2006). Cube law, condition factor and weight-length relationships: history, meta-analysis and recommendations. *Journal of Applied Ichthyology* **22**, 241–253.

Fu, C., J. M. Wilson, P. J. Rombough and C. J. Brauner (2014). Ions first: Na^+ uptake shifts from the skin to the gills before O_2 uptake in developing rainbow trout, *Oncorhynchus mykiss*. *Proceedings of the Royal Society* doi: 10.1098/rspb.2009.1545, 8 pp.

Fulton, T. W. (1904). The rate of growth of fishes. In *Twenty-second Annual Report of the Fishery Board for Scotland, Being for the Year 1903. Part III – Scientific Investigations*. Fishery Board for Scotland, Glasgow, pp. 141–241, 7 plates.

Fulton, T. W. (1911). *The Sovereignty of the Sea: An Historical Account of the Claims of England to the Dominion of the British Seas, and the Evolution of the Territorial Waters: With Special Reference to the Rights of Fishing and the Naval Salute*. William Blackwood and Sons, Edinburgh, xxvi + 799 pp.

Fyhn, H. J., R. N. Finn, M. Reith and B. Norberg (1999). Yolk protein hydrolysis and oocyte free amino acids as key features in the adaptive evolution of teleost fishes to seawater. *Sarsia* **84**, 451–456.

Galloway, J. C. (1941). Lethal effect of the cold winter of 1939–40 on marine fishes at Key West, Florida. *Copeia* **1941**, 118–119.

García, C. B. and O. D. Solano (1995). *Tarpon atlanticus* in Colombia: a big fish in trouble. *Naga (ICLARM Quarterly)* **18**(3), pp. 47–49.

García de León, F. J., C. D. Acuña Leal, I. R. Blandon and R. Ward (2002). Studies in conservation genetics in tarpon (*Megalops atlanticus*). II. Population structure of tarpon in the western Gulf of Mexico. *Contributions in Marine Science* **35**, 18–33.

Garcia-Moliner, G., I. Mateo, S. Maidment-Caseau, W. J. Tobias and B. Kojis (2002). Recreational chartered fishing activity in the U.S. Caribbean. *Proceedings of the Fifty-Third Annual Gulf and Caribbean Fisheries Institute* **53**, 307–317.

Gee, J. H. and J. B. Graham (1978). Respiratory and hydrostatic functions of the intestine of the catfishes *Hoplosternum thoracatum* and *Brochis splendens* (Callichthyidae). *Journal of Experimental Biology* **74**, 1–16.

Gehringer, J. W. (1958). Leptocephalus of the Atlantic tarpon, *Megalops atlanticus* Valenciennes, from offshore waters. *Quarterly Journal of the Florida Academy of Sciences* **21**, 235–240.

Gehringer, J. W. (1959). Early development and metamorphosis of the ten-pounder, *Elops saurus* Linnaeus. *Fishery Bulletin (U.S. Fish and Wildlife Service)* **59**, 619–647.

Geiger, S. P., J. J. Torres and R. E. Crabtree (2000). Air breathing and gill ventilation frequencies in juvenile tarpon, *Megalops atlanticus*: responses to changes in dissolved oxygen, temperature, hydrogen sulfide, and pH. *Environmental Biology of Fishes* **59**, 181–190.

Giacometti Mai, A. C., T. Fernandes A. Silva and J. Francisco A. Legat (2012). Assessment of the fish-weir fishery off the coast of Piauí State, Brazil. *Arquivos de Ciências do Mar, Fortaleza* **45**(2), 40–48.

Gill, T. (1907). The tarpon and lady-fish and their relatives. *Smithsonian Miscellaneous Collections* **48**(3), 31–46.

Gilmore, R. G. Jr. (1977). Fishes of the Indian River lagoon and adjacent waters, Florida. *Bulletin of the Florida State Museum (Biological Sciences)* **22**(3), 101–147. [p. 111]

Gilmore, R. G. Jr., C. J. Donohoe and D. W. Cooke (1981). *Fishes of the Indian River Lagoon and adjacent waters, Florida*. Technical Report No. 41, Harbor Branch Foundation, Fort Pierce, 28 pp.

Girard, C. (1858). Ichthyological notices. *Proceedings of the Academy of Natural Sciences of Philadelphia* **10**, 223–225. [p. 224]

Girling, P., J. Purser and B. Nowak (2003). Effects of acute salinity and water quality changes on juvenile greenback flounder, *Rhombosolea tapirina* (Günther, 1862). *Acta Ichthyologica et Piscatoria* **38**, 1–16.

Gonzalez, B. J. (2013). *Field Guide to Coastal Fishes of Palawan.* USAID Project Number 486-A-00-08-00042-10. World Wildlife Fund and Coral Triangle Initiative, Manila, 208 pp. [p. 27]

Goode, G. B. (1876). Catalogue of the Fishes of the Bermudas: Based Chiefly Upon the Collections of the United States National Museum. *Bulletin of the U.S. National Museum* (**5**), 1–82. [pp. 68–69]

Goode, G. B. (1887). *American Fishes, A Popular Treatise Upon the Game and Food Fishes of North America with Especial Reference to Habits and Methods of Capture.* Estes and Lauriat, Boston, xv + 496 pp. [pp. 406–407]

Gopinath, K. (1946). Notes on the larval and post-larval stages of fishes found along the Trivandrum coast. *Proceedings of the National Institute of Science, India* **12**, 7–21, 7 figs.

Gordon, B. L. (1974). *The Marine Fishes of Rhode Island*, 2nd edition. The Book and Tackle Shop, Watch Hill, 136 pp. [pp. 21–22]

Graham, J. B. (1997). *Air-breathing Fishes: Evolution, Diversity, and Adaptation.* Academic Press, San Diego, CA, xi + 299 pp. [pp. 28–29]

Greenwood, P. H., D. E. Rosen, S. H. Weitzman and G. S. Myers (1966). Phyletic studies of teleostean fishes, with a provisional classification of the living forms. *Bulletin of the American Museum of Natural History* **131**(Article 4), 341–455. [pp.355, 358–359, 362–364, 380]

Grey, Z. (1919). *Tales of Fishes.* Grosset and Dunlap, New York, 267 pp. [pp. 1–7]

Griswold, F. G. (1913). *Sport on Land and Water: Recollections of Frank Gray Griswold.* Plimpton Press, Norwood, Frontispiece + 163 pp. [pp. 91–108]

Griswold, F. G. (1921). *Some Fish and Some Fishing.* John Lane, New York, 251 pp. [pp. 23–24, 171]

Griswold, F. G. (1922). *The Tarpon.* Privately printed, place and publisher unknown, 35 pp. [8–10, 12–15]

Grosell, M. (2006). Intestinal anion exchange in marine fish osmoregulation. *Journal of Experimental Biology* **209**, 2813–2827.

Grubich, J. R. (2001). Prey capture in actinopterygian fishes: a review of suction feeding motor patterns with new evidence from an elopomorph fish, *Megalops atlanticus. American Zoologist* **41**, 1258–1265.

Gudger, E. W. (1937). An albino tarpon, *Tarpon atlanticus*, the only known specimen. American Museum Novitates (**944**), 4 pp.

Guggino, W. B. (1980a). Water balance in embryos of *Fundulus heteroclitus* and *F. bermudae* in seawater. *American Journal of Physiology* **238**, R36–R41.

Guggino, W. B. (1980b). Salt balance in embryos of *Fundulus heteroclitus* and *F. bermudae* adapted to seawater. *American Journal of Physiology* **238**, R42–R49.

Guigand, C. M. and R. G. Turingan (2002). Feeding behavior and prey capture kinematics of juvenile tarpon (*Megalops atlanticus*). *Contributions in Marine Science* **35**, 43–54.

Guindon, K. G. (2011). *Evaluating Lethal and Sub-lethal effects of Catch-and-release Angling in Florida's Central Gulf Coast Recreational Atlantic Tarpon* (Megalops atlanticus) *Fishery.* Doctoral Dissertation, University of South Florida, Tampa, 50 pp.

Gunter, G. (1938). Notes on invasion of fresh water by fishes of the Gulf of Mexico, with special reference to the Mississippi-Atchafalaya River system. *Copeia* **1938**, 69–72.

Gunter, G. (1941). Death of fishes due to cold on the Texas coast, January 1940. *Ecology* **22**, 203–208.

Gunter, G. (1942). A list of the fishes of the mainland of North and Middle America recorded from both freshwater and sea water. *American Midland Naturalist* **28**, 305–326.

Haldeman, W. N. (1892). The tarpon, or Silver King. In *American Game Fishes: Their Habits, Habitat, and Peculiarities; How, When, and Where to Angle for Them*, E. O. Shields (editor). Rand, McNally, Chicago, pp. 111–129.

Halkett, A. (1913). *Check List of the Fishes of the Dominion of Canada and Newfoundland.* C. H. Parmelee, Ottawa, 138 pp., 14 plates. [p. 45]

Hammerschlag, N., A. B. Morgan and J. E. Serafy (2010). Relative predation risk for fishes along a subtropical mangrove-seagrass ecotone. *Marine Ecology Progress Series* **401**, 259–267.

Hammerschlag, N., J. Luo, D. J. Irschick and J. S. Ault (2012). A comparison of spatial and movement patterns between sympatric predators: bull sharks (*Carcharhinus leucas*) and Atlantic tarpon (*Megalops atlanticus*). *PLoS One* **7**(9). doi: 10.1371/journal.pone.0045958, 14 pp.

Hammond, D. L. (2005). Tarpon, *Megalops atlanticus*. South Carolina Department of Natural Resources, Columbia, 5 pp.

Hansson Mild, K. and S. Løvtrup (1985). Movement and structure of water in animal cells. Ideas and experiments. *Biochimica et Biophysica Acta* **822**, 155–167.

Hare, J. A. and R. K. Cowen (1997). Size, growth, development, and survival of the planktonic larvae of *Pomatomus saltatrix* (Pisces: Pomatomidae). *Ecology* **78**, 2415–2431.

Hare, J. A., S. Thorrold, H. Walsh, C. Reiss, A. Valle-Levinson and C. Jones (2005). Biophysical mechanisms of larval fish ingress into Chesapeake Bay. *Marine Ecology Progress Series* **303**, 295–310.

Harris, S. A. and D. P. Cyrus (1995). Occurrence of fish larvae in the St. Lucia estuary, KwaZulu, Natal, South Africa. *South African Journal of Marine Science* **16**, 333–350.

Harrington, R. W. Jr. (1958). Morphometry and ecology of small tarpon, *Megalops atlantica* Valenciennes from the transitional stage through onset of scale formation. *Copeia* **1958**, 1–10, 2 plates.

Harrington, R. W. (1966). Changes through one year in the growth rates of tarpon, *Megalops atlanticus* Valenciennes, reared from mid-metamorphosis. *Bulletin of Marine Science* **16**, 863–883.

Harrington, R. W. Jr. and E. S. Harrington (1960). Food of larval and young tarpon, *Megalops atlantica*. *Copeia* **1960**, 311–319.

Harrington, R. W. Jr. and E. S. Harrington (1961). Food selection among fishes invading a high subtropical salt marsh: from onset of flooding through the progress of a mosquito brood. *Ecology* **42**, 646–666.

Harrington, R. W. Jr. and E. S. Harrington (1982). Effects on fishes and their forage organisms of impounding a Florida salt marsh to prevent breeding by salt marsh mosquitoes. *Bulletin of Marine Science* **32**, 523–531.

Harrison, T. D. (2001). Length-weight relationships of fishes from South African estuaries. *Journal of Applied Ichthyology* **17**, 46–48.

Harvey, B., C. Ross, D. Greer and J. Carolsfeld (editors) (1998). *Action Before Extinction: An International Conference on Conservation of Fish Genetic Diversity*. World Fisheries Trust, Victoria, vii + 259 pp. [p. 217 Table 1]

Hasler, A. D., A. T. Scholz and R. M. Horrall (1978). Olfactory imprinting and homing in salmon. *American Scientist* **66**, 347–355.

Hastie, G. D., B. Wilson, L. H. Tufft and P. M. Thompson (2003). Bottlenose dolphins increase breathing synchrony in response to boat traffic. *Marine Mammal Science* **19**, 74–84.

Hayes, D. B., J. K. T. Brodziak and J. B. O'Gorman (1995). Efficiency and bias of estimators and sampling designs for determining length-weight relationships of fish. *Canadian Journal of Fisheries and Aquatic Sciences* **52**, 84–92.

Hedrick, M. S. and D. R. Jones (1999). Control of gill ventilation and air-breathing in the bowfin *Amia calva*. *Journal of Experimental Biology* **202**, 87–94.

Heilner, V. C. (1953). *Salt Water Fishing*, Second Edition Revised. Alfred A. Knopf, New York, xviii + 329 + xxiv pp.

Heisler, N. (1993). Respiratory and ionic regulation in fish with changes of the environment. In *Aquaculture: Fundamental and Applied Research*, Coastal and Estuarine Studies Vol. 43, B. Lahlou and P. Vitiello (editors). American Geophysical Union, Washington, pp. 15–29.

Helfman, G. S., B. B. Collette and D. E. Facey (1997). *The Diversity of Fishes*. Blackwell Science, Malden, xii + 528 pp.

Herazo C., D., A. Torres P. and E. Olsen V. (2006). Análisis de la Composición y abundancia de la ictiofauna presente en la pesca del camarón rosado (*Penaeus notialis*) en el Golfo de Morrosquillo, Caribe Colombiano. *Revista MVZ Córdoba* **11**, 47–61.

Herbert, B. and J. Peeters (1995). Freshwater fishes of far north Queensland. Information Series Q195018, Department of Primary Industries, Brisbane, 74 pp. [p. 21]

Hernández, L. P., M. J. F. Barresi and S. H. Devoto (2002). Functional morphology and developmental biology of zebrafish: reciprocal illumination from an unlikely couple. *Integrative and Comparative Biology* **42**, 222–231.

Herre, A. W. (1933a). A *Check List* of fishes from Sandakan, British North Borneo. *Journal of the Pan-Pacific Research Institution* **8**(4), 2–5. [p. 2]

Herre, A. W. (1933b). A *Check List* of fishes from Dumaguete, Oriental Negros, P. I., and its immediate vicinity. *Journal of the Pan-Pacific Research Institution* **8**(4), 6–11.

Herre, A. W. (1953). *Check List of Philippine Fishes*. Research Report 20, Fish and Wildlife Service, U.S. Department of the Interior, Washington, 977 pp. [pp. 55–56]

Hickey, C. R. Jr., B. H. Young and J. W. Lester (1976). Tarpon from Montauk, New York. *New York Fish and Game Journal* **23**(2), 186–187.

Hildebrand, S. F. (1934). The capture of a young tarpon, *Tarpon atlanticus*, at Beaufort, North Carolina. *Copeia* **1934**, 45–46.

Hildebrand, S. F. (1937). The tarpon in the Panama Canal. *Scientific Monthly* **44**, 239–248.

Hildebrand, S. F. (1938). *A New Catalogue of the Fresh-water Fishes of Panama*. Publication 425, Zoological Series, Field Museum of Natural History, 22(4), 219–359. [p. 220]

Hildebrand, S. F. (1939). The Panama Canal as a passageway for fishes, with lists and remarks on the fishes and invertebrates observed. *Zoologica (NY)* **24**(3), 15–45.

Hildebrand, S. F. (1963). *Family* Elopidae. In *Fishes of the Western North Atlantic*, H. B. Bigelow (editor-in-chief). Mem oir, Sears Foundation for Marine Research, Yale University, New Haven (1)(Part 3), 111–131. [pp. 111–123]

Hildebrand, S. F. and W. C. Schroeder (1928). *Fishes of Chesapeake Bay*. *Bulletin of the U.S. Bureau of Fisheries* **43**(Part 1), 388 pp. [pp. 79–80]

Hirose, S., T. Kaneko, N. Naito and Y. Takei (2003). Molecular biology of major components of chloride cells. *Comparative Biochemistry and Physiology* **136B**, 593–620.

Hirst, A. G. (2012). Intraspecific scaling of mass to length in pelagic animals: ontogenetic shape change and its implications. *Limnology and Oceanography* **57**, 1579–1590.

Hitchcock G. (2002). Fish fauna of the Bensbach River, southwest Papua New Guinea. *Memoirs of the Queensland Museum* **48**, 119–122.

Hitchcock, G., M. A. Finn, D. W. Burrows and J. W. Johnson (2012). Fishes from fresh and brackish waters of islands in Torres Strait, far north Queensland. *Memoirs of the Queensland Museum* **56**, 13–24.

Hoenig, J. M. (1983). Empirical use of longevity data to estimate mortality rates. *Fishery Bulletin (NOAA)* **81**, 898–903.

Hoese, H. D. (1958). A partially annotated checklist of the marine fishes of Texas. *Publications of the Institute of Marine Science, University of Texas* **5**, 312–352. [p. 320]

Holeton, G. F. and N. Heisler (1978). Acid-base regulation by bicarbonate exchange in the gills after exhausting activity in the larger spotted dogfish *Scyliorhinus stellaris*. *Physiologist* **21**, 56. [Abstract]

Holeton, G. F. and N. Heisler (1983). Contribution of net ion transfer mechanisms to acid-base regulation after exhausting activity in the larger spotted dogfish (*Scyliorhinus stellaris*). *Journal of Experimental Biology* **103**, 31–46.

Hollister, G. (1939). Young *Megalops cyprinoides* from Batavia, Dutch East Indies, including a study of the caudal skeleton and a comparison with the Atlantic species *Tarpon atlanticus*. *Zoologica (NY)* **24**(28), 449–475.

Holstvoogd, C. (1936). *The Development of the Mesonephros in Some Teleosts*. Part I. Meijer's Boek en Handelsdrukkerij. Wormerveer, Netherlands, xvi + 98 pp., Literature Cited (2 pp.), 35 photographs. [pp. 3–5]

Holt, G. J., S. A. Holt and K. T. Frank (2005). What can historic tarpon scales tell us about the tarpon fishery collapse in Texas? *Contributions in Marine Science* **37**, 66–77.

Holzman, R., D. C. Collar, R. S. Mehta and P. C. Wainwright (2012). An integrative modeling approach to elucidate suction-feeding performance. *Journal of Experimental Biology* **215**, 1–13.

Horng, J-L., P-P. Hwang, T-H. Shih, Z-H. Wen, C-S. Lin and L-Y. Lin (2009). Chloride transport in mitochondrion-rich cells of euryhaline tilapia (*Oreochromis mossambicus*) larvae. *American Journal of Physiology – Cell Physiology* **297**, C845–C854.

Hossler, F. E. (1980). Gill arch of the mullet, *Mugil cephalus*. III: Rate of response to salinity change. *American Journal of Physiology* **238**, R160–R164.

Howells, R. G. (1985). Cold tolerance of juvenile tarpon in fresh water. *Annual Proceedings, Texas Chapter, American Fisheries Society* **8**, 26–34.

Howells, R. G. and G. P. Garrett (1992). Status of some exotic sport fishes in Texas waters. *Texas Journal of Science* **44**, 317–324.

Hsia, C. C. W., A. Schmitz, M. Lambertz, S. F. Perry and J. N. Maina (2013). Evolution of air breathing: oxygen homeostasis and the transitions from water to land and sky. *Comprehensive Physiology* **3**, 849–915.

Huang, C-Y., W. Lee and H-C. Lin (2008). Functional differentiation in the anterior gills of the aquatic air-breathing fish, *Trichogaster leeri*. *Journal of Comparative Physiology* **178B**, 111–121.

Huebert, K. B. and M. A. Peck (2014). A day in the life of fish larvae: modeling foraging and growth using quirks. *PLoS One* **9**(6), e98205, 17 pp.

Hughes, G. M. (1960). A comparative study of gill ventilation in marine teleosts. *Journal of Experimental Biology* **37**, 28–45.

Hughes, G. M. (1961). How a fish extracts oxygen from water. *New Scientist* **11**, 346–348.

Hulet, W. H. and R. C. Robins (1989). The evolutionary significance of the leptocephalus larva. In *Fishes of the Western North Atlantic*, Memoir No. 1, Part 9, Vol. 2, E. B. Böhlke, (editor-in-chief). Memoirs, Sears Foundation for Marine Research, Yale University, New Haven, pp. 669–677.

Hwang, P-P., Y-N. Tsai and Y-C. Tung (1994). Calcium balance in embryos and larvae of the freshwater-adapted teleost, *Oreochromis mossambicus*. *Fish Physiology and Biochemistry* **13**, 325–333.

Hwang, P-P., T-H. Lee and L-Y. Lin (2011). Ion regulation in fish gills: recent progress in the cellular and molecular mechanisms. *American Journal of Physiology – Regulatory, Integrative and Comparative Physiology* **301**, R28–R47.

Hyslop, N. L., D. J. Stevenson, J. N. Macey, L. D. Carlile, C. L. Jenkins and J. A. Hostetler *et al.* (2011). Survival and population growth of a long-lived threatened snake species, *Drymarchon couperi* (eastern indigo snake). *Population Ecology* doi 10:1007/s10144-011-0292-3, 12 pp.

Icardo, J. M. (2012). The teleost heart: a morphological approach. In *Ontogeny and Phylogeny of the Vertebrate Heart*, D. Sedmera and T. Wang (editors). Springer Science + Business Media, New York, pp. 35–53.

Ikoma, T., H. Kobayashi, J. Tanaka, D. Walsh and S. Mann (2003). Microstructure, mechanical, and biomimetic properties of fish scales from *Pagurus major*. *Journal of Structural Biology* **142**, 327–333.

Inoue, J. G., M. Miya, K. Tsukamoto and M. Nishida (2004). Mitogenomic evidence for the monophyly of elopomorph fishes (Teleostei) and the evolutionary origin of the leptocephalus larva. *Molecular Phylogenetics and Evolution* **32**, 274–286.

Ip, Y. K. and S. F. Chew (2010). Ammonia production, excretion, toxicity, and defense in fish: a review. *Frontiers in Physiology* **1**(134). doi: 10.3389/fphys.2010.00134, 20 pp.

Job, T. J. and P. I. Chacko (1947). Rearing of saltwater fish in freshwaters of Madras. *Indian Ecologist* **2**(1), 12–20, Plate 3.

Johansen, K., C. P. Mangum and G. Lykkeboe (1978). Respiratory properties of the blood of Amazon fishes. *Canadian Journal of Zoology* **56**, 898–906.

Jones, P. W., F. D. Martin and J. D. Hardy Jr. (1978). *Development of Fishes of the Mid-Atlantic Bight: An Atlas of Egg, Larval and Juvenile Stages, Vol. 1: Acipenseridae through Ictaluridae*. Biological Service Program FWS/OBS-78/12, U.S. Fish and Wildlife Service, Chesapeake Biological Laboratory, Solomons, 366 pp. [pp. 53–62]

Jones, R. E., R. J. Petrell and D. Pauly (1999). Using modified length-weight relationships to assess the condition of fish. *Aquacultural Engineering* **20**, 261–276.

Jordan, D. S. (1884). List of fishes collected at Key West, Florida, with notes and descriptions. *Proceedings of the U.S. National Museum* **7**, 103–150. [p. 107]

Jordan, D. S. and B. W. Evermann (1896). *The Fishes of North and Middle America: A Descriptive Catalog*, Part I. *Bulletin of the U.S. National Museum* **47**(Part 1), lx + 1240 pp. [pp. 408–410]

Jordan, D. S. and B. W. Evermann (1904). *American Food and Game Fishes: A Popular Account of all the Species Found in America North of the Equator, with Keys for Ready Identification, Life Histories and Methods of Capture*. The Nature Library, Vol. V. Doubleday, Page, New York, xlix + 572 pp. [pp. 84–86]

Jordan, D. S. and R. E. Richardson (1909). *A Catalog of the Fishes of the Island of Formosa, or Taiwan, Based on the Collections of Dr. Hans Sauter. Memoirs of the Carnegie Museum* **4**(4), 159–204, Plates LXIII–LXXIV. [p. 165]

Jucá-Chagas, R. and L. Boccardo (2006). The air-breathing cycle of *Hoplosternum littorale* (Hancock, 1828) (Siluriformes: Callichthyidae). *Neotropical Ichthyology* **4**, 371–373.

Junk, W. J. (1984). Ecology of the *várzea*, floodplain of Amazonian white-water rivers. In *The Amazon: Limnology and Landscape Ecology of a Mighty Tropical River and Its Basin*, H. Sioli (editor). Dr W. Junk, Dordrecht, Netherlands, pp. 215–243.

Ka Fai Tse, W. (2014). The role of osmotic stress transcription factor 1 in fishes. *Frontiers in Zoology* **11**(86), 6 pp.

Kang, C-K. and T-H. Lee (2014). Medaka villin 1-like protein (VILL) is associated with the formation of microvilli induced by decreasing salinities in the absorptive ionocytes. *Frontiers in Zoology* **11**(2), 14 pp.

Kang, C-K., W-K. Yang, S-T. Lin, C-C. Liu, H-M. Lin and H-H. Chen *et al.* (2013). The acute and regulatory phases of time-course changes in gill mitochondrion-rich cells of seawater-acclimated medaka (*Oryzias dancena*) when exposed to hypoosmotic environments. *Comparative Biochemistry and Physiology* **164A**, 181–191.

Kaplan, M. N. (1937). *Big Game Anglers' Paradise: A Complete, Non-technical Narrative-treatise on Salt Water Gamefishes and Angling in Florida and Elsewhere*. Liveright Publishing, New York, xiv + 400 pp. [pp. 89–136]

Kaufmann, R. (1990). Respiratory cost of swimming in larval and juvenile cyprinids. *Journal of Experimental Biology* **150**, 343–366.

Kawakami, Y., H. Oku, K. Nomura, S. Gorie and H. Ohta (2009). Metabolism of a glycosaminoglycan during metamorphosis in the Japanese conger eel, *Conger myriaster*. *Research Letters in Biochemistry*: doi 10.1155/209/251731, 5 pp.

Khodabandeh, S., M. S. Moghaddam and B. Abtahi (2009). Changes in chloride cell abundance, Na^+, K^+-ATPase immunolocalization and activity in the gills of golden grey mullet, *Liza aurata*, fry during adaptation to differend [*sic*] salinities. *Vakhteh Medical Journal* **11**, 49–54.

Killam, K. A., R. J. Hochberg and E. C. Rzemien (1992). *Synthesis of Basic Life Histories of Tampa Bay Species*. Technical Publication No. 10-92, Tampa Bay National Estuary Program, xvi + 239 pp. [pp. ix, 3–1 to 3–8 + table following p. 3–7]

Koffi, B. K., B. R. D. Aboua, T. Koné and M. Bamba (2014). Fish distribution in relation to environmental characteristics in the Aby-Tendo-Ehy lagoon system (southeastern Côte d'Ivoire). *African Journal of Environmental Science and Technology* **8**, 407–415.

Kottelat, M. (2013). *The Fishes of the Inland Waters of Southeast Asia: A Catalogue and Core Bibliography of the Fishes Known to Occur in Freshwaters, Mangroves and Estuaries.* Raffles Bulletin of Zoology 2013 (Supplement 27), 663 pp. [pp. 33–34]

Kowarsky, J. and A. H. Ross (1981). Fish movement upstream through a central Queensland (Fitzroy River) coastal fishway. *Australian Journal of Marine and Freshwater Research* **32**, 93–109.

Kramer, D. L. and J. B. Graham (1976). Synchronous air breathing, a social component of respiration in fishes. *Copeia* **1976**, 689–697.

Kramer, D. L. and M. McClure (1982). Aquatic surface respiration, a widespread adaptation to hypoxia in tropical freshwater fishes. *Environmental Biology of Fishes* **7**, 47–55.

Kramer, D. L., C. C. Lindsey, G. E. E. Moodie and E. D. Stevens (1978). The fishes and the aquatic environment of the central Amazon basin, with particular reference to respiratory patterns. *Canadian Journal of Zoology* **56**, 717–729.

Kulbicki, M., N. Guillemot and M. Amand (2005). A general approach to length-weight relationships for New Caledonian lagoon fishes. *Cybium* **29**, 235–252.

Kulkarni, C. V. (1983). Longevity of fish *Megalops cyprinoides* (Brouss). *Journal of the Bombay Natural History Society* **80**, 230–232.

Kulkarni, C. B. (1992). Recent observations on the longevity of *Megalops cyprinoides* (Brouss.). *Journal of the Bombay Natural History Society* **89**, 384.

Kushlan, J. A. and T. E. Lodge (1974). Ecological and distributional notes on the freshwater fish of southern Florida. *Florida Scientist* **37**(2), 110–127.

Kuthalingam, M. D. K. (1958). Studies on post larvae and feeding habits of some fishes found in the Madras plankton. *Journal of Madras University* **28B**, 1–11.

Lalèyè, P., A. Chikou, J-C. Philippart, G. Teugels and P. Vandewalle (2004). Étude de la diversité ichtyologique du basin du flueve Ouémé au Bénin (Afrique de l'Ouest). *Cybium* **28**, 329–339. [p. 334 Table 1]

Landry Jr., A. M. (2002). Historical overview of tarpon in Texas. *Contributions in Marine Science* **35**, 103. [Abstract]

Larson, H. K. (1999). Report to Parks Australia North, on the estuarine fish inventory of Kakadu National Park, Northern Territory, Australia. MAGNT Research Report No. 5, Museums and Art Galleries of the Northern Territory, Darwin, 50 pp. [p. 26]

Lauder, G. V. (1982). Patterns of evolution in the feeding mechanism of actinopterygian fishes. *American Zoologist* **22**, 275–285.

Layman, C. A. and B. R. Silliman (2002). Preliminary survey and diet analysis of juvenile fishes of an estuarine creek on Andros Island, Bahamas. *Bulletin of Marine Science* **70**, 199–210.

Leal-Flórez, J., M. Rueda and M. Wolff (2008). Role of the non-native fish *Oreochromis niloticus* in the long-term variations of abundance and species composition of the native ichthyofauna in a Caribbean estuary. *Bulletin of Marine Science* **82**, 365–380.

Lee, D. S., C. R. Gilbert, C. H. Hocutt, R. E. Jenkins, D. E. McAllister and J. R. Stauffer Jr. (1980). *Atlas of North American Freshwater Fishes.* Publication No. 1980-12, North Carolina Biological Survey. North Carolina State Museum of Natural History, Raleigh, x + 867 pp. [p. 57]

Lee, T-H., P-P. Hwang, H-C. Lin and F-L. Huang (1996). Mitochondria-rich cells in the branchial epithelium of the teleost, *Oreochromis mossambicus,* acclimated to various hypotonic environments. *Fish Physiology and Biochemistry* **15**, 513–523.

Lee, T-H., S-H. Feng, C-H. Lin and Y-H. Hwang (2003). Ambient salinity modulates the expression of sodium pumps in branchial mitochondria-rich cells of Mozambique tilapia, *Oreochromis mossambicus. Zoological Science* **20**, 29–36.

Leim, A. H. and W. B. Scott (1966). *Fishes of the Atlantic Coast of Canada.* Bulletin No. 155, Fisheries Research Board of Canada, Ottawa, 485 pp. [p. 85]

Leis, J. M. (2007). Behaviour as input for modelling dispersal of fish larvae: behaviour, biogeography, hydrodynamics, ontogeny, physiology and phylogeny meet hydrography. *Marine Ecology Progress Series* **347**, 185–193.

Leis, J. M., R. F. Piola, A. C. Hay, C. Wen and K-P. Kan (2009). Ontogeny of behaviour relevant to dispersal and connectivity in the larvae of two non-reef demersal, tropical fish species. *Marine and Freshwater Research* **60**, 211–223.

Lema, S. C. and G. A. Nevitt (2004). Evidence that thyroid hormone induces olfactory cellular proliferation in salmon during a sensitive period for imprinting. *Journal of Experimental Biology* **207**, 3317–3327.

Le Menn, M. (2011). About uncertainties in practical salinity calculations. *Ocean Science* **7**, 651–659.

Letcher, B. H., J. A. Rice, L. B. Crowder and K. A. Rose (1996). Variability in survival of larval fish: disentangling components with a generalized individual-based model. *Canadian Journal of Fisheries and Aquatic Sciences* **53**, 787–801.

Lewis, E. L. and R. G. Perkin (1981). The practical salinity scale 1978: conversion of existing data. *Deep-Sea Research* **28A**, 307–328.

Lewis, W. M. Jr. (1970). Morphological adaptations of cyprinodontoids for inhabiting oxygen deficient waters. *Copeia* **1970**, 319–326.

Ley, J. A. (2008). Indo-Pacific tarpon *Megalops cyprinoides*: a review and ecological assessment. In *Biology and Management of the World Tarpon and Bonefish Fisheries*, J. S. Ault (editor). CRC Press, Boca Raton, pp. 3–26.

Li, J., J. Eygensteyn, R. A. C. Lock, P. M. Verbost, A. J. H. van der Heijden and S. E. Wendelaar Bonga *et al.* (1995). Branchial chloride cells in larvae and juveniles of freshwater tilapia *Oreochromis mossambicus*. *Journal of Experimental Biology* **198**, 2177–2184.

Liem, K. F. (1989). Respiratory gas bladders in teleosts: functional conservatism and morphological diversity. *American Zoologist* **29**, 333–352.

Lin, C-H., C-L. Huang, C-H. Yang, T-H. Lee and P-P. Hwang (2004). Time-course changes in the expression of Na, K-ATPase and the morphometry of mitochondrion-rich cells in gills of euryhaline tilapia (*Oreochromis mossambicus*) during freshwater acclimation. *Journal of Experimental Zoology* **301A**, 85–96.

Lin, H-C. and W-T. Sung (2003). The distribution of mitochondria-rich cells in the gills of air-breathing fishes. *Physiological and Biochemical Zoology* **76**, 215–228.

Lin, L-Y., C-F. Weng and P-P. Hwang (2000). Effects of cortisol and salinity challenge on water balance in developing larvae of tilapia (*Oreochromis mossambicus*). *Physiological and Biochemical Zoology* **73**, 283–289.

Linares, C., D. F. Doak, R. Coma, D. Diaz and M. Zabala (2007). Life history and viability of a long-lived marine invertebrate: the octocorals *Paramuricea clavata*. *Ecology* **88**, 918–928.

Liu, S-f, W-f. Peng, P. Gao, M-j. Fu, H-z. Wu and M-k. Lu *et al.* (2010). Digenean parasites of Chinese marine fishes: a list of species, hosts and geographical distribution. *Systematic Parasitology* **75**, 1–52. [pp. 5, 28]

Loftus, W. F. and J. A. Kushlan (1987). *Freshwater Fishes of Southern Florida. Bulletin of the Florida State Museum* **31**(4), 147–334. [pp. 183, 185, 280, 284, 288, 291, 296]

Lorin-Nebel, V. Boulo, C. Bodinier and G. Charmantier (2006). The $Na^+/K^+/2Cl^-$ cotransporter in the sea bass *Dicentrarchus labrax* during ontogeny: involvement in osmoregulation. *Journal of Experimental Biology* **209**, 4908–4922.

Losse, G. F. (1968). The elopoid and clupeoid fishes of east African coastal waters. *Journal of the East Africa Natural History Society and National Museum* **27**, 77–115. [pp. 80–81]

Lowe (McConnell), R. H. (1962). The fishes of the British Guiana continental shelf, Atlantic coast of South America, with notes on their natural history. *Journal of the Linnean Society (Zoology)* **44**(301), 667–700. [pp. 678, 693]

Lugger, O. (1878). Addition to the list of fishes of Maryland previously published. See Reports, January 1st, 1876, and January, 1877. In *Report of the Commissioner of Fisheries of Maryland, January, 1878*. George Colton, Annapolis, pp. 107–125. [pp. 121–122]

Luo, J., J. S. Ault, M. F. Larkin, R. Humston and D. B. Olson (2008b). Seasonal migratory patterns and vertical habitat utilization of Atlantic tarpon (*Megalops atlanticus*) from satellite PAT tags. In *Biology and Management of the World Tarpon and Bonefish Fisheries*, J. S. Ault (editor). CRC Press, Boca Raton, pp. 275–299.

MacIntyre, F. (1976). Concentration scales: a plea for physico-chemical data. *Marine Chemistry* **4**, 205–224.

MacKenzie, D. S. and C. Chavez (2002). The challenges of spawning tarpon in captivity. *Contributions in Marine Science* **35**, 104. [Abstract]

MacPherson, G. A. H. (1935). A record tarpon from Nigeria. *The Field* **CLXV** (4288), 447.

Mamonekene, V., S. Lavoué, O. S. G. Pauwels, J. Hervé Mve Beh, J-E. Mackayah and L. Tchignoumba (2006). Fish diversity at Rabi and Gamba, Ogooué-Maritime Province, Gabon. *Bulletin of the Biological Society of Washington* (**12**), 285–296. [p. 287 Table 2]

Mancera, J. M. and S. D. McCormick (2000). Rapid activation of gill Na^{+}+ K^{+}-ATPase in the euryhaline teleost *Fundulus heteroclitus*. *Journal of Experimental Zoology* **287**, 263–274.

Mancera, J. M., J. M. Pérez-Fígares and P. Fernández-Llebrez (1994). Effect of cortisol on brackish water adaptation in the euryhaline gilthead sea bream (*Sparus aurata* L.). *Comparative Biochemistry and Physiology* **107A**, 397–402.

Mann, B. Q. (editor) (2000). *South African Linefish Status Reports*. Oceanographic Special Publication No. 7. Research Institute, Durban, iii + 260 pp. [pp. 66–67]

Marín, G. (2000). Ichthyofauna and fisheries of the Unare Lagoon, Estado Anzoátegui, Venezuela. *Acta Biologica Venezuelica* **20**(3), 61–93. [p. 64]

Marín Erausquin, G. C. (2007). *Propuesta para la Ordenacion de la Pesca artisanal en el Parque nacional Jaragua, Republica Dominicana*. Fundación Interuniversitaria, Madrid, 94 pp. [p. 15]

Marshall, W. S. (2003). Rapid regulation of NaCl secretion by estuarine teleost fish: coping strategies for short-duration freshwater exposures. *Biochimica et Biophysica Acta* **1618**, 95–105.

Martin, W. R. (1949). *The mechanics of environmental control of body form in fishes*. University of Toronto Studies, Biological Series No. 58. Publications of the Ontario Fisheries Research Laboratory (70), 72 pp. + appendices.

Marwitz, S. R. (1986). Young tarpon in a roadside ditch near Matagorda Bay in Calhoun County, Texas. Management Data Series No. 100, Texas Parks and Wildlife Department, Coastal Fisheries Branch, 8 pp.

Matamoros, W. A., J. F. Schaefer and B. R. Kreiser (2009). Annotated checklist of the freshwater fishes of continental and insular Honduras. *Zootaxa* **2307**, 38 pp. [pp. 7, 33]

Mateos-Molina, D., M. Schärer-Umpierre, M. Nemeth, H. Ruiz, I. Ruiz-Valentín and J. Vargas-Santiago (2013). Ecology and distribution of tarpons (*Megalops atlanticus*) at the Boquerón Wildlife Refuge, Puerto Rico. Proceedings of the 65th Gulf and Caribbean Fisheries Institute, 5–9 November 2012, Santa Marta, pp. 262–265.

Matey, V., J. G. Richards, Y. Wang, C. M. Wood, J. Rogers and R. Davies *et al.* (2008). The effect of hypoxia on gill morphology and ionoregulatory status in the Lake Qinghai scaleless carp, *Gymnocypris przewalskii*. *Journal of Experimental Biology* **211**, 1063–1074.

McAllister, D. E. (1990). *A List of the Fishes of Canada. Syllogeus* (**64**), 310 pp. [p. 44]

McCormick, S. D. (2001). Endocrine control of osmoregulation in teleost fish. *American Zoologist* **41**, 781–794.

McCulloch, A. R. (1929). *A Check-list of the Fishes Recorded from Australia*. Memoir V. The Australian Museum, Sydney, 436 pp. [p. 34]

McMillen-Jackson, A. L., T. M. Bert, S. Seyoum, T. Orsoy, H. Cruz-Lopez and R. E. Crabtree (2002). Geographic variation in molecular and morphometric characters of tarpon (*Megalops atlanticus* Valenciennes) in the North Atlantic Ocean. *Contributions in Marine Science* **35**, 104. [Abstract]

McMillen-Jackson, A. L., T. M. Bert, H. Cruz-Lopez, S. Seyoum, T. Orsoy and R. E. Crabtree (2005). Molecular genetic variation in tarpon (*Megalops atlanticus* Valenciennes) in the northern Atlantic Ocean. *Marine Biology* **146**, 253–262.

Medillín-Mora, J., A. Polanco F. and G. R. Navas S. (2013). Inventario de larvas de peces registradas para el Caribe colombiano. *Boletín de Investigaciones Marinas y Costeras* **42**, 233–253. [pp. 234, 236]

Meek, A. (1916). *The Migrations of Fish*. Edward Arnold, London, xviii + 427 pp. [pp. 64–65]

Meek, S. E. and S. F. Hildebrand (1923). *The Marine Fishes of Panama*. Publication No. 215, Zoological Series Vol. 15 (Part I). Field Museum of Natural History, Chicago, xi + 330 pp. [pp. 173–175].

Mello Lopes, J. (2013). Ensaio preliminary sobre o cultivo do camurupim (*Megalops atlanticus*) em diferentes densidades de estocagem. In *Pesquisadores do Maranhão*. Fundação de Amparo à Pesquisa e ao Desenvolvimento Científico e Tecnológico do Maranhão (FAPEMA), São Luís, p. 43.

Mercado Silgado, J. E. (1969). La cría de sábalo *Megalops atlanticus* (Pisces Megalopidae) desde leptocefalo hasta juvenil en acuario y lagunas naturales. Centro de Investigaciones de Ciencias Marinas INDERENA, Cartagena, 20 pp.

Mercado Silgado, J. E. (1971). Notas sobre los estados larvales del sábalo *Megalops atlanticus* Valenciennes, con comentarios sobre su importancia comercial. Boletín Museo del Mar (Universidad del Bogotá*)* 2, 1–28.

Mercado-Silgado, J. E. (1981). Los peces comerciales del Golfo de Morrosquillo. *Divulgación Pesquera (INDERENA-Rev., Bogotá)* **15**(2), 1–9.

Mercado-Silgado, J. E. and A. Ciardelli (1972). Contribución a la morfología y organogenésis de los leptocéfalos del sábalo *Megalops atlanticus* (Pisces: Megalopidae). *Bulletin of Marine Science* **22**, 153–184.

Mercado Silgado, J. (2002). Culture of leptocephali and juveniles of tarpon (*Megalops atlanticus*) collected from wild stocks off the Caribbean coast of Colombia. *Contributions in Marine Science* **35**, 96–99.

Merrick, J. R. and G. E. Schmida (1984). *Australian Freshwater Fishes: Biology and Management*. Griffin Press, Netley, 409 pp. [pp. 52–54]

Miller, M. J., T. Otake, J. Aoyama, S. Wouthuyzen, S. Suharti and H. Y. Sugeha *et al.* (2011). Observations of gut contents of leptocephali in the North Equatorial Current and Tomini Bay, Indonesia. *Coastal Marine Science* **35**, 277–288.

Miller, M. J., Y. Chikaraishi, N. O. Ogawa, Y. Yamada, K. Tsukamoto and N. Ohkouchi (2013). A low trophic position of Japanese eel larvae indicates feeding on marine snow. *Biology Letters*: doi 10.1098/rsbl.2012.0826, 5 pp.

Millero, F. J. (1993). What is psu? *Oceanography* **6**(3), 67.

Millward, D. J. (1995). A protein-stat mechanism for regulation of growth and maintenance of the lean body mass. *Nutrition Research Reviews* **8**, 93–120.

Mir, I. H., A. Cana and S. Nabi (2011). Chloride and pavement cells in the gills of snow trout, *Schizothorax curvifrons* Heckel (Teleost [*sic*], Cypriniformes, Cyprinidae). *Cytologia* **76**, 439–444.

Moffett, A. W. and J. E. Randall (1957). The Roger Firestone tarpon investigation. University of Miami Marine Laboratory Progress Report 57–22, 23 pp.

Mol, J. H., D. Resida, J. S. Ramlal and C. R. Becker (2000). Effects of El Niño-related drought on freshwater and brackish-water fishes in Suriname, South America. *Environmental Biology of Fishes* **59**, 429–440.

Moloney, C. A. (1883). *West African Fisheries: With Particular Reference to the Gold Coast Colony.* William Clowes and Sons, London, 79 pp. [p. 7]

Monod, T. (1927). Contribution à l'étude de la faune du Cameroun. In *Faune des colonies françaises,* Vol.1, A. Gruvel (editor). Société d'éditions géographiques, maritimes et colonials, Paris, pp. 643–742. [p. 654]

Monteiro, S. M., A. Fontainhas-Fernandes and M. Sousa (2010). An immunohistochemical study of gill epithelium cells in the Nile tilapia, *Oreochromis niloticus. Folia Histochemica et Cytobiologica* **48**, 112–121.

Moore, G. I., S. M. Morrison, J. B. Hutchins, G. R. Allen and A. Sampey (2014). Kimberley marine biota. Historical data: fishes. *Records of the Western Australian Museum* (Supplement **84**), 161–206. [p. 174]

Moore, R. H. (1975). Occurrence of tropical marine fishes at Port Aransas, Texas 1967–1973, related to sea temperatures. *Copeia* **1975**, 170–172.

Morgan, D. L., M. G. Allen, P. Bedford and M. Horstman (2004). Fish fauna of the Fitzroy River in the Kimberley region of Western Australia – including the Bunuba, Gooniyandi, Ngarinyin, Nyikina and Walmajarri aboriginal names. *Records of the Western Australian Museum* **22**, 147–161.

Moseley, A. (1877). *Annual Report of the Fish Commissioner of the State of Virginia for the Year 1877.* R. F. Walker, Richmond, 60 pp. [p. 9]

Moses, S. T. (1923). A statistical account of the fish supply of Madras. In *Bulletin No. XV, Fishery Reports for 1922,* J. Hornell (editor). Madras Fisheries Department, Superintendent, Government Press, Madras, pp. 131–166. [pp. 134, 138, 159]

Movahedinia, A. A., A. Savari, H. Morovvati, P. Koochanin, J. G. Marammazi and M. Nafisi (2009). The effects of changes in salinity on gill mitochondria-rich cells of juvenile yellowfin seabream, *Acanthopagrus latus. Journal of Biological Sciences* **9**, 710–720.

Moyle, P. B. and J. J. Cech Jr. (1982). *Fishes: An Introduction to Ichthyology.* Prentice-Hall, Englewood Cliffs, xiv + 593 pp.

Munro, I. S R. (1955). *The Marine and Fresh Water Fishes of Ceylon.* Department of External Affairs, Canberra, xvi + 351 pp., 56 plates. [p. 22–23]

Munro, I. S. R. (1967). *The Fishes of New Guinea.* Department of Agriculture, Stock and Fisheries, Port Moresby, xxxvii + 650 pp, 78 plates. [p. 41]

Murdy, E. O., R. S. Birdsong and J. A. Musick (1997). *Fishes of Chesapeake Bay.* Smithsonian Institution Press, xi + 324 pp., 49 plates. [p. 60]

Nakada, T., C. M. Westhoff, A. Kato and S. Hirose (2007). Ammonia secretion from fish gill [*sic*] depends on a set of Rh glycoproteins. *FASEB Journal* **21**, 1067–1074.

Nash, R. D. M., A. H. Valencia and A. J. Geffen (2006). The origin of Fulton's condition factor – setting the record straight. *Fisheries* **31**(5), 236–238.

Neal, J. W., C. G. Lilyestrom and T. J. Kwak (2009). Factors influencing tropical island freshwater fishes: species, status, and management implications in Puerto Rico. *Fisheries* **34**(11), 546–554.

Nelson, D. M. (editor), M. E. Monaco, C. D. Williams, T. E. Czapla, M. E. Pattillo and L. Coston-Clements *et al.* (1992). *Distribution and Abundance of Fishes and Invertebrates in Gulf of Mexico Estuaries, Vol. I: Data Summaries.* ELMR Report Number 10. NOAA/NOS Strategic Environmental Assessments Division, Rockville, 273 pp. [pp. 9, 10, 61, 84–94, 151, 215]

Nelson, J. A. and A. M. Dehn (2011). The GI tract in air breathing. In *Fish Physiology,* Vol. 30, M. Grosell, A. P. Farrell and C. J. Brauner (editors). Academic Press, London, pp. 395–433.

Nevitt, G. and A. Dittman (1999). A new model for olfactory imprinting in salmon. *Integrative Biology* **1**, 215–223.

Nichols, J. T. (1912). Notes on Cuban fishes. *Bulletin of the American Museum of Natural History* **31**(18), 179–194. [p. 181]

Nichols, J. T. (1921). A list of Turk Islands fishes, with a description of a new flatfish. *American Museum Novitates* **44**(3), 21–24 + 1 plate.

Nichols, J. T. (1929). *The Fishes of Porto Rico and the Virgin Islands*. In *Scientific Survey of Porto Rico and the Virgin Islands*. New York Academy of Sciences, New York, pp. 161–535. [pp. 198–199]

Nichols, J. T. (1943). *The Fresh-Water Fishes of China*. American Museum of Natural History, New York, xxxvi + 322 pp. [pp. 17–18]

Nichols, J. T. and C. M. Breder Jr. (1927). *The Marine Fishes of New York and Southern New England*. *Zoologica (NY)* IX (1), 192 pp. [p. 33]

Nichols, J. T. and C. H. Pope (1927). The fishes of Hainan. *Bulletin of the American Museum of Natural History* **54**(2), 321–394. [p. 325]

Nielsen, R. and P. Munk (2004). Growth pattern and growth dependent mortality of larval and pelagic juvenile North Sea cod *Gadus morhua*. *Marine Ecology Progress Series* **278**, 261–270.

Nietschmann, B. (1973). *Between Land and Water: The Subsistence Ecology of the Miskito Indians, Eastern Nicaragua*. Seminar Press, New York, vii + 279 pp.

Noble, A. (1973). Food and feeding of the post-larvae and juveniles of *Megalops cyprinoides* (Brouss.). *Indian Journal of Fisheries* **20**, 203–204.

Nordlie, F. G. and D. Kelso (1975). Trophic relationships in a tropical estuary. *Revista de Biología Tropical* **23**, 77–99.

Norman, J. R. and F. C. Fraser (1938). *Giant Fishes Whales and Dolphins*. W. W. Norton, New York, xxvii + 361 pp. [pp. 93–97]

Norton, S. F. and E. L. Brainerd (1993). Convergence in the feeding mechanics of ecomorphologically similar species in the Centrarchidae and Cichilidae. *Journal of Experimental Biology* **176**, 11–29.

O'Bryan, P. D., D. B. Carlson and R. G. Gilmore (1990). Salt marsh mitigation: an example of the process of balancing mosquito control, natural resource, and development interests. *Florida Scientist* **53**, 189–203.

Ohata, R., R. Masuda, M. Ueno, Y. Fukunishi and Y. Yamashita (2011). Effects of turbidity on survival of larval ayu and red sea bream exposed to predation by jack mackerel and moon jellyfish. *Fisheries Science* **77**, 207–215.

Okamura, A., Y. Yamada, N. Mikawa, N. Horie and K. Tsukamoto (2012). Effect of starvation, body size, and temperature on the onset of metamorphosis in Japanese eel (*Anguilla japonica*). *Canadian Journal of Zoology* **90**, 1378–1385.

Okoro, C. B., P. E. Anyanwu, G. Oladosu, O. A. Okunade, K. A. Musa-Agemo and E. P. Anwa *et al.* (2009). Culture trial of *Megalops atlanticus* in plastic tanks pond [*sic*]. In *2009 Annual Report*. Nigerian Institute for Oceanography and Marine Research, Victoria Island, Lagos, pp. 46–50.

Okoro, B. C., P. E. Anyanwu, G. Oladosu and K. A. Musa-Agemo (2010). Research on growth, reproduction and nutritional requirements of *Tarpon atlanticus*. In *2010 Annual Report*. Nigerian Institute for Oceanography and Marine Research, Victoria Island, Lagos, pp. 17–20.

Otake, T. K. Nogami and K. Maruyaka (1990). Possible food sources for eel leptocephali. *La Mer* **28**, 218–224.

Otake, T., K. Nogami and K. Maruyama (1993). Dissolved and particulate organic matter as possible food sources for eel leptocephali. *Marine Ecology Progress Series* **92**, 27–34.

Ouattara, N'G., S. Ouattara, Y. Bamba and K. Yao (2014). Influence de la salinité sur la structure des branchies et l'ultrastructure des ionocytes chez le tilapia *Sarotherodon melanotheron heudelotii* provenant d'un estuaire hypersalé (Saloum, Sénégal). *Journal of Applied Biosciences* **79**, 6808–6817.

Oufiero, C. E., R. A. Holzman, F. A. Young and P. C. Wainwright (2012). New insights from serranid fishes on the role of trade-offs in suction-feeding diversification. *Journal of Experimental Biology* **215**, 3845–3855.

Overstreet, R. M. (1974). An estuarine low-temperature fish-kill in Mississippi, with remarks on restricted necropsies. *Gulf Coast Research Reports* **4**, 328–350.

Oviedo Perez, J. L., L. Gonzales Ocaranza, E. Varga Molinar, R. Morales Hernandez, J. A. Hernandez Valencia and L. Dominguez Trejo *et al.* (2002). Sport fishing of tarpon (*Megalops atlanticus*) along the Mexican coast of the Gulf of Mexico. *Contributions in Marine Science* **35**, 105–106. [Abstract]

Packard, G. C. (1974). The evolution of air breathing in Paleozoic gnathostome fishes. *Evolution* **28**, 320–325.

Padmaja, G. and L. M. Rao (2001). Breeding biology of *Megalops cyprinoides* from Visakhapatnam coast. *Journal of Environmental Biology* **2**, 91–99.

Palumbi, S. R. (2004). Why mothers matter. *Nature* **430**, 621–622.

Pandian, T. J. (1968). Feeding habits of the fish *Megalops cyprinoides* Broussonet, in the Cooum backwaters, Madras. *Journal of the Bombay Natural History Society* **65**, 569–580.

Pankhurst, N. W. (2011). The endocrinology of stress in fish: an environmental perspective. *General and Comparative Endocrinology* **170**, 265–275.

Pankhurst, N. W. and D. F. Sharples (1992). Effects of capture and confinement on plasma cortisol concentrations in the snapper, *Pagrus auratus*. *Australian Journal of Marine and Freshwater Research* **43**, 345–356.

Pardo, S. A., A. B. Cooper and N. K. Dulvy (2013). Avoiding fishy growth curves. *Methods in Ecology and Evolution*. doi: 10.1111/2041-210x.12020, 8 pp.

Paris, C. B. and R. K. Cowen (2004). Direct evidence of a biophysical retention mechanism for coral reef fish larvae. *Limnology and Oceanography* **49**, 1964–1979.

Park, K., W. Kim and H-Y. Kim (2014). Optimal lamellar arrangement in fish gills. *Proceedings of the National Academy of Sciences* **111**, 8067–8070.

Parrish, J. K., S. V. Viscido and D. Grünbaum (2002). Self-organized fish schools: an examination of emergent properties. *Biological Bulletin* **202**, 296–305.

Pellegrin, J. (1923). *Poissons des Eaux Douces de l'Afrique Occidentale (du Sénégal au Niger)*. Émile Larose, Paris, 373 pp. [44–45]

Pelletier, C., K. C. Hanson and S. J. Cooke (2007). Do catch-and-release guidelines from state and provincial *Fisheries* agencies in North America conform to scientifically based best practice? *Environmental Management* **39**, 760–773.

Penrith, M. J. (1976). Distribution of shallow water marine fishes around southern Africa. *Cimbebasia (Journal of the State Museum, Windhoek), Series A* **4**(7), 137–154.

Penrith, M. J. (1978). An annotated check-list of the inshore fishes of southern Angola. *Cimbebasia (Journal of the State Museum, Windhoek), Series A* **4**(11), 179–190.

Perna, C. N., M. Cappo, B. J. Pusey, D. W. Burrows and R. G. Pearson (2011). Removal of aquatic weeds greatly enhances fish community richness and diversity: an example from the Burdekin River floodplain, tropical Australia. *River Research and Applications*. doi: 10.1002/rra.1505, 12 pp.

Perry, S. F. and G. McDonald (1993). Gas exchange. In *The Physiology of Fishes*, D. H. Evans (editor). CRC Press, Boca Raton, pp. 251–278.

Perry, S. F., C. Daxboeck, B. Emmett, P. W. Hochachka and R. W. Brill (1985). Effects of exhausting exercise on acid-base regulation in skipjack tuna (*Katsuwonus pelamis*) blood. *Physiological Zoology* **58**, 421–429.

Pew, P. (1971). *Food and Game Fishes of the Texas Coast*, 6th Printing. Bulletin 33, Texas Parks and Wildlife Department, Austin, 70 pp. [pp. 18–19]

Pfeiler, E. (1984). Glycosaminoglycan breakdown during metamorphosis of larval bonefish *Albula*. *Marine Biology* Letters **5**, 241–249.

Pfeiler, E. (1986). Towards an explanation of the developmental strategy in leptocephalous [*sic*] larvae of marine teleost fishes. *Environmental Biology of Fishes* **5**, 3–13.

Pfeiler, E. (1996). Energetics of metamorphosis in bonefish (*Albula* sp.) leptocephali: role of keratan sulfate glycosaminoglycan. *Fish Physiology and Biochemistry* **15**, 359–362.

Pfeiler, E. (1999). Developmental physiology of elopomorph leptocephali. *Comparative Biochemistry and Physiology* **123A**, 113–128.

Pfeiler, E., J. Donnelly, J. J. Torres and R. E. Crabtree (1991). Glycosaminoglycan composition of tarpon (*Megalops atlanticus*) and ladyfish (*Elops saurus*) leptocephali. *Journal of Fish Biology* **39**, 613–615.

Pfeiler, E. and J. J. Govoni (1993). Metabolic rates in early life history stages of elopomorph fishes. *Biological Bulletin* **185**, 277–283.

Phillip, D. A. T., D. C. Taphorn, E. Holm, J. F. Gilliam, B. A. Lamphere and H. López-Fernández (2013). Annotated list and key to the stream fishes of Trinidad & Tobago. *Zootaxa* **3711** (1), 64 pp. [pp. 7, 12, 14, 27]

Pinto Paiva, M. and M. Ferreira de Menezes (1963). Estudo biométrico de alevinos do camurupim, *Megalops atlanticus* Val., 1846. *Arquivos da Estação de Biologia Marinha da Universidade Federal do Ceará* **3**(2), 57–64.

Planquette, P., P. Keith and P-Y. Le Bail (1996). *Atlas des Poissons d'Eau Douce de Guyane*, Tome 1. Muséum National d'Histoire Naturelle, Paris, 429 pp. [pp. 54–55]

Playfair, R. L. (1866). *The Fishes of Zanzibar*. John van Voorst, London, xiv + 153 pp. [p. 122]

Pollard, D. A. (1980). Family Megalopidae. Oxeye herring. In *Freshwater Fishes of South-Eastern Australia (New South Wales, Victoria and Tasmania)*, R. M. McDowall (editor). A. H. and A. W. Reed Pty., Sydney, pp. 53–54.

Pope, K. L. and C. G. Kruse (2007). Condition. In *Analysis and Interpretation of Freshwater Fisheries Data*, C. S. Guy and M. L. Brown (editors). American Fisheries Society, Bethesda, pp. 423–471.

Poulakis, G. R., J. M. Shenker and D. S. Taylor (2002). Habitat use by fishes after tidal reconnection of an impounded estuarine wetland in the Indian River Lagoon, Florida (USA). *Wetlands Ecology and Management* **10**, 51–69.

Powell, J. H. and R. E. Powell (1999). The freshwater ichthyofauna of Bougainville Island, Papua New Guinea. *Pacific Science* **53**, 346–356.

Power, J. H. and P. J. Walsh (1992). Metabolic scaling, buoyancy, and growth in larval Atlantic menhaden *Brevoortia tyrannus*. *Marine Biology* **112**, 17–22.

Power, T. and T. Marsden (2007). Tedlands Fish Fauna Survey: Tedlands Wetland. Queensland Department of Primary Industries and Fisheries, Brisbane, ii + 38 pp. [pp. 10, 12, 27]

Prince, P. A., P. Rothery, J. P. Croxall and A. G. Wood (1994). Population dynamics of black-browed and grey-headed albatrosses, *Diomedea melanophris* and *D. chrystoma* at Bird Island, South Georgia. *Ibis* **136**, 50–71.

Pritchett, J. T. (2008). Tarpon and bonefish fishery on Turneffe Atoll, Belize. In *Biology and Management of the World Tarpon and Bonefish Fisheries*, J. S. Ault (editor). CRC Press, Boca Raton, pp. 99–102.

Pusey, B. J. and A. H. Arthington (2003). Importance of the riparian zone to the conservation and management of freshwater fish: a review. *Marine and Freshwater Research* **54**, 1–16.

Pusey, B., M. Kennard and A. Arthington (2004). *Freshwater Fishes of North-Eastern Australia*. CSIRO Publishing, Collingwood, Victoria, xiv + 684 pp. [pp. 64–70].

Qualia, S. (2002). Tarpon tagging in the Gulf of Mexico. *Contributions in Marine Science* **35**, 106–107. [Abstract]

Quero, J. C. (1998). Changes in the Euro-Atlantic fish species composition resulting from fishing and ocean warming. *Italian Journal of Zoology* **65**(Supplement), 493–499.

Quero, J-C. and G. Delmas (1982). Captures au large de la Côte Basque du tarpon *Tarpon atlanticus* (Valenciennes, 1847) (Pisces, Elopiformes, Megalopidae), espèce nouvelle pour la faune française. *Cybium* **6**(3), 34.

Quero, J-C., P. Decamps, G. Delmas, M. Duron and J. Fonteneau (1982). Observations ichtyologiques effectuées en 1981). *Annales de la societe des Sciences naturelles de la Charente-Maritime, La Rochelle* **6**(9), 1021–1028.

Radcliffe, L. (1916). An extension of the recorded range of three species of fishes in New England waters. *Copeia* **26**, 2–3.

Rahman, A. K. A. (1989). *Freshwater Fishes of Bangladesh*. Zoological Society of Bangladesh, Department of Zoology, University of Dhaka, Dhaka, xvii + 364 pp. [pp. 234–236].

Raimundo Bastos, J., T. Teixeira Alves, C. A. Esteves Araripe and F. J. Siqueira Telles (1973). Sobre a elaboração de conservas de pescado em leite de côco e em óleos de algodão e de babaçu. *Arquivos de Ciências do Mar* **13**(1), 25–29.

Raj, B. S. (1916). *Notes on the freshwater fish of Madras. Records of the Indian Museum* 12, 249–294, Plates XXV–XXIX. [pp.252–253]

Ramsundar, H. (2005). The distribution and abundance of wetland ichthyofauna, and exploitation of the *Fisheries* in the Godineau Swamp, Trinidad – case study. *Revista de Biología Tropical* **53** (Supplement 1), 11–23.

Randall, D. (1982). The control of respiration and circulation in fish during exercise and hypoxia. *Journal of Experimental Biology* **100**, 275–288.

Randall, D. J. and H. Lin (1993). Effects of variations in water pH on fish. In *Aquaculture: Fundamental and Applied Research*, Coastal and Estuarine Studies Vol. 43, B. Lahlou and P. Vitiello (editors). American Geophysical Union, Washington, pp. 31–45.

Randall, J. E. (1967). Food habits of reef fishes of the West Indies. *Studies in Tropical Oceanography, Miami* **5**, 665–847. [p. 684?]

Randall, J. E. and A. W. Moffett (1958). The tarpon has many secrets. *Sea Frontiers* **4**(3), 136–146.

Rantin, F. T. and A. L. Kalinin (1996). Aquatic surface respiration (ASR) and cardiorespiratory responses of *Piaractus mesopotamicus* (Teleostei, Serrasalmidae), during graded and acute hypoxia. In *International Congress on the Biology of Fish, 1996*. The Physiology of Tropical Fishes Symposium Proceedings. American *Fisheries* Society, San Francisco, pp. 59–67.

Reddy, P., N. C. James, A. K. Whitfield and P. D. Cowley (2011). Marine connectivity and fish length frequencies of selected species in two adjacent temporarily open/closed estuaries in the eastern Cape Province, South Africa. *African Zoology* **46**, 239–245.

Richards, J. G. (2011). Physiological, behavioral and biochemical adaptations of intertidal fishes to hypoxia. *Journal of Experimental Biology* **214**, 191–199.

Richards, W. J. (1969). Elopoid leptocephali from Angolan waters. *Copeia* **1969**, 515–518.

Richards, W. J. (2006). Introduction. In *Early Stages of Atlantic Fishes: An Identification Guide for the Western Central North Atlantic*, Vol. I, W. J. Richards (editor). Taylor and Francis, Boca Raton, pp. 1–14.

Richardson, J. (1846). *Report on the Ichthyology of the Seas of China and Japan*. Richard and John E. Taylor, London, pp. 187–320. [p. 310–311]

Rickards, W. L. (1968). Ecology and growth of juvenile tarpon, *Megalops atlanticus*, in a Georgia salt marsh. *Bulletin of Marine Science* **18**, 220–239.

Roberts, T. R. (1972). Ecology of fishes in the Amazon and Congo basins. *Bulletin of the Museum of Comparative Zoology* **143**, 117–147.

Roberts, T. R. (1978). An ichthyological survey of the Fly River in Papua New Guinea with descriptions of new species. *Smithsonian Contributions to Zoology* **281**, vi + 72 pp.

Rocha, L. A. and I. L. Rosa (2001). Baseline assessment of reef fish assemblages of Parcel Manuel Luiz Marine State Park, Maranhão, north-east Brazil. *Journal of Fish Biology* **58**, 985–998.

Röhl, E. (1956). *Fauna Descriptiva de Venezuela (Vertebrados)*, tercera edition, (Corregida y Augmentata). Nuevas Graficas, Madrid, xxxvi + 516 pp. [pp.455–456]

Rombough, P. J. (1999). The gill of fish larvae. Is it primarily a respiratory or an ionoregulatory structure? *Journal of Fish Biology* **55**(Supplement A), 186–204.

Rombough, P. (2007). The functional ontogeny of the teleost gill: which comes first, gas or ion exchange? *Comparative Biochemistry and Physiology* **148A**, 732–742.

Ross, S. T. (2001). *The Inland Fishes of Mississippi.* Mississippi Department of Wildlife, Fisheries and Parks, Jackson, xx + 624 pp. [p. 47]

Roughley, T. C. (1953). *Fish and Fisheries of Australia* (revised edition). Angus and Robertson, Sydney, xv + 343 pp. [p. 7, Plate 1]

Roux, C. (1960). Note sur le tarpon (*Megalops atlanticus* C. et V.) des côtes de la République du Congo. *Bulletin du Muséum National d'Histoire Naturelle* **32**(4), 314–319.

Rueda, M. and O. Defeo (2003). Spatial structure of fish assemblages in a tropical estuarine lagoon: combining multivariate and geostatistical techniques. *Journal of Experimental Marine Biology and Ecology* **296**, 93–112.

Rueda, M., J. Blanco, J. C. Naváez, E. Viloria and C. S. Beltrán (2011). Coastal Fisheries of Colombia. In *Coastal Fisheries of Latin America and the Caribbean.* Fisheries and Aquaculture Technical Paper 544. FAO, Rome, pp. 117–136. [pp. 122–123]

Rundle, K. R., D. C. Jackson, E. D. Dibble and O. Ferrer (2002). Atlantic tarpon distribution in brackish-water lagoons, Humacao Natural Reserve, Puerto Rico. *Proceedings of the Annual Conference of the Southeastern Association of Fish and Wildlife Agencies* **56**, 86–94.

Russell, B. C. and W. Houston (1989). Offshore fishes of the Arafura Sea. *The Beagle* **6**, 69–84.

Russell, D. J. and R. N. Garrett (1983). Use by juvenile barramundi *Lates calcarifer* (Bloch), and other fishes of temporary supralittoral habitats in a tropical estuary in northern Australia. *Australian Journal of Marine and Freshwater Research* **34**, 805–811.

Saadatfar, Z. and D. Shahsavani (2008). Structure of chloride cell in *Telaji* [*sic*] (Cyprinidae, Teleost [*sic*]) of Caspian Sea. *American Eurasian Journal of Agricultural and Environmental Sciences* **4**, 599–602.

Saadatfar, Z. and D. Shahsavani (2011). Structure of lamellae and chloride cells in the gill of *Alosa Caspio* [*sic*] *Caspio* [*sic*] (Clupeidae, Teleostei). *American Journal of Applied Sciences* **8**, 535–539.

Sadowsky, V. (1958). Ocorrência do "camarupim" – *Megalops atlanticus* Val., na região lagunar de Cananéia. *Boletim do Instituto Oceanográfico, São Paulo* **9**(1–2), 61–63.

Safran, P. (1992). Theoretical analysis of the weight-length relationship in fish juveniles. *Marine Biology* **112**, 545–551.

Saint-Paul, U. (1984). Physiological adaptation to hypoxia of a neotropical characoid fish *Colossoma macropomum*, Serrasalmidae. *Environmental Biology of Fishes* **11**, 53–62.

Sakamoto, T., K. Uchida and S. Yokota (2001). Regulation of the ion-transporting mitochondria-rich cell during adaptation of teleost fishes to different salinities. *Zoological Science* **18**, 1163–1174.

Salama, A., I. J. Morgan and C. M. Wood (1999). The linkage between Na^+ uptake and ammonia excretion in rainbow trout: kinetic analysis, the effects of $(NH_4)_2SO_4$ and NH_4HCO_3 infusion and the influence of gill boundary layer pH. *Journal of Experimental Biology* **202**, 697–709.

Santer, R. M. and M. Greer Walker (1980). Morphological studies on the ventricle of teleost and elasmobranch hearts. *Journal of Zoology (London)* **190**, 259–272.

Santhanam, R. N. Ramanathan and G. Jegatheesan (1990). *Coastal Aquaculture in India.* CBS Publishers and Distributors, Delhi, 180 pp. [pp. 8–9, 28–29, 81, 140]

Santoro, A. E. (2016). The do-it-all nitrifier. *Science* **351**, 342–343.

Sanvicente-Añorve, L., M. Sánchez-Ramírez, A. Ocaña-Luna, C. Flores-Coto and U. Ordóñez-López (2011). Metacommunity structure of estuarine fish larvae: the role of regional and local processes. *Journal of Plankton Research* **33**, 179–194. [p. 186 Table II]

Sardella, B. A., V. Matey, J. Cooper, R. J. Gonzalez and C. J. Brauner (2004). Physiological, biochemical and morphological indicators of osmoregulatory stress in "California" Mozambique tilapia (*Oreochromis mossambicus* x *O. urolepis hornorum*) exposed to hypersaline water. *Journal of Experimental Biology* **207**, 1399–1413.

Saroglia, M., G. Caricato, F. Frittella, F. Brambilla and G. Terova (2010). Dissolved oxygen regimen (PO_2) may affect osmo-respiratory compromise in European sea bass (*Dicentrarchus labrax*, L.). *Italian Journal of Animal Science*. doi: 10.4081/ijas.2010.e15, 7 pp.

Scarabotti, P. A., J. A. López, R. Ghirardi and M. J. Parma (2011). Morphological plasticity associated with environmental hypoxia in characiform fishes from neotropical floodplain lakes. *Environmental Biology of Fishes* **92**, 391–402.

Schmitter-Soto, J. J., L. Vásquez-Yeomans, A. Aguilar-Perera, C. Curiel-Mondragón and J. A. Caballero Vázquez (2000). Lista de peces marinos del Caribe mexicano. *Anales del Instituto de Biología, Serie Zoología* **71**(2), 143–177. [p. 147]

Schmitter-Soto, J. J., A. M. Arce-Ibarra and L. Vásquez-Yeomans (2002). Records of *Megalops atlanticus* on the Mexican Caribbean coast. *Contributions in Marine Science* **35**, 34–42.

Schreiber, A. M. and J. L. Specker (1999). Metamorphosis in the summer flounder *Paralichthys dentatus*: changes in gill mitochondria-rich cells. *Journal of Experimental Biology* **202**, 2475–2484.

Schultz, L. P. (1952). A further contribution to the ichthyology of Venezuela. *Proceedings of the U.S. National Museum* **99**(3235), 1–211 + 3 plates. [p. 33]

Seeto, J. and W. J. Baldwin (2010). *A Checklist of the Fishes of Fiji and a Bibliography of Fijian Fishes*. Technical Report 1/2010. Division of Marine Studies, University of the South Pacific, Suva, 102 pp. [p. 60]

Seidelin, M. and S. S. Madsen (1997). Prolactin antagonizes the seawater – adaptive effect of cortisol and growth hormone in anadromous brown trout (*Salmo trutta*). *Zoological Science* **14**, 249–256.

Senigaglia, V. and H. Whitehead (2012). Synchronous breathing by pilot whales. *Marine Mammal Science* **28**, 213–219.

Senior, H. D. (1907a). The conus arteriosus in *Tarpon atlanticus* (Cuvier & Valenciennes). *Biological Bulletin* **12**, 146–151.

Senior, H. D. (1907b). Note on the conus arteriosus of *Megalops cyprinoides* (Broussonet). *Biological Bulletin* **12**, 378–379.

Serafy, J. E., S. J. Cooke, G. A. Diaz, J. E. Graves, M. Hall and M. Shivji *et al.* (2012). Circle hooks in commercial, recreational, and artisanal fisheries: research status and needs for improved conservation and management. *Bulletin of Marine Science* **88**, 371–391.

Seraine, F. (1958). Curral-de-pesca no litoral cearense. *Boletim de Antropologia, Fortaleza* **2**(1), 21–44 + 12 figs.

Sérat, B. (1981, re-issued 1990). *Poissons de Mer de l'Ouest Africain Tropical*. Initiations – Documentations Techniques No. 49. ORSTOM, Paris, viii + 450 pp. [pp. 84–85]

Serrano, X., M. Grosell and J. E. Serafy (2010). Salinity selection and preference of the grey [*sic*] snapper *Lutjanus griseus*: field and laboratory observations. *Journal of Fish Biology* **76**, 1592–1608.

Serrão Santos, R., F. Mora Porteiro and J. Pedro Barreiros (1997). *Marine Fishes of the Azores: An Annotated Checklist and Bibliography*. Boletim da Universidade dos Açores, Arquipélago: Ciências Biológicas e Marinhas (Supplemento 1), xxiii + 242 pp. [p. 120]

Seymour, R. S., K. Christian, M. B. Bennett, J. Baldwin, R. M. G. Wells and R. V. Baudinette (2004). Partitioning of respiration between the gills and air-breathing organ in response to aquatic hypoxia and exercise in the Pacific tarpon, *Megalops cyprinoides. Physiological and Biochemical Zoology* **77**, 760–767.

Seymour, R. S., A. P. Farrell, K. Christian, T. D. Clark, M. B. Bennett and R. M. G. Wells *et al.* (2007). Continuous measurement of oxygen tensions in the air-breathing organ of Pacific tarpon (*Megalops cyprinoides*) in relation to aquatic hypoxia and exercise. *Journal of Comparative Physiology* **177B**, 579–587.

Seymour, R. S., N. C. Wegner and J. B. Graham (2008). Body size and the air-breathing organ of the Atlantic tarpon *Megalops atlanticus. Comparative Biochemistry and Physiology* **150A**, 282–287.

Shanks, E. (2014). Calcasieu River, Louisiana: Lake History and Management Issues, Part VI-A, Waterbody Management Plan Series. Office of Fisheries, Inland Fisheries Section, Louisiana Department of Wildlife and Fisheries, Baton Rouge, 30 pp. [p. 19]

Sharaf, M. M., S. M. Sharaf and H. I. El-Marakby (2004). The effect of acclimatization of fresh water red hybrid tilapia in marine water. *Pakistan Journal of Biological Sciences* **7**, 628–632.

Shen, J. W. (1982). Three new species of genus *Bivesicula* Yamaguti, 1934 (Trematode: Bivesiculidae) from marine fishes of China. *Oceanologia et Limnologia Sinica* **13**, 570–576. [Chinese]

Shen, K. N., C. W. Chang, Y. Iizuka and W. N. Tzeng (2009). Facultative habitat selection in Pacific tarpon *Megalops cyprinoides* as revealed by otolith Sr:Ca ratios. *Marine Ecology Progress Series* **387**, 255–263.

Shenker, J. M., R. Crabtree and G. Zarillo (1995). Recruitment of larval tarpon and other fishes into the Indian River Lagoon. *Bulletin of Marine Science* **57**, 284. [Abstract]

Shenker, J. M., E. Cowie-Mojica, R. E. Crabtree, H. M. Patterson, C. Stevens and K. Yakubik (2002). Recruitment of tarpon (*Megalops atlanticus*) leptocephali into the Indian River Lagoon, Florida. *Contributions in Marine Science* **35**, 55–69.

Shiao, J-C. and P-P. Hwang (2004). Thyroid hormones are necessary for teleostean otolith growth. *Marine Ecology Progress Series* **278**, 271–278.

Shiao, J-C. and P-P. Hwang (2006). Thyroid hormones are necessary for the metamorphosis of tarpon *Megalops cyprinoides* leptocephali. *Journal of Experimental Marine Biology and Ecology* **2**, 121–132.

Shiffman, D. S., A. J. Gallagher, J. Wester, C. C. Macdonald, A. D. Thaler and S. J. Cooke *et al.* (2014). Trophy fishing for species threatened with extinction: a way forward building on a history of conservation. *Marine Policy* **50**, 318–322.

Shih, H-H., W. Liu and Z-Z. Qiu (2004). Digenean fauna in marine fishes from Taiwanese water with the description of a new species, *Lecithochirium tetraorchis* sp. nov. *Zoological Studies* **43**, 671–676.

Shih, T-H., J-L. Horng, P-P. Hwang and L-Y. Lin (2008). Ammonia excretion by the skin of zebrafish (*Danio rerio*) larvae. *American Journal of Physiology – Cell Physiology* **295**, C1625–C1632.

Shih, T-H., J-L. Horng, Y-T. Lai and L-Y. Lin (2013). Rhcg1 and Rhbg mediate ammonia excretion by ionocytes and keratinocytes in the skin of zebrafish larvae: H^+-ATPase-linked active ammonia excretion by ionocytes. *American Journal of Physiology: Regulatory, Integrative and Comparative Physiology* **304**, R1130–R1138.

Shine, R. (1978). Propagule size and parental care: the "safe harbor" hypothesis. *Journal of Theoretical Biology* **75**, 417–424.

Shlaifer, A. (1941). Additional social and physiological aspects of respiratory behavior in small tarpon. *Zoologica (NY)* **26**(11), 55–60.

Shlaifer, A. and C. M. Breder Jr. (1940). Social and respiratory behavior of small tarpon. *Zoologica (NY)* **25**(30), 493–512, 2 plates.

Shreves, M. L. (1959). Tarpon fishing in Virginia: a new development of importance to Virginia's salt water sport fishing program. In *Fourth International Game Fish Conference*, Nassau, pp. 1–4.

Shultz, A. D., K. J. Murchie, C. Griffith, S. J. Cooke, A. J. Danylchuk and T. L. Goldberg *et al.* (2011). Impacts of dissolved oxygen on the behavior and physiology of bonefish: implications for live-release angling tournaments. *Journal of Experimental Marine Biology and Ecology* **402**, 19–26.

Siewert, H. F. and J. B. Cave (1990). Survival of released bluegill, *Lepomis macrochirus*, caught on artificial flies, worms, and spinner lures. *Journal of Freshwater Ecology* **5**, 407–411.

Simmons, W. E. (1900). *The Nicaragua Canal*. Harper and Brothers, New York, 335 pp. [pp. 329–330]

Simpson, D. G. (1954). Two small tarpon from Texas. *Copeia* **1954**, 71–72.

Sirois, P. and J. J. Dodson (2000). Critical periods and growth-dependent survival of larvae of an estuarine fish, the rainbow smelt *Osmerus mordax*. *Marine Ecology Progress Series* **203**, 233–245.

Skomal, G. B. (2007). Evaluating the physiological and physical consequences of capture on post-release survivorship in large pelagic fishes. *Fisheries Management and Ecology* **14**, 81–89.

Skomal, G. and B. Chase (2002). The physiological effects of angling on post-release survivorship in tunas, sharks, and marlin. In *Catch and Release in Marine Recreational Fisheries*, J. A. Lucy and A. L. Studholme (editors). American Fisheries Society, Bethesda, pp. 135–138.

Skomal, G. B., B. C. Chase and E. D. Prince (2002). A comparison of circle hook and straight hook performance in recreational Fisheries for juvenile Atlantic bluefin tuna. In *Catch and Release in Marine Recreational Fisheries*, J. A. Lucy and A. L. Studholme (editors). American Fisheries Society, Bethesda, pp. 57–65.

Sloman, K. A., R. D. Sloman, G. De Boeck, G. R. Scott, F. I. Iftikar and C. M. Wood *et al.* (2009). The role of size in synchronous air breathing of *Hoplosternum littorale*. *Physiological and Biochemical Zoology* **82**, 625–634.

Smith, C. L., J. C. Tyler, W. P. Davis, R. S. Jones, D. G. Smith and C. C. Baldwin (2003). Fishes of the Pelican Cays, Belize. *Atoll Research Bulletin* **497**, ii + 88 pp. [p. 7]

Smith, D. G. (1980). Early larvae of the tarpon, *Megalops atlantica* Valenciennes (Pisces: Elopidae), with notes on spawning in the Gulf of Mexico and the Yucatan Channel. *Bulletin of Marine Science* **30**, 136–141.

Smith, D. G. (1989). *Order* Elopiformes *Families* Elopidae, Megalopidae, and Albulidae: leptocephali In *Fishes of the Western North Atlantic*, Memoir No. I, Part 9, Vol. 2, E. B. Böhlke (editor-in-chief). Memoirs, Sears Foundation for Marine Research, Yale University, New Haven, pp. 961–972. [pp. 968–970]

Smith, H. M. (1896). Notes on Biscayne Bay, Florida, with reference to its adaptability as a site of a marine hatching and experiment station. In *Report of the Commissioner for the Year Ending June 30, 1895*. U.S. Commission of Fish and Fisheries, Part XXI. Government Printing Office, Washington, pp. 169–191. [p. 173]

Smith, H. M. (1907). *North Carolina Geological and Economic Survey, Vol. II: The Fishes of North Carolina*. E. M. Uzzell, Raleigh, xi + 431 pp. [pp. 114–115]

Smith, J. L. B. (1965). *The Sea Fishes of Southern Africa*. Central News Agency, Cape Town, xvi + 580 pp. [p. 86]

Smith, R. S. and D. L. Kramer (1986). The effect of apparent predation risk on the respiratory behavior of the Florida gar (*Lepisosteus platyrhincus*). *Canadian Journal of Zoology* **64**, 2133–2136.

Smith-Vaniz, W. F. and H. L. Jelks (2014). Marine and inland fishes of St. Croix, U.S. Virgin Islands: an annotated checklist. *Zootaxa* **3803**(1), 120 pp. [p. 22]

Smith-Vaniz, W. F., B. B. Collette and B. E. Luckhurst (1999). *Fishes of Bermuda: History, Zoogeography, Annotated Checklist, and Identification Keys*. Special Publication No. 4, American Society of Ichthyologists and Herpetologists, Lawrence, KS, x + 424 pp., 12 plates. [pp. 126–127]

Snelson, F. F. Jr. (1983). Ichthyofauna of the northern part of the Indian River Lagoon System, Florida. *Florida Scientist* **46**, 187–206.

Snelson, F. F. Jr. and W. K. Bradley Jr. (1978). Mortality of fishes due to cold on the east coast of Florida, January, 1977. *Florida Scientist* **41**, 1–12.

Solarin, B. B., A. Oresegun, A. B. Dunsin, M. A. Matanmi, D. Adetoye and F. Aniebona *et al.* (2010). Preliminary investigation into community-based fish cage and pen culture in Ikate Elegushi water front, Lekki, Lagos State. *2010 Annual Report*. Nigerian Institute for Oceanography and Marine Research, Victoria Island, Lagos, pp. 121–124.

Sollid, J., P. De Angelis, K. Gundersen and G. E. Nilsson (2003). Hypoxia induces adaptive and reversible gross morphological changes in crucian carp gills. *Journal of Experimental Biology* **206**, 3667–3673.

Sollid, J., R. E. Weber and G. E. Nilsson (2005). Temperature alters the respiratory surface area of crucian carp *Carassius carassius* and goldfish *Carassius auratus*. *Journal of Experimental Biology* **208**, 1109–1116.

Southworth, A. S. (1888). The silver king. *Frank Leslie's Popular Monthly* **27**(6), unpaginated. [9 pp.]

Spotte, S. (1992). *Captive Seawater Fishes: Science and Technology*. Wiley, New York, xxii + 942 pp. [pp. 699–701, 850, 851]

Spotte, S. (2014). *Free-ranging Cats: Behavior, Ecology, Management*. Wiley Blackwell, Chichester, xix + 296 pp.

Spotte, S. and G. Adams (1983). Estimation of the allowable upper limit of ammonia in saline waters. *Marine Ecology Progress Series* **10**, 207–210.

Spotte, S. and G. Anderson (1989). Plasma cortisol changes in seawater-adapted mummichogs (*Fundulus heteroclitus*) during exposure to ammonia. *Canadian Journal of Fisheries and Aquatic Sciences* **46**, 2065–2069.

Spotte, S., P. M. Bubucis and G. Anderson (1991). Plasma cortisol response of seawater-adapted mummichogs (*Fundulus heteroclitus*) during deep MS-222 anesthesia. *Zoo Biology* **10**, 75–79.

Springer, V. G. and K. D. Woodburn (1960). *An ecological study of the fishes of the Tampa Bay area*. Professional Papers Series, No. 1. Florida State Board of Conservation, St. Petersburg, 104 pp. [p. 16]

Stamatopoulos, C. (1993). *Trends in Catches and Landings Atlantic Fisheries: 1970–1991*. FAO, Rome, vii + 223 pp. [pp. 102, 202]

Steffe, A. S. and M. Westoby (1991). Tidal currents, eddies and orientated behaviour by larval fishes. *Bureau of Rural Resources Proceedings* (**15**), 110–115.

Stein, W. III., N. J. Brown-Peterson, J. S. Franks and M. T. O'Connell (2012). Evidence of spawning capable tarpon (*Megalops atlanticus*) off the Louisiana coast. *Gulf and Caribbean Research* **24**, 73–74.

Stewart, K. R. and J. Wyneken (2004). Predation risk to loggerhead hatchlings at a high-density nesting beach in southeast Florida. *Bulletin of Marine Science* **74**, 325–335.

Stiassny, M. L. J., G. G. Teugels and C. D. Hopkins (2007). *Poissons d'eaux douces et saumatres de basse Guinée, ouest de l'Afrique centrale*, Vol. 1. Collection Faune et Flore tropicales 42, IRD, Muséum national d'histoire Naturelle, Paris, 800 pp. [pp. 54-55]

Stickney, R. R. (1986). Tilapia tolerance of saline waters: a review. *Progressive Fish-culturist* **48**, 161–167.

Storer, T. I. and R. L. Usinger (1957). *General Zoology*, 3rd edition. McGraw-Hill, New York, 664 pp.

Storey, M. (1937). The relation between normal range and mortality of fishes due to cold at Sanibel Island, Florida. *Ecology* **18**, 10–26.

Storey, M. and W. W. Gudger (1936). Mortality of fishes due to cold at Sanibel Island, Florida, 1886–1936. *Ecology* **17**, 640–648.

Storey, M. and L. M. Perry (1933). A record of young tarpon at Sanibel Island, Lee County, Florida. *Science* **78**, 284–285.

Sucré, E., M. Charmantier-Daures, E. Grousset and P. Cucchi-Mouillot (2011). Embryonic ionocytes in the European sea bass (*Dicentrarchus labrax*): structure and functionality. *Development, Growth and Differentiation* **53**, 26–36.

Sunier, A. L. J. (1922). Contributions to the knowledge of the natural history of the marine fishponds of Batavia. *Treubia* **2**(2–4), 159–405. [p. 226]

Suski, C. D., S. J. Cooke, A. J. Danylchuk, C. M. O'Connor, M-A. Gravel and T. Redpath *et al.* (2007). Physiological disturbance and recovery dynamics of bonefish (*Albula vulpes*), a tropical marine fish, in response to variable exercise and exposure to air. *Comparative Biochemistry and Physiology* **148A**, 664–673.

Swanson, P. L. (1946). Tarpon in the Pacific. Copeia **1946**, 175.

Tabb, D. C. and R. B. Manning (1961). A checklist of the flora and fauna of northern Florida Bay and adjacent brackish waters of the Florida mainland collected during the period of July, 1957 through September, 1960. *Bulletin of Marine Science of the Gulf and Caribbean* **11**, 552–649. [pp. 606–607]

Tagatz, M. E. (1968). Fishes of the St. Johns River, Florida. *Quarterly Journal of the Florida Academy of Sciences* **30**, 25–50. [p. 33]

Tagatz, M. E. (1973). A larval tarpon, *Megalops atlanticus*, from Pensacola, Florida. *Copeia* **1973**, 140–141.

Takasuka, A., I. Aoki and Y. Oozeki (2007). Predator-specific growth-selective predation on larval Japanese anchovy *Engraulis japonicus*. *Marine Ecology Progress Series* **350**, 99–107.

Tang, C-H. and T-H. Lee (2011). Morphological and ion-transporting plasticity of branchial mitochondrion-rich cells in the euryhaline spotted green pufferfish, *Tetraodon nigroviridis*. *Zoological Studies* **50**, 31–42.

Tano de la Hoz, M. F., A. M. García, M. González Castro and A. Ofelia Díaz (2014). Histochemical and scanning electron microscopic approaches to gills in juveniles of *Odontesthes argentinensis* (Actinopterygii, Atherinopsidae). *International Journal of Aquatic Science* **5**, 154–166.

Taylor, H. F. (1919). Mortality of fishes on the west coast of Florida. In *Report of the United States Commissioner of Fisheries for the Fiscal Year 1917 with Appendixes*. Bureau of Fisheries Document No. 848, Appendix III, pp. 1–24.

Taylor, R. G., J. A. Wittington and D. E. Haymans (2001). Catch-and-release mortality rates of common snook in Florida. *North American Journal of Fisheries Management* **21**, 70–75.

Taylor, W. R. (1964). Fishes of Arnhem Land. In *Records of the American-Australian Scientific Expedition to Arnhem Land* 4, 45–307, 68 plates, map. [pp. 59–60]

Teixeira Alves, T., M. Ogawa, M. da Conceição Caland-Noronha and C. A. Esteves Araripe (1972). Industrialização da ova do camurupim *Tarpon atlanticus* (Valenciennes). *Arquivos de Ciências do Mar, Fortaleza* **12**(2), 151–154.

Thomas, H. S. (1887). *Tank Angling in India*. Hamilton Adams, London, xx + 190 pp. [pp. xx, 124, 168, Plate XIII]

Titus, R. G. and H. Mosegaard (1992). Fluctuating recruitment and variable life history of migratory brown trout, *Salmo trutta* L., in a small, unstable stream. *Journal of Fish Biology* **41**, 239–255.

Torelli, J., I. L. Rosa and T. Watanabe (1997). Ictiofauna do rio Gramame, Paraíba, Brasil. *Iheringia (Série Zoologica)* **82**, 67–73.

Tota, B., V. Cimini, G. Salvatore and G. Zummo (1983). Comparative study of the arterial and lacunary systems of the ventricular myocardium of elasmobranch and teleost fishes. *American Journal of Anatomy* **167**, 15–32.

Townsend, S. (1992). The occurrence of fish kills, and their causes, in the Darwin-Katherine-Jabiru region of the Northern Territory. Report 36/92. Water Resources Division, Power and Water Authority, Darwin, 28 pp.

Townsend, S. A., K. T. Boland and T. J. Wrigley (1992). Factors contributing to a fish kill in the Australian wet/dry tropics. *Water Research* **8**, 1039–1044.

Tran, H. Q., R. S. Mehta and P. C. Wainwright (2010). Effects of ram speed on prey capture kinematics of juvenile Indo-Pacific tarpon, *Megalops cyprinoides*. *Zoology (Jena)* **113**, 75–84.

Tripathi, R. K., V. Mohindra, A. Singh, R. Kumar, R. M. Mishra and J. K. Jena (2013). Physiological responses to acute experimental hypoxia in the air-breathing Indian catfish, *Clarias batrachus* (Linnaeus, 1758). *Journal of Bioscience* **38**, 373–383.

Tsukamoto, Y. and M. Okiyama (1993). Growth during the early life history of the Pacific tarpon, *Megalops cyprinoides*. *Japanese Journal of Ichthyology* **39**, 379–386.

Tsukamoto, Y. and M. Okiyama (1997). Metamorphosis of the Pacific tarpon, *Megalops cyprinoides* (Elopiformes, Megalopidae) with remarks on development patterns in the Elopomorpha. *Bulletin of Marine Science* **60**, 23–26.

Tucker, J. W. and R. G. Hodson (1976). Early and mid-metamorphic larvae of the tarpon, *Megalops atlantica*, from the Cape Fear River estuary, North Carolina, 1973–1974. *Chesapeake Science* **17**, 123–125.

Tuma, D. (1994). *Sea Catch: Identifying and Handling Fish and Shellfish from Sydney to Shark Bay*. Department of Primary Industries. Brisbane, 117 pp. [unnumbered page]

Twomey, E. and P. Byrne (1985). A new record for the tarpon, *Tarpon atlanticus* Valenciennes (Osteichthyes-Elopiformes-Elopidae), in the eastern North Atlantic. *Journal of Fish Biology* **26**, 359–362.

Tzeng, W-N., C-E. Wu and Y-T. Wang (1998). Age of Pacific tarpon, *Megalops cyprinoides*, at estuarine arrival and growth during metamorphosis. *Zoological Studies* **37**, 177–183.

Tzeng, W-N., Y-T. Wang and C-W. Chang (2002). Spatial and temporal variations of the estuarine larval fish community on the west coast of Taiwan. *Marine and Freshwater Research* **53**, 419–430.

van den Thillart, G., D. Randall and L. Hoa-Ren (1983). CO_2 and H^+ excretion by swimming coho salmon, *Oncorhynchus kisutch*. *Journal of Experimental Biology* **107**, 169–180.

van der Heijden, A. J. H., P. M. Verbost, J. Eygensteyn, J. Li, S. E. Wendelaar Bonda and G. Flik (1997). Mitochondria-rich cells in gills of tilapia (*Oreochromis mossambicus*) adapted to fresh water or sea water: quantification by confocal laser scanning microscopy. *Journal of Experimental Biology* **200**, 55–64.

van der Heijden, P. G. M. (2002). The artisanal fishery in the Sepik-Ramu catchment area, Papua New Guinea. *Science in New Guinea* **27**, 101–119.

van Kampen, P. N. (1909). Kurze Notezin über Fische des Java-Meeres. III. Die Larvae von *Megalops cyprinoides* Brouss. *Bulletin du Département de l'Agriculture aux Indes Néerlandaises* **20**, 10–12.

Varsamos, S., C. Nebel and G. Charmantier (2005). Ontogeny of osmoregulation in postembryonic fish: a review. *Comparative Biochemistry and Physiology* **141A**, 401–429.

Vasilakopoulos, P., F. G. O'Neill and C. T. Marshall (2011). Misspent youth: does catching immature fish affect *Fisheries* sustainability? *ICES Journal of Marine Science* **68**, 1525–1534.

Vásquez-Yeomans, L., U. Ordónez-López and E. Sosa-Cordero (1998). Fish larvae adjacent to a coral reef in the western Caribbean Sea off Mahahual, Mexico. *Bulletin of Marine Science* **62**, 229–245.

Vega-Cendejas, M. E. and M. Hernández (2002). Isla Contoy – a Mexican Caribbean ecosystem used by tarpon, *Megalops atlanticus* as a feeding area. *Contributions in Marine Science* **35**, 70–80.

Vikebø, F., S. Sundby, B. Ådlandsvik and Ø. Fiksen (2005). The combined effect of transport and temperature on distribution and growth of larvae and pelagic juveniles of Arcto-Norwegian cod. *ICES Journal of Marine Science* **62**, 1375–1386.

Vikebø, F., C. Jørgensen, T. Kristiansen and Ø. Fiksen (2007). Drift, growth, and survival of larvae of northeast Arctic cod with simple rules of behaviour. *Marine Ecology Progress Series* **347**, 207–219.

Vilaró-Diaz, J. (1893). *Algo Sobre Peces de Cuba con Cierta Extension a los de Puerto Rico y los Estados Unidos*. Imprenta de A. Alvarez y Compañia, Havana, 176 pp. [pp. 12, 148–149]

Villa, J. (1982). *Peces Nicaragüenses de Agua Dulce*. Unión Cardoza, Managua, xiv + 253 pp. [pp. 77–78]

Viñola Valdez, L. Cotayo Cedeño and N. Zurcher (2008). Coastal ecosystem management to support bonefish and tarpon sportfishing in Peninsula de Zapata National Park, Cuba. In *Biology and Management of the World Tarpon and Bonefish Fisheries*, J. S. Ault (editor). CRC Press, Boca Raton, pp. 93–98.

Vladykov, V. D. and R. A. McKenzie (1935). *The Marine Fishes of Nova Scotia*. Proceedings of the Nova Scotian Institute of *Science* XIX(part 1), 113 pp. [p. 53]

von Bertalanffy, L. (1938). A quantitative theory of organic growth (inquiries on growth laws, II). *Human Biology, A Record of Research* **10**, 181–218.

Vyas, P. (2012). *Biodiversity Conservation in Indian Sundarban in the Context of Anthropogenic Pressures and Strategies for Impact Mitigation*. Doctoral dissertation, Saurashtra University, Gujrat, Rajkot, xxix + 320 pp., 50 plates. [p. 308]

Wade, R. A. (1962). The biology of the tarpon, *Megalops atlanticus*, and the ox-eye, *Megalops cyprinoides*, with emphasis on larval development. *Bulletin of Marine Science of the Gulf and Caribbean* **12**, 545–622.

Wang, C-H., C-H. Kuo, H-K. Mok and S-C. Lee (2003). Molecular phylogeny of elopomorph fishes inferred from mitochondrial 12S ribosomal RNA sequences. *Zoological Scripta* **32**, 231–241.

Ward, R., I. R. Blandon, F. J. Garcia de Leon, A. M. Landry, W. Dailey and C. D. Acuña Leal (2004). Studies in conservation genetics of tarpon (*Megalops atlanticus*) – IV. Population structure among Gulf of Mexico collection sites inferred from variation in restriction site polymorphisms of tarpon mitochondrial DNA. *Proceedings of the Gulf and Caribbean Fisheries Institute* **55**, 373–383.

Ward, R., I. R. Blandon, R. Vegal, A. M. Landry, W. Dailey and F. J. García de León *et al.* (2005). Studies in conservation genetics of tarpon (*Megalops atlanticus*) – III. Variation across the Gulf of Mexico in the nucleotide sequence of a 12S mitochondrial rRNA gene fragment. *Contributions in Marine Science* **37**, 46–60.

Ward, R., I. R. Blandon, F. García de León, S. J. Robertson, A. M. Landry and A. O. Anyanwu *et al.* (2008). Studies in conservation genetics of tarpon (*Megalops atlanticus*): microsatellite variation across the distribution of the species. In *Biology and Management of the World Tarpon and Bonefish Fisheries*, J. S. Ault (editor). CRC Press, Boca Raton, pp. 131–145.

Weihrauch, D., M. P. Wilkie and P. J. Walsh (2009). Ammonia and urea transporters in gills of fish and aquatic crustaceans. *Journal of Experimental Biology* **212**, 1716–1730.

Wells, R. M. G. and J. Baldwin (2006). Plasma lactate and glucose flushes following burst swimming in silver trevally (*Pseudocaranx dentex*: Carangidae) support the "releaser" hypothesis. *Comparative Biochemistry and Physiology* **143A**, 347–352.

Wells, R. M. G., J. Baldwin, R. S. Seymour and R. E. Weber (1997). Blood oxygen transport and hemoglobin function in three tropical fish species from northern Australian freshwater billabongs. *Fish Physiology and Biochemistry* **16**, 247–258.

Wells, R. M. G., J. Baldwin, R. S. Seymour, R. V. Baudinette, K. Christian and M. B. Bennett (2003). Oxygen transport capacity in the air-breathing fish, *Megalops cyprinoides*: compensations for strenuous exercise. *Comparative Biochemistry and Physiology* **134A**, 45–53.

Wells, R. M. G., J. Baldwin, R. S. Seymour, K. Christian and T. Brittain (2005). Red blood cell function and haematology in two tropical freshwater fishes from Australia. *Comparative Biochemistry and Physiology* **141A**, 87–93.

Wells, R. M. G., J. Baldwin, R. S. Seymour, K. A. Christian and P. A. Farrell (2007). Air breathing minimizes post-exercise lactate load in the tropical Pacific tarpon, *Megalops cyprinoides*

Broussonet 1782 but oxygen debt is repaid by aquatic breathing. *Journal of Fish Biology* **71**, 1649–1661.

Wendelaar Bonga, S. E. (1997). The stress response in fish. *Physiological Reviews* **77**, 591–625.

Weng, C-F., C-C. Chiang, H-Y. Gong, M. H-C. Chen, C. J-F. Lin and W-T. Huang *et al.* (2002) Acute changes in gill Na^{+}-K^{+}-ATPase and creatine kinase in response to salinity changes in the euryhaline teleost, tilapia (*Oreochromis mossambicus*). *Physiological and Biochemical Zoology* **75**, 29–36.

Wheeler, A. (1992). A list of the common and scientific names of fishes of the British Isles. *Journal of Fish Biology* **41**, 1–37. [pp. 7, 17]

Whitehead, C. E. (1891/1991). *The Camp-Fires of the Everglades or Wild Sports in the South.* University of Florida Press, Gainesville, xvi + 281 pp. [pp. (198–203] [Reprint of 1891 Edinburgh edition]

Whiton, H. D. and C. H. Townsend (1928). Tarpon behavior in Calcasieu Pass. *Bulletin of the New York Zoological Society* **31**(5), 170–172.

Wiggers, R. K. (2010). South Carolina marine game fish tagging program 1978–2009. Marine Resources Division, South Carolina Department of Natural Resources, Charleston, iv + 55 pp. [p. 42]

Wilkie, M. P. (2002). Ammonia excretion and urea handling by fish gills: present understanding and future research challenges. *Journal of Experimental Zoology* **293**, 284–301.

Willcox, J. (1887). Fish killed by cold along the Gulf of Mexico and coast of Florida. *Bulletin of the U.S. Fish Commission* **VI**, 123.

Wilcox, W. A. (1897). Commercial Fisheries of Indian River, Florida. In *Report on the Fisheries of Indian River, Florida*, J. J. Brice (Commissioner). U.S. Government Printing Office, Washington, pp. 27–40.

Wilson, J. M. and P. Laurent (2002). Fish gill morphology: inside out. *Journal of Experimental Zoology* **293**, 192–123.

Winemiller, K. O. (1989). Development of dermal lip protuberances for aquatic surface respiration in South American characid fishes. *Copeia* **1989**, 382–390.

Winemiller, K. O. and W. H. Dailey (2002). Life history strategies, population dynamics, and consequences for supplemental stocking of tarpon. *Contributions in Marine Science* **35**, 81–94.

Wirtz, P., R. Fricke and M. J. Biscoito (2008). The coastal fishes of Madeira Island – new records and an annotated check-list. *Zootaxa* **1715**, 26 pp. [p. 1]

Wirtz, P., A. Brito, J. M. Falcón, R. Freitas, R. Fricke and V. Monteiro *et al.* (2013). The coastal fishes of the Cape Verde Islands – new records and an annotated check-list. *Spixiana (München)* **36**(1), 113–142. [pp. 113, 117, 138]

Woodcock, S. H. and B. D. Walther (2014). Trace elements and stable isotopes in Atlantic tarpon scales reveal movements across estuarine gradients. *Fisheries Research* **153**, 9–17.

Worm, B. (2015). A most unusual (super) predator. *Science* **349**, 784–785.

Wright, D. G., P. Pawlowicz, T. J. McDougall, R. Feistel and G. M. Marion (2011). absolute salinity, "density salinity" and the reference-composition salinity scale: present and future use in the seawater standard TEOS-10. *Ocean Science* **7**, 1–26.

Wright, P. A. (1995). Nitrogen excretion: three end products, many physiological roles. *Journal of Experimental Biology* **198**, 273–281.

Wright, P. A. and C. M. Wood (2009). A new paradigm for ammonia excretion in aquatic animals: role of Rhesus (Rh) glycoproteins. *Journal of Experimental Biology* **212**, 2303–2312.

Zale, A. V. and S. G. Merrifield (1989). Species profiles: life histories and environmental requirements of coastal fishes and invertebrates (South Florida) – ladyfish and tarpon. TR EL-82-4. U.S. Fish and Wildlife Service Biological Report 82 (11.104) and U.S. Army Corps of Engineers, Vicksburg and Washington, vi + 17 pp.

Zaneveld, J. S. (1962). The fishery resources and fishery industries of the Netherlands Antilles. *Proceedings of the Gulf and Caribbean Fisheries Institute* **14**, 137–171. [p. 144 Table 2, 161 Table 9]

Zappler, G. (editor) (1993). *Saltwater Fishes of Texas: A Guide to Knowing and Catching Bay and Gulf Fishes,* revised edition. Texas Parks and Wildlife Department, Austin, 41 pp. [p. 14]

Zerbi, A. and C. Aliaume (2002). Effects of depth and plant cover on juvenile sportfish distribution in a Puerto Rican mudflat: an experimental field study. *Contributions in Marine Science* **35**, 107. [Abstract]

Zerbi, A. J., C. Aliaume and J. M. Miller (1999). A comparison between two tagging techniques with notes on juvenile tarpon ecology in Puerto Rico. *Bulletin of Marine Science* **64**, 9–19.

Zerbi, A., C. Aliaume and J. Joyeux (2001). Growth of juvenile tarpon in Puerto Rican estuaries. *ICES Journal of Marine Science* **58**, 87–95.

Zerbi, A., C. Aliaume, J. Miller and J-C. Joyeux (2005). Contributions to the ecology of juvenile tarpon *Megalops atlanticus* in an impoundment in southwestern Puerto Rico. *Proceedings of the Forty-Seventh Annual Gulf and Caribbean Fisheries Institute,* Fort Pierce, pp. 478–498.

Index

Page numbers in *italics* refer to illustrations; those in **bold** refer to tables.

Tarpons: Biology, Ecology, Fisheries, First Edition. Stephen Spotte.